AutoCAD® for Architects and Engineers

A Practical Guide to Design, Presentation, and Production

By John M. Albright and Elizabeth H. Schaeffer

with B. Dwayne Loftice, Scot Woodard and J. Wise Smith

 New Riders Publishing, Thousand Oaks, California

AutoCAD® for Architects and Engineers

The Complete Guide to Design, Presentation and Production

By John M. Albright and Elizabeth Schaeffer
with D. Loftice, S. Woodard and J. Wise Smith

Published by:

New Riders Publishing
Post Office Box 4846
Thousand Oaks, CA 91360, U.S.A.

First Edition, 1989

Printed in the United States of America

Library of Congress Cataloging-in-Publication Data

Albright, John M., 1949-
    AutoCAD for architects and engineers.

    1. AutoCAD (Computer program) I. Schaeffer,
Elizabeth, 1957-   . II. Title
T385.A436 1989   620'.00425'02855369   89-14489
ISBN 0-934035-53-9

About the Authors

John M. Albright

John M. Albright is System Manager and Project Architect for J. Wise Smith Associates, Inc., Memphis Tennessee. He has used AutoCAD on a day-to-day bases since version 1.3. Before joining J. Wise Smith, he set up, operated and managed the first inhouse AutoCAD workstations devoted to facility management and later expanded the system to include workstations in all four regional offices of a large corporation. Mr. Albright has consulted, lectured and instructed other architects on the use and management of AutoCAD. He attended State Technical Institute at Memphis and graduated with an Associated Degree of Architecture and returned to add computer science to his education background.

Elizabeth Schaeffer

Elizabeth Schaeffer holds a Batchelors of Science degree in Biology from Rhodes College in Memphis, Tennessee. She began her post-college career as a Senior Research Technician in the field of Cellular Immunology at the University of Tennessee Center for the Health Sciences. Her interest turn to computers through her work with statistical analyses and she joined the Data Processing Division of the IBM Corporation in Memphis in 1981. She worked as a Systems Engineer at IBM for four years and specialized in Intermediate Systems. A direct descendant of Nathaniel Hawthorne, Ms. Schaeffer enjoys the challenge of rendering technical subject matter into readable material.

B. Dwayne Loftice

B. Dwayne Loftice is a licensed and practicing Architect of 23 years. After graduation from the College of Architecture at Texas A&M University, he became interested in the various forms of systems drafting and production. When he started his own practice in 1977, he guided his production staff into computer aided drafting to reduce the overall time expended on mundane production procedures. This enabled him to devote more time on quality design ad the marketing of his small firm's capabilities. Recently, he has relocated to Memphis where he manages an architectural firm's design and production staff in large scale health, and industrial type projects produced on the AutoCAD system.

Scot Woodard

Scot Woodard holds a Bachelors of Science degree in Engineering Technology from Heriff College of Engineering, Memphis State University.   From 1984 to 1987 he managed the CAD system and supplied training to all personnel of the Architecture and Planning firm of Bologna and Associates, Inc. He is past chairman of the Research and Development Committee for the Memphis AutoCAD User's Group, which boasted international acceptance of its monthly newsletter. He is currently responsible for CAD production at J. Wise Smith Associates, Inc.

J. Wise Smith

J. Wise Smith is Founder and President of J. Wise Smith Associates, Inc., a Multi-disciplined Architectural and Planning firm based in Memphis, Tennessee. The firm limits itself to commercial, institutional, educational and business occupancies. Because of the complexity and size of their projects, Mr. Smith tailored his Architectural and Planning firm to take advantage of all available computer technology. Mr. Smith earned his Bachelors of Architecture degree from Auburn University in 1968 and has been in self practice since 1973.

Acknowledgments

The author's wish to thank the following companies for their assistance. Autodesk, Inc. supplied AutoCAD, AutoShade and AutoFlix for the book. Symsoft provided a copy of Hotshot Plus and KETIV Technologies supplied their Menu System ARCH-T. AC3 loaned us hardware and supplied us with technical support.

John Albright wishes to acknowledge: "Jesus Christ, the Son of God, as my savior. I would also like to give a special thank-you, with love, to my wife Marjorie and our son John David. If it were not for them I would have never exhausted the amount of energy, effort and worry this project took. And finally, to my parents, who through out my life have always encouraged and supported my different ventures."

"I would also like to acknowledge the following people, without whose friendship and assistance I would probably either, still be writing, or would have long since given up: Beth Schaeffer, my co-author, for putting up with me in general. Alfred Heyman of HHS Software Solutions for his custom CADLOG program that exported the AutoCAD command sequence into our text editor, and for all of the other programs he gave to us. Page and Tammy, for their untiring efforts in obtaining the hardware I needed. Mark Stockton of AC3, who kept that hardware running. Todd Summer who mostly just listened. Joe Waston, for his insight in advising me to go back to college and learn about computers a long time ago. Dan Griffin, for his advice and assistance. And finally to Harbert Rice, for giving Beth and me the opportunity to write and complete this book. To all of you, THANK YOU."

Beth Schaeffer would like to thank John Albright for "leaning over my fence one day and asking me if I could write (prosaically, that is). Also, I'd like to thank Harbert Rice for backing this project and for all his helpful comments and direction. A very special thanks to my parents, for their support through the years and for the tremendous contribution they made to this book by spending time with their grandchildren — above and beyond the call of duty — when I had to work. And last but not least, my thanks and love to my husband, Sandy, for his almost infinite patience and support, and to Emily and Michael, for helping me see what's really important.".

Dwayne Loftice acknowledges that since most of the time spent writing his portion of this book infringed with his family life, he would like: "to dedicate my contributions to my wife Kelly, my son Jason, and my two daughters Sarah and Maggie. Their patience and understanding during my time at this endeavor was truly remarkable. Likewise, I want to acknowledge my mother and father, Bill and Jean, for their understanding for not being as much help to them in the orchard this year during this undertaking."

Scot Woodard wishes: "to thank you Lord, for the opportunity, and all my love to Melinda, Nealy and Mallory for their love and patience. Much respect and gratitude to the crew at Wise Smith for their diligence."

Production

Director of Production: Carolyn Porter
Graphics Art Director: Todd Meisler
Cover Design: Jill Casty and Associates

The production staff wishes to thank Beany, Orville and Cecil Dekeles for their unflagging support.

Table of Contents

Introduction

Introduction	I-1
Your Basic AutoCAD	I-2
Getting the Best Out of AutoCAD	I-2
The Book's Project	I-3
Book Organization	I-4
How to Use the Book	I-5
Setting Up the Drawings for the Chapters	I-7
The AE-SYSTEM Disk	I-8
Items to Check When Doing the Exercises	I-9
Prerequisites	I-10
Start with Planning	I-10

Section 1-Project Planning and System Setup

CHAPTER 1 *Planning and Production*

Planning, or the Lack Thereof	1-1
Production Concepts	1-2
Production Issues	1-5
Organizing with a Project Work Book	1-7
Hardware and Project Planning	1-14
System Development and Standards	1-15
Getting Started with the Simulation Project	1-16
Naming Convention for the Project	1-18
Creating and Using Prototype Drawings	1-20
Technique for Creating Prototype Drawings	1-22
Technique for Using and Creating Border Drawings	1-27
Technique for Creating AEBORDER Drawing	1-29
Summary	1-30

CHAPTER 2 *Symbol Libraries*

Introduction	2-1
Drawing Setup for Chapter Two	2-2

Symbol Types   2-3
Rules for Creating Symbols   2-3
Attributes and Symbols   2-5
  Technique for Creating Symbols   2-6
Real World Symbols   2-18
AE SYSTEM Setup   2-23
AE-MENU Setup and Installation   2-26
A Little AutoLISP   2-29
Summary   2-39

CHAPTER 3 — *Menu Routines and AutoLISP*

Introduction   3-1
Menu System Concepts   3-2
AutoCAD's Standard Menu Areas   3-3
Macro Command Structure   3-11
The AE-MENU System   3-12
AE-MENU Tablet Area One   3-16
Using the AE MENU System Setup   3-16
AutoLISP and Productivity   3-24
AutoLISP Utilities   3-25
Summary   3-33

SECTION II — DEVELOPING THE DESIGN

CHAPTER 4 — *Design Drawings with AutoCAD*

Design Drawings   4-1
Advantages of the AE-MENU System in Design Drawings   4-2
Project Introduction   4-3
Site Plan Design Drawing Sequence   4-7
  Technique for Property Line Input   4-7
  Technique for Creating Setback Lines   4-9
  Technique for Creating Setback Block   4-12
  Technique for Calculating Areas   4-13
Floor Plan Design Drawing Project Sequence   4-14
  Technique for Floor Plan Area Layout   4-16
  Technique for Adjusting the Areas   4-20
  Technique for Defining Wall Thicknesses   4-22
Creating the Base Plan   4-24
  Technique for Defining Mullions   4-33

Project Sequence for Completing the Site Plan   4-38
     Technique for Adding the North Parking Area Striping   4-52
Summary   4-56

CHAPTER 5   *3D Presentations and Animation*

Introduction   5-1
Advantages to Using AutoShade and AutoFlix   5-1
Using Release 10's 3D Features for Presentation   5-2
AutoShade Basics   5-3
AutoFlix Basics   5-4
Presentation Drawings and Movie Introduction   5-4
Drawing Setup for Chapter Five   5-5
AutoLISP Drawing Utility   5-7
Presentation Drawings Project Sequence   5-8
     Technique for Enhancing and Correcting Exterior Entities   5-10
     Technique For Adding Columns   5-14
     Technique for Adding 3D Curbs   5-17
     Technique for Adding 3D Recessed Areas   5-23
     Technique for Outlining 3D Roof Curves   5-26
Introduction to 3D Surfaces   5-28
RULESURF and EDGESURF Surfaces   5-29
     Technique for Defining 3D Vertical Ruled Surfaces   5-31
     Technique for Adding Remaining 3D Curved Surfaces   5-33
     Technique for Defining 3D Horizontal Surfaces   5-36
     Technique for 3D Ceiling Surfaces   5-38
Procedure for Setting Up AutoShade   5-45
Procedure for Loading and Operating AutoShade   5-48
Procedure for Setting Up and Operating AutoFlix   5-49
Generating A Movie   5-54
Summary   5-54

SECTION III — PROJECT PRODUCTION

CHAPTER 6   *Plans*

Introduction   6-1
Drawing Setup for Chapter Six   6-3
Site Plan Project Sequence   6-4
     Technique for Adding Concrete Walk Construction Joints   6-7
     Technique for Drawing New Contours   6-9

Technique for Labeling Contour Lines 6-12

Technique for Adding Landscaping 6-14

Techniques for Dimensioning 6-19

Floor Plan Project Sequence 6-24

Techniques for Adding Room, Door, and Window Annotations 6-27

Technique for Wall Poucheing 6-33

Ceiling Plan Project Sequence 6-37

Technique for Creating Ceiling Grid System 6-40

Technique for Inserting Light Fixtures 6-45

Summary 6-49

CHAPTER 7   *Elevations*

Introduction 7-1

Building on Existing Data 7-1

Exterior Elevations 7-1

Drawing Setup for Chapter Seven 7-3

AutoLISP Drawing Utilities 7-4

Exterior Elevation Project Sequence 7-6

East Elevation 7-7

Technique for Defining Major Components 7-7

Technique for Defining Sloped Roof Lines 7-11

Technique for Adding the Grade Lines and Curbs 7-14

Technique for Creating and Placing Windows 7-16

Technique for Creating Base Drawing for West Elevation 7-20

West Elevation 7-21

North Elevation 7-26

Technique For Creating Store Front and Doors 7-32

South Elevation 7-39

Technique for Placing Elevation and Annotation Symbols 7-41

Summary 7-43

CHAPTER 8   *Building Sections*

Introduction 8-1

Building Sections and Details 8-1

Drawing Setup for Chapter Eight 8-2

AutoLISP Drawing Utilities 8-4

Building Section Project Sequence 8-7

Technique for Defining Foundation, Floor Slab and Wall Components 8-10

Technique for Creating Joist Web Members ... 8-24
Technique for Creating the Other Half of the Section ... 8-28
Creating a Detail ... 8-33
Technique for Creating the Detail Base Drawing ... 8-33
Summary ... 8-39

CHAPTER 9   *Engineering Drawings*

Introduction ... 9-1
Engineering Drawings ... 9-1
Drawing Setup for Chapter Nine ... 9-3
AutoLISP Drawing Utilities ... 9-4
Roof Framing Plan Project Sequence ... 9-6
Technique for BASEPLAN Framing Modifications ... 9-7
Mechanical Plan Project Sequence ... 9-15
Technique for Supply and Return Grilles ... 9-16
Plumbing Plan Project Sequence ... 9-22
Technique for BASEPLAN Plumbing Modifications ... 9-23
Electrical Lighting Plan Project Sequence ... 9-29
Summary ... 9-35

CHAPTER 10   *Project Completion and Revisions*

Introduction ... 10-1
Plotter History ... 10-1
Plotter Technologies ... 10-2
Plotter Bottlenecks ... 10-4
Bottleneck Solutions ... 10-5
Standards for Plotting ... 10-6
Creating a Title Block and Completing a Border Drawing ... 10-7
Technique for Creating the Title Block ... 10-8
Line Weights, Views and Scale ... 10-17
Technique for Using Macros and Script Files for Plotting ... 10-18
Plotting the Project Drawings ... 10-21
Plot Logs ... 10-23
Adding Revisions ... 10-23
Technique for Adding Revisions ... 10-24
Alternate Output Methods ... 10-25
Summary ... 10-26

SECTION IV — ADVANCED TOPICS

CHAPTER 11   *Data Extraction, Specifications and Reports*

Introduction	11-1
Concepts of the Reporting System	11-1
System Software Components	11-4
The Key to Extracting AutoCAD Data	11-7
Technique for Data Extraction	11-7
Setting Up the Software Components (Optional)	11-10
Framework Database Setup	11-11
Methods of System Operation	11-24
Technique for Generating Base Data from AutoCAD	11-24
Generating a Specification Reference Report	11-25
Generating a Drawing Number Log	11-27
Generating the Project Number Log	11-29
Summary	11-30

CHAPTER 12   *Systems Management*

Introduction	12-1
Getting the Best from AutoCAD and the People Who Use It	12-1
Project Planning	12-2
AutoCAD vs Manual and AutoCAD with Manual	12-3
Production Management	12-4
Project Work Book	12-5
System Standards Manual	12-6
Managing Drawing Setup	12-7
Managing Symbol Libraries	12-8
Managing Menu Systems	12-12
Managing Design Development	12-16
Managing Project Production	12-20
Managing Communications	12-25
Managing the System	12-28
Summary	12-31

APPENDIX A   *Techniques and AutoLISP Routines*

Techniques	A-1
AutoLISP Routines	A-3

APPENDIX B   *AE-MENU System*

Introduction	B-1
AE DISK Installation	B-1
ACAD.LSP Installation	B-4
AE-MENU Installation	B-7
Bonus AutoLISP Routines	B-7

APPENDIX C   *Authors' Comments*

	C-1

APPENDIX D   *Our System's Hardware*

Hardware	D-2

Index

	Index-1

AE-MENU Tablet Menu

Introduction

Just a decade ago, only the largest professional design firms could afford the handful of Computer Aided Design and Drafting (CADD) systems that were available. These early and highly complicated *mainframe* systems, whose roots were planted in the aerospace and automobile manufacturing industries, were priced in the six figures per workstation. It was not hard to see why the majority of architectural and engineering firms chose to stick with their more familiar production methods, and maintain the labor-intensive approach to designing and producing technical drawings that had changed little since the days of Sir Christopher Wren.

When mainframe CADD systems gave way to systems running on more affordable and in many ways more *friendly* mini-computers, more professionals took the plunge and converted their production methods to CADD. While some of the available software applications were modified to better suit the needs of architects and related professionals, the systems were still generally geared for use in other industries.

When CADD systems designed to run on personal computers were introduced, many of the last *holdouts* on the decision to automate were finally persuaded — the cost in terms of not keeping up with their CADD-using competitors far exceeded the actual cost of converting.

Trading One Set of Management Problems for Another

While today's Architecture/Engineering (A/E) professionals are educated on PC applications, hardware comparisons, and computer lingo, many still face the lessons learned the hard way by those who chose to automate earlier in the evolution of CADD systems:

❑ Purchasing a CADD system will not solve all your problems. Drawings are not created *at the touch of a button*, as many vendors would like you to believe.

❑ Just as you had to manage the old system of doing things, you have to manage the new. A new sense of discipline is required to keep a CADD system functioning to your benefit. Unfortunately, the mismanagement of CADD systems is often one of the biggest problems that design firms face.

❑ You'll probably never find a piece of software that exactly fits your requirements. No matter how friendly the developers make it, it still won't come out of the box doing things exactly the way you'd like them done.

How AutoCAD FOR ARCHITECTS AND ENGINEERS Can Help

Most of the current publications about AutoCAD take a *how to use* approach to describe the various menus, features, and commands of the software. While many of these books have proven useful in helping their readers get established with the fundamentals of AutoCAD, we felt that it was important to give architects and engineers a way to focus their knowledge and experience more directly on *how to apply* AutoCAD to their practices.

Within the chapters of this book is a comprehensive, step-by-step tutorial for taking a simulated project from the design concept to the actual working drawings. Our goals in mapping the development of this project are not only to broaden your understanding of AutoCAD, but to give you the tools to apply that understanding to the design and production facets of your practice or almost any other project. Our production techniques include suggestions on how to set up layer naming conventions, how to manage the phases of production under AutoCAD, and how to organize and document the flow of work.

Your Basic AutoCAD

AutoCAD is the most popular PC-based CADD system on the market today. It is economical — workstations cost little more than you would have paid for one or two electric typewriters ten years ago — and it successfully combines considerable power with ease of use. What benefits AutoCAD users the most is that it gives us the ability to easily incorporate changes and modifications into our drawings. This, in turn, enhances the design process by giving us more opportunities to explore alternative design solutions.

Getting the Best Out of AutoCAD

There's more to getting the best out of AutoCAD than just using the system, though. To avoid some of the pitfalls of mis-management, you must carefully plan your approach in implementing and using the system. Understanding AutoCAD and its capabilities, organizing the system around the requirements of the project, and managing the processes with a controlled and consistent approach are essential for successfully implementing AutoCAD into professional A/E practices.

We show you how to use the native commands of AutoCAD to accomplish specific tasks and how to take advantage of some of the advanced features of AutoCAD to perform those same tasks more efficiently. We also show you how to begin customizing AutoCAD, using a working menu system as an example, to perform more closely to your own needs. Included in the discussions and exercises are the features and capabilities of AutoCAD Release 10, AutoLISP, AutoShade, and AutoFlix.

The Book's Project

The simulated project we chose for this book is a branch bank with drive-through facilities. We selected it because we felt its relative simplicity would be a good case study to incorporate AutoCAD with *real life* design and production requirements. In addition, our particular design solution includes a diversity of building products, systems, and features. Representing these features in drawing form will challenge both you — the user — and the native and expanded features of AutoCAD. We present techniques for optimizing the use of AutoCAD in all phases of design and drawing production, including the following processes and drawings:

Project Planning
Three-dimensional Drawing Presentations
Working Drawing Setup
Plans
Elevations
Sections
Details
Structural roof framing
Mechanical plan
Plumbing plan
Electrical lighting plan

We also cover project completion, how to plot drawings, how to extract the data they contain for specification and report writing and for use with other software packages, and how to manage AutoCAD resources.

The techniques and sequences illustrated in the book are intended to address the majority of the production issues that you will encounter — for this simulated project, and almost any other project type.

➡ *NOTE! While we do not intend to represent the project drawings as "complete", we do intend to use them to illustrate specific techniques and planning processes which will help you simplify your own production challenges.*

The AE-MENU System

We have customized a subset of the KETIV Technologies menu system, called ARCH-T, for use with this book.[1] We integrated custom symbols libraries, prototype drawings, and AutoLISP routines to streamline tasks, make AutoCAD easier to apply, and save you time in the process. While you do not have to use our menu system to benefit from the book, we hope to accomplish three goals in presenting it to you:

1. To show you how to organize and customize the features of AutoCAD to take the greatest advantage of its capabilities in the A/E environment.

2. To give you some *ready-made* tools to experiment with, hone your production skills on, and implement with your own projects.

3. To provide you with a good example on which to base the customization of your own menu system.

A companion diskette to the book is available to provide you direct access to the programs and files of this AE-MENU System. The menu system is shown within appropriate chapters of the book.

Book Organization

AutoCAD FOR ARCHITECTS AND ENGINEERS is organized into four sections. Each section revolves around specific phases or tasks necessary to develop the drawings for the simulated project. The individual sections represent portions and combinations of the A/E's traditional phases of services as they pertain to project design and production processes. Icons are indicated within the margins to provide you with a quick, graphic reference to specific production techniques and related management production issues.

Section I: Project Planning and System Setup

Chapters 1 — 3

In the first section, we illustrate how to plan projects around the use of AutoCAD and how to set up and use custom prototype drawings and

---

1   The complete KETIV Technologies Menu is available through KETIV Technologies, Inc., Portland, OR. See the information on KETIV's menu system in the back of this book.

Symbols Libraries. We also show you how we organized our menu system and we present techniques for customizing the existing AutoCAD menu system for A/E use. This section is devoted to the planning steps and production setup for the actual branch bank demonstration project.

Section II: Developing the Design

Chapters 4 and 5

In this section, we implement the AE-MENU System and present techniques for using AutoCAD, AutoLISP, AutoShade, and AutoFlix as tools for design development and 3D drawing presentations. This section begins the actual A/E service phases that would be performed in an office environment for a project such as a branch bank.

Section III: Project Production

Chapters 6 — 10

Building upon the design that we created in Section II, we use the AE-MENU System and present techniques to enhance producing and plotting the working drawings for our project. This section's chapters break the working drawings into the various drawing descriptions such as floor plans, elevations, building sections and engineering drawings.

Section IV: Advanced Topics

Chapters 11 and 12

In this last section, we illustrate how to extract the project data and use it for writing specifications and reports. We also show how can you use AutoCAD data with some third-party networking and integrated applications packages currently available.

How to Use the Book

We designed this book to serve not only as a tutorial, but also as a reference for all aspects of integrating and using AutoCAD in the architectural and engineering environment.

If you are an AutoCAD instructor or student, use the book as a training manual — you can read about and complete the exercises to develop your AutoCAD skills. We have provided you with the data you'll need to complete each drawing, but you will need the companion diskette set for the complete AE-MENU System and certain drawings.

If you are an experienced AutoCAD user, you do not need to complete each drawing exercise — just pick and choose among the techniques or specific areas of production that interest you to expand your current production skills. You can also use the AE-MENU System, detailed in the book, for your own production work. You may add to and modify it as you refine your methods, or use it to model your own menu system.

We suggest that you follow the sequence of the chapter exercises since they build upon each other. Here is a Quick Reference Guide to what each chapter contains, should you want to skip right to a specific topic.

QUICK REFERENCE GUIDE

If You Want Information On	Turn To This Chapter
Naming Conventions, Prototype and Border Drawings	Chapter 1
Symbols, Attributes, AutoLISP Insertion Routines	Chapters 2, 6
Menu Areas, Macro Syntax, AutoLISP Menu Routines	Chapter 3
Design Drawings, Techniques and Routines	Chapter 4
Adding 3D Data and Surfaces, Defining 3D Curves	Chapter 5
Using AutoShade and AutoFlix	Chapter 5
Developing Working Drawings	Chapter 6
Site, Floor and Ceiling Plans	Chapter 6
Exterior Elevations	Chapter 7
Building Sections and Details	Chapter 8
Developing Engineering Drawings	Chapter 9
Structural, Mechanical, Plumbing, Electrical	Chapter 9
Plotting and Revisions	Chapter 10
Extracting and Exporting Project Data	Chapter 11
Automating Specification Document Creation	Chapter 11
Managing AutoCAD Systems and Resources	Chapter 12

The Book's Icons

We use the following Icons in the left-hand margin to give you a quick, visual way to find the different categories of information defined in the text.

SYSTEM STANDARDS — Indicates items that should be included in your System Standards of Operation.

 MANAGEMENT ISSUES — Indicates management problems and solutions that relate to AutoCAD's operation.

AE-MENU ROUTINES — Indicates menu items and AutoLISP routines that are included in the AE-MENU System.

 TECHNIQUES — Indicates a special way to use AutoCAD commands to perform a defined production task.

If you thumb through the book now, you will notice how easy it is to find the different categories these icons represent. We describe how we use several of these icons in more detail later.

Project Sequences and Techniques

At the beginning of each of the chapters that contain project drawings and drawing exercises, we include a section listing a *Project Sequence*. These outlines of project steps correspond to the sequences and titles of the actual drawing tasks contained in the exercises. These sequences duplicate key tasks in the design development, presentation, and working drawing phases of an actual project produced with AutoCAD. We provide you with notations, illustrations, and tutorial information for each sequence. We also provide the information needed to complete the entire drawing for each phase of the project.

Techniques occur within project sequence steps that illustrate special uses of AutoCAD commands, AutoLISP routines, or special features of the AE-MENU System. We describe both how to get started with a technique as well as the format used for the actual techniques.

Setting Up the Drawings for the Chapters

To help you practice each chapter's concepts and techniques, we have set up two approaches. First, you can follow each project sequence and complete the drawings as if they were part of a real project. Second, you can read along with the project sequence and use a defined scratch drawing to experiment with the techniques. Either way, you will benefit by discovering new ways to increase your productivity and by learning how to apply AutoCAD more effectively in your own work environment. Let's take a closer look at these methods.

Method One

If you plan to complete the drawings from chapter to chapter, you will need the menu system and drawings from the companion disk, the AE DISK. You will begin a new drawing, or use one of the drawings we have provided on the diskette. You will then be able to select a command from the AE-MENU and move to a scratch area within the drawing editor (defined with our custom setup routine) to experiment with the techniques before you apply them to the drawing.

Method Two

If you plan to just practice the techniques in the chapters, but not complete the drawings, you will be able to experiment using an AE-SCRATCH drawing, which we define with setup information in a table specific to each chapter. You can use AutoCAD's setup routine to get this drawing started.

Format for Techniques

We illustrate each project sequence's steps by calling them to your attention with a heading. Then, tell you how we are going to accomplish the particular drawing task and provide you with a drawing to refer to. The instructions that follow are annotated with a numbered bubble that corresponds to numbered areas on the drawing.

Techniques are steps that we think merit greater detail or show a special way of doing things, we separate them out from the project sequence flow, calling them to your attention with the technique icon. We give each technique a lengthier explanation, a technique — specific illustration, and show you the actual AutoCAD commands necessary to complete the task. Data that you need to input is shown in bold type.

The AE-SYSTEM Diskette Set

The AE DISK set contains the AutoLISP routines and menu files that we use to produce the book's project, as well as all the completed project working drawings. We provide instructions for using files from the disk set via the disk icons. In general, any time we refer to a file from the disk, you will see instructions like these:

Disk Icons

    Enter selection: Do "this" if you have the AE DISK set.

    Enter selection: Do "this" if you don't have the AE DISK set.

Installation instructions for the AE DISK are given in Chapters One and Two, and Appendix B.

AutoLISP Routines

Listings of the AutoLISP routines that are part of the AE-MENU System are listed in each chapter. (Some longer routines are listed in Appendix B.) Each routine performs a function that speeds up, simplifies, or enhances the production of drawings. These routines are contained on the AE DISK. We would rather you spent your time *using* these routines than typing them in, so we suggest that you purchase the disk set to save yourself a lot of time and effort. You will find an order form at the back of the book.

Menu and Drawing Files

The AE DISK contains the custom menu .MNU files, the symbol library drawing .DWG files, and other support files that you can use. We show you how we developed much of this information in Section I of the book; the AE DISK contains the complete AE-MENU System for your use.

Items to Check When Doing the Exercises

The routines and techniques that we developed for this book were tested by us and by our reviewers. There are a few things we suggest you be careful of when doing the exercises.

Because the printing font does not distinguish clearly between the number zero and the letter "O", nor between the number one and the lower case letter "l", you should watch these closely:

```
0   this is a zero
O   this is an upper case letter O
1   this is the number one
l   this is a lower case letter L
```

The steps within the project sequences and techniques show the most important, but not all, of the command prompts and tasks necessary to complete the drawings.

Prerequisites

To get the most from this book, you should be familiar with common AutoCAD commands and be comfortable with their use. If you have worked through INSIDE AUTOCAD (New Riders Publishing), or have gained a similar skill level elsewhere, you will have the experience to do our exercises.

If you encounter unfamiliar commands, refer to the above book, The AUTOCAD REFERENCE GUIDE (New Riders Publishing), your AutoCAD Reference Manual, or use the HELP command.

We assume that you have the following hardware and software available:

■ An 80286 or 80386 workstation, or an equivalent type workstation (Sun, Dec, Macintosh), with a math co-processor and an EGA or VGA (or equivalent) monitor.

■ MS or PC-DOS release 3.0 or later.

■ AutoCAD Release 10, with AutoShade and AutoFlix as options. You'll miss out on some important techniques if you are at Release 9 or farther back.

The AutoShade and AutoFlix exercises of Chapter Five were done on an IBM PS/2 Model 70 and an IBM 8514 monitor and graphics adapter.

The Framework III software used in Chapter 11, part of the Advanced Topics section, is an option. If you don't have Framework, you can still read along and get the basic ideas you need to automate your data reporting with other database and spreadsheet software.

Start with Planning

Now that we've told you why we created this book and have taken care of the housekeeping by explaining how the material will be presented, we hope you're ready to join us on our quest to increase productivity using AutoCAD in the architectural and engineering design and production processes. Let's begin where all worthwhile endeavors begin — with Planning.

PROJECT NUMBER LOG

JOB NUMBER	NAME OF PROJECT	DATE	PROJECT MANAGER

DRAWING NUMBER LOG

PROJECT NUMBER: PROJECT NAME:

DRAWING NUMBER	DRAWING TITLE	DATE	DRAWN BY	REVIS. DATE	CHECK BY	REMARKS

DRAWING SPECIFICATION LOG

PROJECT NUMBER: FILE NAME:
PROJECT NAME: DIRECTORY NAME:
DRAWING NUMBER: AutoCAD REL:
DRAWING TITLE: DATE:

TEXT STYLE: SCALE:
TEXT STYLE: LTSCALE:
TEXT STYLE: DIMSCALE:
DRAWN BY: SHEET SIZE:

LAYER AND PLOT DATA

No.	LAYER NAME	COLOR	LTYPE	PEN No.	PEN WID.	REMARKS
1						
2						
3						
4						
5						
6						
7						
8						
9						
10						
11						
12						
13						
14						
15						

PROJECT SPECIFICATION LOG

PROJECT NUMBER: DRAWING NUMBER:
PROJECT NAME: DRAWING TITLE: SHEET 1 of

CSI NUMBER	DESCRIPTION	LAYER NAME	Y/N	REMARKS
02275	SITE – GEOTEXTILES/GEOGRID			
02276	EARTH RETAINAGE			
02280	SOIL TREATMENT			
02350	PLIES & CAISSONS			
02488	DOCKS & BOAT FACILITIES			
02510	WALK, ROAD & PARKING			
02535	ATHLETIC SURFACING			
02710	SUBSURFACE DRAINAGE MATL			
02712	SUBSURFACE DRAINAGE PIPE			
02722	DRAINS & INLETS			
02725	PRECAST TRENCH DRAIN SYS.			
02770	PONDS & RESERVOIRS			
02820	FOUNTAINS & POOLS			
02830	FENCES & GATES			
02835	GATE OPERATORS			
02842	BICYCLE RACKS/LOCKERS			
02860	RECREATIONAL FACILITY			
02870	SITE, STREET & MALL FUR.			
02875	SITE & STREET SHELTERS			
02880	OUTDOOR SCULPTURE			
02890	FOOTBRIDGE			
02980	LANDSCAPE ACCESSORIES			

Log Forms

CHAPTER 1

Planning and Production

It's important to take time to plan your approach to integrating and using AutoCAD on a project by project basis. As you develop each project, you will not only increase your skill as a system user, but you will generate more symbols, define more productive procedures or refine some standards that will save you time and effort on future projects. As you establish a system for evaluating the applicability of your current AutoCAD procedures to each project, and integrate new information in the process, you will gradually build a smooth-running and adaptive AutoCAD system. You'll be able to develop each new project more easily than the project that preceded it.

Planning, or the Lack Thereof

Although the construction document phase is the single most labor and resource intensive part of an A/E firm's services, planning for time and resources during this phase is frequently overlooked or discounted by CAD users. Many CAD users in A/E firms mistakenly believe, at least early in their CAD experience, that an increase in production speed will make planning time *unnecessary*. As A/E firms develop an expanded utilization of CAD, production planning becomes a more important issue in meeting deadlines, particularly deadlines set by the client.

"There will be plenty of time later to correct problems and still meet schedule." Right? Wrong!

Next to the lack of good training, miscalculating the importance of planning is the greatest cause of systems application failure. SPEED NEVER COMPENSATES FOR POOR PLANNING. In fact, this maxim holds true for all aspects of computer usage in the A/E firm. Planning, in system integration, if not production procedures, is an on-going learning process. Planning should never be considered *complete*; and the planning process should be continually revised and refined as new techniques are integrated for system usage. Unfortunately, the lack of planning merely leads to making mistakes more quickly, and then requires additional time for making corrections.

Production Concepts

In the early days, projects were often selected for AutoCAD production on the basis of how many entities were repeatable. Hotels, motels, and apartment projects were good project candidates because the basic components were repeated throughout the project and required little individualization. Although this criterion has not changed completely, today almost any project (or at least a major portion of it) is a good candidate for production under AutoCAD. The determining factors are now more likely to be:

❑ How advanced the system's data base is

❑ How proficient the users are with various aspects of the system

❑ If the client has special requirements which require or preclude the use of CAD

❑ If your consultants are using CAD

An AutoCAD system that contains an extensive symbol library, that is well-integrated, well-organized and well-managed, can be used to produce a project in less time, with more accuracy, than using conventional hand-drafting methods. An additional inducement to producing projects under AutoCAD is the growing emphasis on facilities management. The spatial needs of many clients change frequently. Keeping this possibility in mind, use AutoCAD if the client's future needs require you to access old data for renovation projects. These renovation tasks are much more easily performed on an AutoCAD system.

Secondly, almost any project is a good candidate for CAD where:

❑ The Symbol library has become extensive enough to literally pull entities from the shelf to complete the necessary drawings.

❑ Continual updating is necessary as in the cases of corporate office facilities, manufacturing centers, or hospital and medical facilities.

❑ The possibility exists where a client's building program requires new additions, and a portion of the original drawings are reusable.

❑ Prototype buildings such as franchise restaurants, supermarkets, and auto repair shops have repeatable floor plans, elevations, and details.

Production Decisions

Now that its been established that almost any project is a good candidate for application under AutoCAD, your first real concern is deciding whether or not the entire project should be produced with the system. This decision is usually predetermined by the following questions:

❑ How definitive is the client about design decisions? Will his or her program develop as an on going procedure during the design and production process?

❑ What are the client's requirements for this project? What will his or her requirements be for future projects?

❑ How extensive is your current symbol library? Can it support the project without a lot of modifications or additions?

❑ Do you have enough workstations and trained personnel to produce the whole project under AutoCAD?

❑ Do you foresee future renovations or additions to the project that the client might require? Would using AutoCAD now help expedite these future changes? If so, what portions of the work will require alteration?

❑ What goals have you set for expanding your professional target markets? If you expand your integration of AutoCAD now, will it help you accomplish these goals?

❑ Which consultants will you require for the production of this project, and are they using AutoCAD? If they are, how proficient are they as AutoCAD users?

Chances are you're not going to decide to do the whole project on AutoCAD if your data base or CAD personnel resources are limited. If this is the case, answers to these questions will help you decide which portions of the project to produce on the system, and which to produce manually. In many instances, this *combined method* of producing some of the drawings with AutoCAD and producing some of them with manual methods may be the best choice. Remember though, the sooner you use the system in an effort to expand your data base, the larger the proportion of projects you can produce under AutoCAD; and the more proficient you'll become.

Combining AutoCAD and Manual Methods

If you combine AutoCAD with manual methods, a good rule of division is this: use AutoCAD to produce drawings such as site plans, floor plans, and reflected ceiling plans because these drawings are generally used as backgrounds for consultants. Your consultants won't have to repeat the production efforts of creating these same drawings. As we all know, we do have to make revisions and modifications to drawings during the production process. Consequently, using AutoCAD to produce this portion of the work also insures that the same modifications to the floor plan, for example, will apply to the electrical floor plan.

If you choose to integrate AutoCAD and manual methods for producing a project, we think it is best to produce the details by hand-drafting methods. Most details, particularly large scale details, are project specific. Granted, it is easier to retrieve and revise old details with AutoCAD once you've created them on the system. However, details determine control dimensions, such as column grid references to the exterior walls, and most of the time you cannot establish these control dimensions until you complete the actual detail. This can cause you to have to manipulate the affected dimensions of the floor plan, and it is much easier and faster to do so with AutoCAD. In addition, you don't generally produce details until later in the construction document phase.

To further support your decision to produce plans on the system, there are always more repeatable entities or blocks on these drawings for doors, millwork, windows, columns, toilet fixtures and furnishings. The occurrence of repeatable items is an excellent basis for determining what you should put on the AutoCAD system.

Using this approach, drawing data that you create under AutoCAD in the design development phase, for example, can be reused to create the working drawings. This approach will also help the engineers and other consultants using AutoCAD to produce their drawings.

Production Variables

Whether you use AutoCAD or a combination of AutoCAD and manual methods, different phases of project development require different amounts of time. For planning purposes, this system of dividing allowable time determines how much of the total time you can allot for each phase of the project. Time constraints in any of these phases raise issues about determining what portions of a project to produce under AutoCAD or manual methods:

❑ What will be the overall time requirements of your staff and your consultants?

❑ What is your present work load? How will your current work load impact our anticipated work load?

❑ Will the engineers or the other consultants you need be available during the production time frame? With which method of production are they most experienced?

❑ How complex is the project? Which portions of the project can be produced the most quickly under which method?

You have to answer these questions before you can effectively plan a project's production under the combined method. In addition to these questions, to determine what portions of the work you will produce manually and which portions you will produce with AutoCAD, you need to analyze the following questions for application to your needs:

❑ What is the staff's expertise in using the system for this type of project, and specifically, the portions of the project that you are thinking of producing under AutoCAD?

❑ How versed are the staff members having experience in predominantly manual production methods with the procedures and requirements of the staff members who have AutoCAD experience, and visa versa?

❑ What future projects would benefit from expanding the symbol library for the specific portions of the project you are thinking about producing under AutoCAD?

Many A/E firms today successfully utilize the combined method for production. Our purpose in posing these questions is to guide you in the planning process to help you make an informed decision that suits your office's and project's needs. We advocate, ultimately, the full and complete utilization of AutoCAD for all drawing needs for all projects. However, until your system's library and staff are ready for this plunge, a combined method of production can yield satisfactory results.

Production Issues

How you answer the questions listed above has great impact on how you produce projects under AutoCAD. To plan the most effective and efficient means of production, you must evaluate the training of your staff and the capabilities of the system. The only thing you have to base these evaluations on are your past production experiences. How accurate you are in predicting the production time for each phase of a new project also depends on your experience with those same phases from projects in the past.

Another production issue you need to resolve is whether to use third party software along with AutoCAD. Third party custom tablet menus, AutoLISP routines and symbol libraries come in handy during some phases of development. In general, a well-organized menu system, like the one you use in this book, can boost production time savings by 10 to 15 percent. Organizing and managing other aspects of the system, like symbols libraries and systems operations, which we will discuss in subsequent chapters, can add an additional 10 to 15 percent gain in productivity.

Production Planning

While all drawings should be considered important and necessary portions of the construction documents, some drawings contain the primary scope of the work and others contain the supportive or secondary information that is required for construction of the project. For planning the production of the branch bank project, we will define the following drawings as primary:

❑ Plans — Site, Floor, and Reflected Ceiling

❑ Exterior Elevations

❑ Building Sections

You will notice that our logic behind this separation is based on using AutoCAD to prepare the design development drawings, converting them to the construction documents and transferring this information to the engineers and consultants. Still a key issue in using AutoCAD is to determine what data you can reuse. This includes not only the data in the architect's working drawings, but the parts of the drawings that an engineer using AutoCAD will need. From an engineering standpoint, the primary drawings are the:

❑ Structural Plan

❑ Mechanical Plan

❑ Plumbing Plan

❑ Lighting/Power Plan

Project Naming Convention

We use a year-project number naming convention. For the bank project, we assigned the project number of "8910." We based this on the contract year, "19<u>89</u>," and the project being the "<u>10</u>"th project of the calendar year. To name our first two design drawings, we will add to this job number the suffixes "SP-D" for site plan design, and "FP-D" for floor plan design. The complete names will be "8910SP-D" and "8910FP-D."

Using this naming convention, let's suppose, that you have a project that has more than one floor and is the 103rd project in your current calendar year. Since DOS limits the file naming to eight characters, you must adjust your file naming convention accordingly. Here are some examples of extending our drawing name conventions:

- Design Development Phase — Site and Floor Plan Examples

Site Plan drawing	89103SPD
First Floor Plan drawing	89103F1D
Second Floor Plan drawing	89103F2D and etc.
Tenth Floor Plan drawing	9103F10D (drop the 8 in 89)

- Working Drawing Phase — Site and Floor Plan Examples

Site Plan drawing	89103SP
First Floor Plan drawing	89103F1
Second Floor Plan drawing	89103F2 and etc.
Tenth Floor Plan drawing	89103F10

- Other Working Drawing Examples

Ceiling Plan-First Floor	89103R01
Ceiling Plan-Second Floor	89103R02
Ceiling Plan-Tenth Floor	89103R10
Building Sections	89103BS1
Construction Details	89103D01

Organizing with a Project Work Book

We use a project work book and advocate that you use one to help make project production run more smoothly. Our work book helps keep track of the project before, during, and after production. It's contents include completed drawings plus the following production records and charts:

❑ Production Flow Chart

❑ Project Drawing Layout

❑ Project Number Log

❑ Drawing Number Log

❑ Drawing Specification Log

❑ Project Specification Log

If you assign a production manager to your projects, we suggest that you include in your production manager's responsibilities the maintenance of the work book throughout the project. At the conclusion of the project, the production manager can place all of the AutoCAD drawing disk files in the back of this book and file it away in a safe place. It's also a good idea to keep an additional copy of the disk files at another location in case one is lost or destroyed.

Let's take a look at some of the practical benefits of maintaining each of these components of the project work book. They also provide some important management benefits, which we'll discuss more thoroughly when we expand the topic of systems management in Chapter Twelve.

Production Flow Chart

The first chart you can use in planning a project's production is a flow chart of the drawings that you anticipate creating. Flow charts are aptly named because they help everyone involved in a project see how the work in the office will *flow* within the project's relative time frame. In a production flow chart, you lay out which areas of the drawings you will produce first, the relationships between drawings, and the order in which you are going to create them. The flow chart below shows these relationships for the Branch Bank project. To create a production flow chart for any project, the different A/E professionals involved have to determine at what stages they will need to exchange information so that they'll be able to meet their respective requirements and the production schedule.

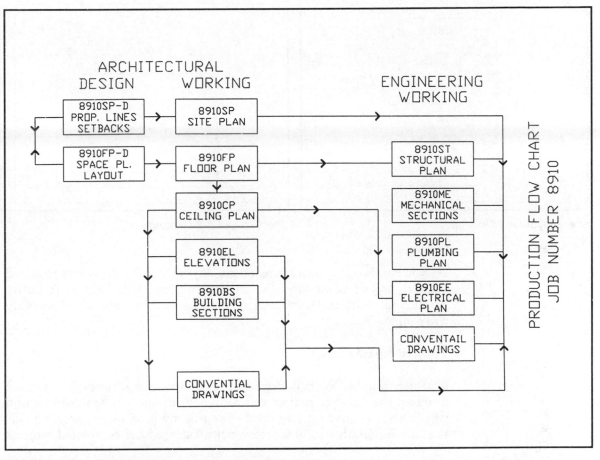

Production Flow Chart Diagram

Project Drawing Layout

Another work book document that we think is very useful is a project drawing layout, detailing the number and types of drawings the project will require. One way to do this is to create mock-up sheets on 8 1/2"x 11" grid paper. You can hand sketch the required areas of each detail of the plan for each sheet, using the grid to approximate the scale of the details.

Creating drawing layouts like this helps you to define which drawings will require additional details for construction information, what the scaling factors for each drawing are, what legends and schedules you will need, and what specification section will be required for the project.

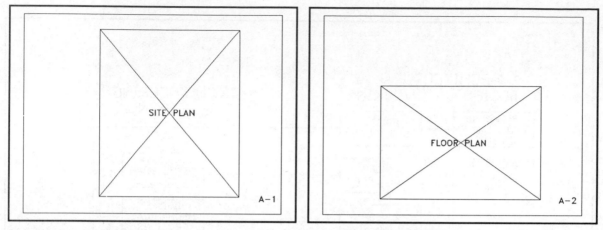

Sheet Layout A-1 Drawing   *Sheet Layout A-2 Drawing*

As you can see in illustrations above, laying out the drawings required for the project in advance is not difficult and will help your entire production team in their quest to produce an accurate set of working drawings.

Project Number Log

A project number log is designed to be an accurate record of the project number, the date the project was assigned, who it was assigned to, and who it was assigned by. Log sheets like this are good record-keeping aids that are especially useful for reference after you have completed projects

PROJECT NUMBER LOG			
JOB NUMBER	NAME OF PROJECT	DATE	PROJECT MANAGER

Project Number Log Form

Drawing Number Log

We recommend that you also keep a drawing number log to keep track of the drawing numbers, titles, dates, sheet sizes, who they were drawn by, and who checked them. You never know when you might need this information, and you can find most of what you need to know about any specific drawing in one document. Data specific to each drawing, such as drawing file name and date, should also be placed on the border of each drawing.

DRAWING NUMBER LOG

PROJECT NUMBER: PROJECT NAME:

DRAWING NUMBER	DRAWING TITLE	DATE	DRAWN BY	REVIS. DATE	CHECK BY	REMARKS

Drawing Number Log Form

Drawing Specification Log

A drawing specification log is designed to contain data about the computer and drawing specifications. Multiple log sheets may be required according to the project.

```
┌─────────────────────────────────────────────────┐
│         DRAWING  SPECIFICATION  LOG             │
├─────────────────────────────────────────────────┤
│ PROJECT NUMBER:        FILE NAME:               │
│ PROJECT NAME:          DIRECTORY NAME:          │
│ DRAWING NUMBER:        AutoCAD REL:             │
│ DRAWING TITLE:         DATE:                    │
├─────────────────────────────────────────────────┤
│ TEXT STYLE:            SCALE:                   │
│ TEXT STYLE:            LTSCALE:                 │
│ TEXT STYLE:            DIMSCALE:                │
│ DRAWN BY:              SHEET SIZE:              │
└─────────────────────────────────────────────────┘
```

No.	LAYER NAME	COLOR	LTYPE	PEN No.	PEN WID.	REMARKS
1						
2						
3						
4						
5						
6						
7						
8						
9						
10						
11						
12						
13						
14						
15						

Drawing Specification Log Form

Project Specification Log

The project specification log contains the CSI (Construction Specifications Institute) format number so that as you develop the project and complete the working drawings, you can place a check mark by the appropriate specification section number. This will give the specification writer a guide to follow. Secondly, if you will maintain copies of the project specification log form at each computer work station, it will serve as a reference for choosing the appropriate CSI division as a part of the naming convention when assigning names to layers and blocks. We will cover this convention later in the chapter.

PROJECT SPECIFICATION LOG					
PROJECT NUMBER:	DRAWING NUMBER:		SHEET 1 of		
PROJECT NAME:	DRAWING TITLE:				
CSI NUMBER	DESCRIPTION	LAYER NAME	Y/N	REMARKS	
02275	SITE — GEOTEXTILES/GEOGRID				
02276	EARTH RETAINAGE				
02280	SOIL TREATMENT				
02350	PLIES & CAISSONS				
02488	DOCKS & BOAT FACILITIES				
02510	WALK, ROAD & PARKING				
02535	ATHLETIC SURFACING				
02710	SUBSURFACE DRAINAGE MATL				
02712	SUBSURFACE DRAINAGE PIPE				
02722	DRAINS & INLETS				
02725	PRECAST TRENCH DRAIN SYS.				
02770	PONDS & RESERVOIRS				
02820	FOUNTAINS & POOLS				
02830	FENCES & GATES				
02835	GATE OPERATORS				
02842	BICYCLE RACKS/LOCKERS				
02860	RECREATIONAL FACILITY				
02870	SITE, STREET & MALL FUR.				
02875	SITE & STREET SHELTERS				
02880	OUTDOOR SCULPTURE				
02890	FOOTBRIDGE				
02980	LANDSCAPE ACCESSORIES				

Project Specification Log Form

You will find project work books even more useful if you include printerplots showing the details that you used. If you also file the details from different projects together, you can develop a very extensive symbol library for everyone to use. You are much more likely to use existing details if you know where to find them and you can find them quickly.

Hardware and Project Planning

Due to the rapid changes occurring in the computer industry, the best workstation to fit your needs today may not necessarily be the best for tomorrow. Being aware of changes in technology, making assessments of how well the current system performs and determining whether and when the benefits of obtaining new hardware outweigh the cost are duties that usually fall on the AutoCAD Manager's shoulders.

Having the right hardware plays a very important role in production planning. If your system can't handle the requirements of the project (for example, CPU speed, storage capacity, etc.), then you should consider up-dating your current system. It would be ideal to be able to do such a good job of planning and purchasing hardware that the capacity of the systems always exceeds the demands of the projects. But computers are like just about everything else you buy — they have limited life spans due to continual changes in both hardware technology and the software application demands placed on them.

The first rule of hardware selection is:
NEVER UNDER-PURCHASE.

Investing in good equipment is well worth the cost. Purchasers of most systems find it easier to stomach a higher price than to watch the productivity of their employees dwindle due to unacceptably long response and *regen* times. Recouping your investment over a longer period of time is a better alternative than finding that your system is undersized and will not perform as you expected. It is also vitally important for you to select a dealer who maintains a current stock of parts and who employs good service technicians to keep your system operating.

The second rule of hardware selection is:
BETTER TOO MUCH THAN TOO LITTLE.

If you always have a little more capacity or *horsepower* than you need, you won't get stuck in the middle of a project with a system that won't do the job.

In Chapter Eleven and Appendix D, we give you some more specific recommendations on the types of hardware devices available and tell you what we used to develop the book's project.

System Development and Standards

Developing system standards is a continuing process. Standards can be maintained, improved, and built upon throughout the life of the system. We suggest that you pay special attention to documenting the basic areas of system operation. The following subjects are particularly important and will contribute tremendously to everyone's productivity if you will define them and keep them well-documented in a book of system standards for every system user's reference.

❑ How to name directories and drawings

❑ How to create and use prototype drawings

❑ How to name layers

❑ How to assign colors and line weights to layers

❑ How to create and use border drawings

❑ How to adopt and use plotting standards

We will illustrate our recommendations for the beginnings of this book by giving you our system standards for the branch bank project.

Getting Started with the Simulation Project

Since the book is built around simulating the bank branch project, you need to set up your AutoCAD environment for the project. To help you set up this environment, you need to create a series of subdirectories. These directories are designed to act as a separate working AutoCAD environment, and to let you work with the companion disk menu without interfering with your AutoCAD set up and any ongoing work on your workstation.

Necessary Directories

The following table shows the directories and subdirectories that you need to create. The table names show the names we prefer for the system directories as they relate to DOS, AutoCAD and the book's project. The book's AutoCAD support files are placed in a directory called \AE-ACAD. The 8910 directory is used for project drawing files. The FW3 directory is an optional directory that is used in the advanced chapter on making reports.

We assume that your DOS files are in a directory called \DOS, and that your standard AutoCAD files are in a directory called \ACAD. We also assume that your hard disk is drive C:.

 IMPORTANT NOTE! To insure proper operation of AutoCAD from the AE-ACAD directory, you must set a path in your AUTOEXEC.BAT file to your ACAD directory. For example, `PATH=C:\ACAD`

Directories	Usage
\ (ROOT)	Used to place all Batch files
\DOS	Used to place all DOS files
\ACAD	Used to place all AutoCAD files
\AE-ACAD	Used for AutoCAD files for book
\8910	Used for Project Drawings
\FW3	Used for Framework files (Optional)

Directory Table

Use the DOS command MD (Make Directory) to create the table's directories and subdirectories.

Creating the Subdirectory Setup

```
C:\> MD \AE-ACAD            Make a directory named AE-ACAD.
C:\> MD \8910               Make a directory called 8910.
```
Continue to make the rest of the directories, if you need to.

When you are done you should have the directories and subdirectories listed in the table (above).

AutoCAD Program and Support Files

We assume that AutoCAD's support files are in your \ACAD subdirectory. Besides support files, AutoCAD also requires device driver files (*.DRV) during configuration. These drivers are used by AutoCAD's configuration overlay files (*.OVL). Configuration also creates a configuration file (ACAD.CFG). To make the book's menu and project environment work, you need to transfer your AutoCAD overlay and configuration files to the AE-ACAD directory as shown below. If your configuration files are not in your \ACAD directory, substitute your directory name (where the files are located) for the \ACAD.

Copying AutoCAD Files to the AE-ACAD Directory

```
C:\> CD\ACAD                          Change to the ACAD directory.
C:\ACAD> COPY ACAD?.OVL \AE-ACAD\*.*
ACADPL.OVL                            Plotter overlay file.
ACADPP.OVL                            Printer/plotter overlay file.
2 File(s) copied

C:\ACAD> COPY ACADD?.OVL \AE-ACAD\*.*
ACADDS.OVL                            Display (video) overlay file.
ACADDG.OVL                            Digitizer (or mouse) overlay file.
2 File(s) copied.

C:\ACAD> COPY ACAD.CFG \AE-ACAD\*.*
1 File(s) copied                      General AutoCAD configuration file.
```

After you have copied these files, test your AutoCAD setup by changing to the AE-ACAD directory, and booting AutoCAD.

Testing Your AutoCAD Setup

`C:\> CD \AE-ACAD`	Change to AE-ACAD.
`C:\AE-ACAD> ACAD`	AutoCAD should boot normally.

When you are done testing your configuration, remain in the \AE-ACAD directory to install the AE DISK files.

Installing the AE DISK

It's time to install the Architects and Engineers optional AE DISK. If you don't have the disk yet, see the order form in the back of the book. We recommend getting the disk. It will save you a lot of typing time and drawing setup time.

You need to copy the AE DISK files into the \AE-ACAD directory and the 8910 directory. To conserve diskette space the disks files are merged into a two files called AE-LOAD.EXE, and DR-LOAD.EXE which automatically copy all files into the current directory.

Installing the AE DISK

Put the AE DISK in your disk drive A:

`C:\AE-ACAD> A:AE-LOAD`	Copies disk files to the directory.
`C:\AE-ACAD> CD \8910`	Change to 8910 directory.
`C:\8910> A:DR-LOAD`	Copies drawing files to directory.
`C:\8910> CD \AE-ACAD`	Change back to AE-ACAD directory.

Now that you have copied the disk files, you have all the starting files that you need to complete this chapter, including the prototype drawings used in the branch bank.

Naming Convention for the Project

Defining standards for naming system directories and drawings, mentioned as components of the system standards manual, is an important task in system setup. If you take the time to set up these standards, and make sure that everyone follows them, you will insure more consistency between your projects and among your users. Secondly, you will find that data retrieval becomes an easier process.

A good method for naming directories is to create a separate directory for each project, naming it by the project number. For our simulated project, we have created a directory named 8910. Using this convention, you can name the drawings for a project by the project number, PLUS an abbreviation for the drawing type. For example, the Site Plan for our project will be named 8910SP.

If you are using MS-DOS, remember that neither DOS (Disk Operating System) nor AutoCAD accepts a number or name that is more than eight characters in length. With only eight characters to work with, choose your drawing names with care so that they are easily recognized.

Another method for naming drawings that we often see is to use standard drawing names across projects. This method works out fine while the project is active — if you still use a separate directory named to include the project number, you will be able to access specific drawings. The major drawback to this method, which causes us to caution against its use, is: after the project is completed and archived, you must depend upon the accuracy of labels (not to mention the labeler) to determine which project is on which diskette or tape. Once you separate like-named drawings from their respective directories, the chances of interchanging them increases. You should always avoid naming drawings as FLOOR1 for the Floor Plan or SITEPL for the Site Plan, for example. These names will create problems when you retrieve archived data. Always try to make it a practice to include the project number in the drawing name to avoid any confusion.

The following table lists the drawing names for the branch bank project, using our drawing naming convention, and the contents of each. You can also use this table as a guide for projects that contain more drawings than our simulated project by adding suffixes to the names (as described earlier in the chapter).

Drawing Name	Description	Sheet Number
8910SP	Site Plan	(A-1)
8910FP	Floor Plan	(A-2)
8910CP	Ceiling Plan	(A-3)
8910EL	Elevations	(A-4)
8910BS	Building Section	(A-5)
8910ST	Structural	(S-1)
8910ME	Mechanical	(M-1)
8910PL	Plumbing	(P-1)
8910EE	Electrical	(E-1)

Drawing Name Table

Now that we've established the directories and drawing names that you will use for the project, let's discuss how to create some prototype drawings to speed up producing the project's drawings.

Creating and Using Prototype Drawings

In the early days of AutoCAD, every time a new drawing was begun, the layers for that drawing had to be created. This took a lot of time and caused problems in maintaining a standard layering convention. The addition of prototype drawings in version 2.1 (or Release 6) of AutoCAD solved this problem to a degree since it allowed you to predefine different items to use in all drawings — kind of like the way you use templates in conventional drawings.

With prototype drawings, you could set your layers, units, grids, axes and limits, among other things, and reuse them every time you began a new drawing. In practice, however, this also created problems — not many drawings maintain the same requirements for all settings. So, you still had to make adjustments for each drawing, and not until AutoLISP was added did this problem become resolved.

When AutoCAD added AutoLISP (in version 2.18), it gave us a way to create flexible prototype drawings that can be configured with AutoLISP Setup routines. Now only the basic items, such as layers, need to be defined in the prototype drawing; the remaining scale-related items can be added very effectively with AutoLISP.

We will show you how to create prototype drawings (in just a few minutes); and we will show you a custom AutoLISP setup routine in Chapter Three to illustrate how these two AutoCAD features give you greater consistency and control over operational standards. In Chapter Twelve, we will discuss the management issues involved in creating separate prototype drawings for different drawing tasks and why we think it's advisable to maintain these different drawings instead of one large prototype drawing file.

We will show you how to use four different prototype drawings for the branch bank project, indicating their type with a suffix of PT:

❏ SITE-PT – for creating all Site Plan drawings

❏ PLAN-PT – for creating all Architectural Plans, Elevations and 3D Presentation drawings

❏ SECT-PT – for creating all Section and Detail drawings

❑ ENGR-PT – for creating all Structural, Mechanical, Electrical and Plumbing drawings

If you have installed the companion AE DISK, you already have these drawings in your AE-ACAD directory.

Naming Layers

Being able to preset layer names and associated options is now the main reason you use prototype drawings. In the layer-naming convention for the project, we combine the CSI numbering format with a three-character descriptive prefix. If you are not familiar with the CSI (Construction Specifications Institute) categorization system, obtain a copy of the number and category listings to refer to so that you will become familiar with the naming layer conventions. As an example of using this system, if you are drawing in a layer to indicate walks, you would choose the CSI number 02510 and combine it with an easily recognized prefix of WALK. Your layer for walks would be named WALK02510.

AutoCAD lets you use layer names up to 31 characters long, but the status line of the Drawing Editor will only display eight. So, be careful in adopting a lengthy naming system.

➡ *NOTE! If you use consistent-length layer names, you can use the wildcard symbol * to specify layer names that you want to turn on or off. This is the reason we place the CSI number last. It gives you the option to name the layers with the full numbering system of CSI, but it also maintain a simple abbreviation in front to help you identify the layers.*

Using Layer Colors

When you plot drawings with AutoCAD, the layer colors you specify when you set up a drawing determine which pens the plotter selects when plots are made. Using layer colors is how you can vary the line weights of entities within your drawings. Entities will be plotted with the pen weights that correspond to their own layer colors.

Take a look at the following table showing AutoCAD's standard layer number/color associations and some suggested line weights for each color. For your simulated project, the titles that you draw on layer TBL-010 (which we will define with a color of WHITE) will be assigned a heavier line weight than the entities that you draw on parking layer PKG-020 (which we will define with a color of YELLOW).

Color	Number	Line Weight	Pen Size
Red	1	x-fine	.18mm
Yellow	2	x-fine	.18mm
Green	3	fine	.25mm
Cyan	4	fine	.25mm
Blue	5	medium	.35mm
Magenta	6	medium	.35mm
White	7	bold	.50mm

Layer Colors and Line Weights Table

Technique for Creating Prototype Drawings

Now that we have discussed the advantages of using prototype drawings, standardizing layer names and colors, let's create some prototypes to use in developing the drawings for the branch bank project. You can follow the instructions, tables, and commands below to specify the layers that you will need for your prototype drawings, which are named SITE-PT, PLAN-PT, SECT-PT, and ENGR-PT. As we mentioned, if you have the AE DISK, you already have the prototypes. Examine the prototypes, and skip the sequences shown below for creating the drawings.

Creating Prototype Drawing Files

Verify that you have the prototype drawing files SITE-PT.DWG, PLAN-PT.DWG, SECT-PT.DWG, and ENGR-PT.DWG in your AE-ACAD directory. Examine them.

Follow the command sequence and tables below to create the four prototype drawings and their layers.

Creating Prototype Drawings

❑ Start by loading AutoCAD and selecting Option 1, Begin a New drawing.

❑ Name the drawing with a prototype drawing name, starting with SITE-PT.DWG.

When the drawing editor appears, use the commands shown after the following tables of layers to create each drawing. Start each prototype drawing with Layer 0.

The following is a list of the layers you'll need for the SITE-PT prototype drawing. Notice that we have included our suggested assignment of layer color, based on the types of entities represented in each layer.

➡ *NOTE! All layers followed by an * are unique to this prototype drawing. Don't type in the asterisks.*

➡ *TIP! If you are going to create several prototype drawings that have layers in common (such as the four drawings we show below), you may want to create a base prototype drawing containing the common members. To create each individual prototype drawing, copy the base and add the remaining layers. For example, to create the SITE-PT prototype drawing using a base drawing called BASE-PT, start by entering SITE-PT=BASE-PT when AutoCAD prompts you for the new drawing name. Then you only need to add the nonbase layers.*

This method is the method we shown below (after the tables).

Layer name	*	State	Color	Linetype
0		On	7 (white)	CONTINUOUS
TXT-010		On	4 (cyan)	CONTINUOUS
TLB-010		On	7 (white)	CONTINUOUS
TLE-010		On	6 (magenta)	CONTINUOUS
DIM-010		On	4 (cyan)	CONTINUOUS
SYM-010		On	4 (cyan)	CONTINUOUS
HAT-010		On	1 (red)	CONTINUOUS
EXT-010		On	1 (red)	CONTINUOUS
BLD-010		On	1 (red)	CONTINUOUS
MIS-010		On	1 (red)	CONTINUOUS
NCT-020	*	On	3 (green)	CONTINUOUS
ECT-020	*	On	2 (yellow)	DASHED
STE-020	*	On	2 (yellow)	CONTINUOUS
FEN-020	*	On	1 (red)	CONTINUOUS
PRL-020	*	On	5 (blue)	PHANTOM
PKG-020		On	2 (yellow)	CONTINUOUS
LAN-020	*	On	3 (green)	CONTINUOUS
EAS-020	*	On	1 (red)	DASHED
CON-030		On	7 (white)	CONTINUOUS
PLU-150		On	2 (yellow)	CONTINUOUS
ELE-160		On	3 (green)	CONTINUOUS

SITE-PT Layers Table

The following is a list of the layers you'll need for the PLAN-PT Prototype Drawing.

Layer name	State	Color	Linetype
0	On	7 (white)	CONTINUOUS
TXT-010	On	4 (cyan)	CONTINUOUS
TLB-010	On	7 (white)	CONTINUOUS
TLE-010	On	6 (magenta)	CONTINUOUS
DIM-010	On	4 (cyan)	CONTINUOUS
SYM-010	On	4 (cyan)	CONTINUOUS
HAT-010	On	1 (red)	CONTINUOUS
EXT-010	On	1 (red)	CONTINUOUS
BLD-010	On	1 (red)	CONTINUOUS
MIS-010	On	1 (red)	CONTINUOUS
GRA-020	On	3 (green)	CONTINUOUS
PKG-020	On	2 (yellow)	CONTINUOUS
CON-030	On	3 (green)	CONTINUOUS
MAS-040	On	2 (yellow)	CONTINUOUS
COL-050	On	5 (blue)	CONTINUOUS
STD-060	On	4 (cyan)	CONTINUOUS
INS-070	On	1 (red)	CONTINUOUS
ROF-070	On	5 (blue)	CONTINUOUS
CAN-070	On	5 (blue)	CONTINUOUS
DRS-080	On	3 (green)	CONTINUOUS
WIN-080	On	1 (red)	CONTINUOUS
GRD-090	On	2 (yellow)	CONTINUOUS
SUR-090	On	5 (blue)	CONTINUOUS
SPL-100	On	1 (red)	CONTINUOUS
FUR-120	On	1 (red)	CONTINUOUS
EQU-150	On	5 (blue)	CONTINUOUS
PLU-150	On	2 (yellow)	CONTINUOUS

PLAN-PT Layers Table

The following is a list of the layers we'll need for the SECT-PT Prototype Drawing.

Layer name	State	Color	Linetype
0	On	7 (white)	CONTINUOUS
TXT-010	On	4 (cyan)	CONTINUOUS
TLB-010	On	7 (white)	CONTINUOUS
TLE-010	On	6 (magenta)	CONTINUOUS
DIM-010	On	4 (cyan)	CONTINUOUS
SYM-010	On	4 (cyan)	CONTINUOUS
HAT-010	On	1 (red)	CONTINUOUS
EXT-010	On	1 (red)	CONTINUOUS
BLD-010	On	1 (red)	CONTINUOUS
MIS-010	On	1 (red)	CONTINUOUS
GRA-020	On	3 (green)	CONTINUOUS
PKG-020	On	2 (yellow)	CONTINUOUS
CON-030	On	3 (green)	CONTINUOUS
MAS-040	On	2 (yellow)	CONTINUOUS
STL-050	On	5 (white)	CONTINUOUS
COL-050	On	5 (blue)	CONTINUOUS
JST-050	On	5 (blue)	CONTINUOUS
BMS-050	On	5 (blue)	CONTINUOUS
FRA-050	On	2 (yellow)	CONTINUOUS
STD-060	On	4 (cyan)	CONTINUOUS
INS-070	On	1 (red)	CONTINUOUS
ROF-070	On	5 (blue)	CONTINUOUS
CAN-070	On	5 (blue)	CONTINUOUS
DRS-080	On	3 (green)	CONTINUOUS
WIN-080	On	1 (red)	CONTINUOUS
GYP-090	On	1 (red)	CONTINUOUS
GRD-090	On	2 (yellow)	CONTINUOUS
SUR-090	On	5 (blue)	CONTINUOUS
SPL-100	On	1 (red)	CONTINUOUS
FUR-120	On	1 (red)	CONTINUOUS

SECT-PT Layers Table

The following is a list of the layers we'll need for the ENGR-PT Prototype Drawing.

```
Layer name    *        State      Color       Linetype

0                      On         7 (white)    CONTINUOUS
TXT-010                On         4 (cyan)     CONTINUOUS
TLB-010                On         7 (white)    CONTINUOUS
TLE-010                On         6 (magenta)  CONTINUOUS
DIM-010                On         4 (cyan)     CONTINUOUS
SYM-010                On         4 (cyan)     CONTINUOUS
HAT-010                On         1 (red)      CONTINUOUS
BLD-010                On         1 (red)      CONTINUOUS
MIS-010                On         1 (red)      CONTINUOUS
STL-050                On         5 (blue)     CONTINUOUS
JST-050                On         5 (blue)     CONTINUOUS
BMS-050                On         5 (blue)     CONTINUOUS
FRA-050                On         2 (yellow)   CONTINUOUS
STD-060                On         4 (cyan)     CONTINUOUS
GRD-090                On         2 (yellow)   CONTINUOUS
EQU-150                On         5 (blue)     CONTINUOUS
PLU-150                On         2 (yellow)   CONTINUOUS
REC-160       *        On         3 (green)    CONTINUOUS
SWT-160       *        On         3 (green)    CONTINUOUS
PAN-160       *        On         3 (green)    CONTINUOUS
ELE-160       *        On         3 (green)    CONTINUOUS
VNT-150       *        On         2 (yellow)   CONTINUOUS
DIF-150       *        On         2 (yellow)   CONTINUOUS
WAS-150       *        On         5 (blue)     CONTINUOUS
EMG-160       *        On         3 (green)    CONTINUOUS
LIG-160       *        On         3 (green)    CONTINUOUS
```

ENGR-PT Layers Table

Creating Prototype Drawings

```
Command: LAYER
?/Make/Set/New/ON/OFF/Color/Ltype/Freeze/Thaw: N          Type NEW or just N.
New layer name(s): TXT-010,                               Repeat for all layers.

Command: LAYER
?/Make/Set/New/ON/OFF/Color/Ltype/Freeze/Thaw: C
Color: 4
Layer name(s) for color 4 (cyan) <0>: TXT-010,           Repeat for appropriate layers.
?/Make/Set/New/ON/OFF/Color/Ltype/Freeze/Thaw: ?         Display Layer information.
?/Make/Set/New/ON/OFF/Color/Ltype/Freeze/Thaw: <RETURN>
```

➡ *TIP! When you specify layer names, type in as many as you want at one time by separating each with a comma. Anticipate the layers that job will require, be sure no spaces are included, and then press Enter.*

Now that you have the prototype drawings that you need for the project, let's talk about the advantages of custom border drawings and show you how to create one which you can insert with a custom setup routine from the AE DISK (described in Chapter Three).

Technique for Using and Creating Border Drawings

AutoCAD's setup routine, which you can pick from AutoCAD's first screen menu, includes the generation of a border drawing. A simple rectangle defining the drawing area, this border drawing is inserted according to the scale and sheet size you select during the execution of the setup routine.

You can gain added benefits by creating your own border drawings. You can use custom border drawings to match different sheet layouts to the requirements of different clients or, as you will employ in the book's project, define a separate area for creating design drawings, experimenting with techniques, and creating blocks without interfering with other drawings.

The following figure illustrates the simulated project's custom border drawing, AEBORDER. The area in which to experiment is called "SCRATCH." Once you have perfected a technique or task, you can move to the "MAIN" drawing area to work on the real drawing. You can set the SCRATCH area up using a User Coordinate System (UCS) so that you can move between it and the MAIN area easily.

Again if you have loaded the AE DISK, you already have the AEBORDER drawing.

Using a UCS for 2D Drawings

You can define a UCS at any location in 2D or 3D space. You will use UCS's in Chapter Four when you generate our 3D presentation drawings of the project. To move around and draw in the 2D SCRATCH area of the proposed border drawing in the same manner that you do in the MAIN area, you can define a UCS for the SCRATCH area, specifying its lower left hand corner as the origin. If you don't do this, your coordinates will all be off when you try to move an entity from the SCRATCH area to the MAIN area.

When you work in an area defined by a UCS, a directional icon appears on the screen and shows how your current orientation relates to the normal or World Coordinate System (WCS).

Creating AEBORDER Drawing

When you create a border drawing, you need to remember that the drawing has to be inserted at the right size no matter what sheet size and scale you select from the Setup routine. You get this size by creating the border drawing on a 1-unit by 1-unit (AutoCAD drawing units) basis. We will discuss this method of creating drawings for insertion more thoroughly in Chapter Two.

If you have the AE DISK, just examine the drawing. If you don't have the disk, use the sequences below as a guide to create the drawing.

AEBORDER Drawing File

  Verify that you have a copy of the AEBORDER.DWG in your AE-ACAD directory. Just examine it.

Follow the command sequence below to create AEBORDER.

Using the following commands, we will show you how to define a UCS for the SCRATCH area and create the AEBORDER drawing which you will use with the custom Setup routine, AE-SETUP, in Chapter Three. Draw the border drawing on Layer 0.

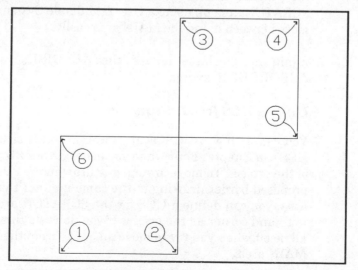

Defined MAIN and SCRATCH Areas

Technique for Creating AEBORDER Drawing

① Begin by loading AutoCAD and select Option 1, Begin a New drawing.

② Enter the drawing name AEBORDER.

Creating Border Drawing

```
Command: LINE
From point: 0,0               Start point ①
To point: @0.500<0            Create line ②
To point: @1.000<90           Create line ③
To point: @0.500<0            Create line ④
To point: @0.500<270          Create line ⑤
To point: @1.000<180          Create line ⑥
To point: C                   Will cause the last point to close to the first.

Command: ZOOM
All/Center/Dynamic/Extents/Left/Previous/Window/ale(X): E
Regenerating drawing.
```

Next, you'll define the UCS, also named SCRATCH, by using the following commands. The origin for the UCS will be defined at a location different than the default 0,0,0.

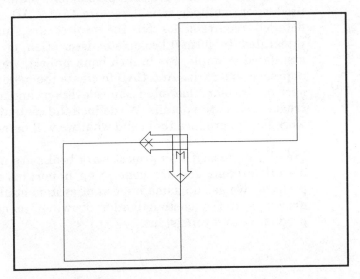

Scratch Area UCS

Creating Scratch Area

```
Command: UCS                    Specify origin at lower left corner.
Origin/ZAxis/3point/Entity/View/X/Y/Z/Prev/Restore/Save/Del/?/<World>: O
Origin point <0,0,0>: 0.5,0.5,0.0

Command: UCS                    Save the UCS.
Origin/ZAxis/3point/Entity/View/X/Y/Z/Prev/Restore/Save/Del/?/<World>: S
?/Name of UCS: SCRATCH

Command: UCS                    To verify, reset to World.
Origin/ZAxis/3point/Entity/View/X/Y/Z/Prev/Restore/Save/Del/?/<World>: <RETURN>
```

To make the custom border drawing even more useful, in Chapter Three we will show you how to create a title drawing to insert into the border before plotting. This drawing, which is called AETITLE, will contain attributes for the sheet titles, the drawing number, revision dates and numbers.

When you are done with the AEBORDER drawing, save it and exit AutoCAD. Let's review our setup progress so far.

Summary

To effectively plan a project's production, you need to determine what production methods it will require (AutoCAD vs Manual), what the detailed requirements for the project are, and how they can be integrated to form a complete description of the project. For the simulated example, the branch bank project, we decided that the best approach was to use AutoCAD to create the major or primary drawings such as site plan, floor plans, and elevations and use manual methods to create some of the details. We defined the contents of the drawings, the work flow to produce them and what we will name each of them.

We also discussed why project work books are important for creating a work process and for generating important references for future projects. We set up some important system standards, like prototype drawings and a custom border drawing, to make our work more productive and consistent.

In addition, the foundations that we have set in this chapter will help you define some additional standards for operation that can make added contributions to your overall productivity. These added standards, symbols libraries and menu routines, are the subjects of our next two chapters.

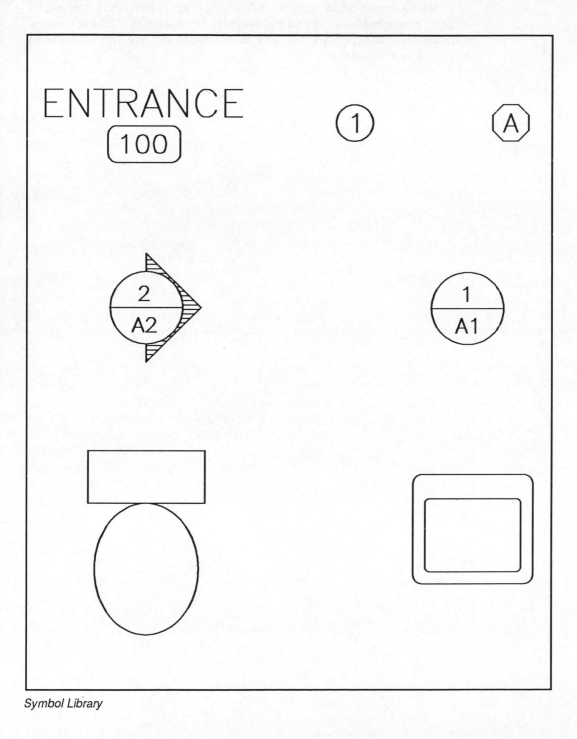

Symbol Library

Symbol Libraries

Introduction

Symbol libraries are critically important to the *health* of your AutoCAD system. The status of this resource can make or break a production system. A well-maintained, well-managed, and well-thought out symbol library is the single most beneficial element that an AutoCAD system can offer to its users. Ideally, every new drawing should make a contribution to the symbol library. Such contributions let you eventually create standard portions of drawings by inserting symbols instead of drawing them over and over again. Without a symbol library, you have to repeatedly create the same symbols with each new drawing. In the end, confusion and repetitious efforts will take the place of productivity.

How you use and manage the library is just as important as building it. Many A/E firms have created extensive symbol libraries. Unfortunately, a common shortcoming is that they forfeit control over how symbols are named and lose track of which directories they use. Then no one knows where to find the information or how to use it. The result is drawing efforts are repeated time and time again because every AutoCAD operator develops his or her own personal library.

The ideal use of AutoCAD is to design and manage your AutoCAD data so that you can use what you created in past projects to enhance your productivity on current and future projects. If you are able to meet this ideal, once you create group of entities and form them into a symbol, you never have to create that same symbol again. Or, where you need slight variations of an entity, you are able to modify the first entity and save the new one with a different name. The underlying process for creating an extensive library is to build the library as the need arises.

Creating the same symbol twice is once too often.

How to Develop Symbol Libraries

In this chapter, we will show you how to develop two types of symbols — annotations and real world symbols. Annotations contain variables, such as dimensions and notations, that you are prompted to enter at the time you insert the symbols. Real world symbols are fixed, defined by predetermined dimensions or characteristics at the time you create them, and inserted into your drawings as you need them. We will discuss these symbol types in more detail in a moment.

Later, in Chapter Three, we will show you how to make symbols easy to use by accessing them and some associated AutoLISP insertion routines through our AE MENU system. These symbols and routines will be used in Chapters Four and Six, when you begin developing the design and working drawings for the branch bank project.

Drawing Setup for Chapter Two

As we discussed in the Introduction, there are two general ways you can approach the drawings in this book. First, you can follow each instruction sequence and complete the drawings as they are presented, or second, you can use the AE-SCRATCH drawings to experiment with the techniques. But this chapter presents a special case because its drawings are not complicated plans, but individual *symbols*. Which approach you take depends on whether you have the AE DISK.

Method One

If you have the AE DISK, this chapter's symbols are already available to you as drawings. If you copied the disk files in Chapter One, you will find the symbol drawings in the AE-ACAD directory. To examine them:

❏ Load AutoCAD and Select Option 2, Edit an Existing drawing.

❏ Specify any of the symbol names that are defined in this chapter.

❏ Examine the symbols as you follow along with the instructions on how to create them.

➡ *NOTE! If you change or revise these symbols to suit your own needs, don't forget to give your own names to your revised symbols. You will need the unrevised symbols in later chapters.*

Method Two

If you do not have the AE DISK, you can create the symbols by following the exercise sequences below. You will be working in the AE-ACAD directory. Use the following sequence as a guide.

❏ Load AutoCAD and Select Option 1, Begin a New drawing.

❏ Specify the symbol names that we define.

❏ Use the default ACAD prototype drawing of full scale, A-8 1/2" x 11" sheet size; set the UNITS to four Architectural (default the other UNITS settings) and use the information in the following setup table.

```
COORDS       GRID        SNAP       ORTHO       UCSICON
ON           OFF         1          ON          WORLD

LIMITS    0,0 to 8 1/2"x11"
ZOOM      Zoom All.

Layer name       State        Color        Linetype
0                On           7 (white)    CONTINUOUS
```

Chapter Two Setup Table

Symbol Types

There are two basic types of symbols or blocks that are used to create architectural drawings. Annotation symbols include symbols such as section and detail marks and elevation flags. They are usually created on a 1-unit by 1-unit basis, and inserted into drawings by using predetermined scale factors.

Real world symbols include entities such as plumbing fixtures and furnishings where their dimensions are fixed. They are created to full size and inserted without scale factors. A subset of the real world type are variation symbols which are entities, such as doors and windows that can require size variations. These are inserted with different X and Y scale factors for the specific location, thus avoiding the need to create multiple symbols to represent different sizes of the same entity.

Rules for Creating Symbols

The rules for creating symbols or blocks are simple and few:

❑ Use a short but descriptive name, preferably eight characters or less.

❑ Create symbols with the appropriate scale factor for that symbol type.

❑ Create symbols on layer 0.

❑ Define a base or insertion point that relates to the symbol's function.

Let's take a look at these rules in more detail.

Naming Symbols

How you name symbols is almost as important as how you create them. If a symbol name is vague or nondescriptive, it will be too difficult to recall and/or revise when you need to do so. It is important to adopt a naming convention for symbols that everyone using the system can easily recognize.

When possible, your symbol names should provide a description of the symbol. For example, a section annotation named SECTION is much easier to find and reuse than one named 123. Keeping track of symbol names that you are currently using is important, since using a second symbol with the same name can over-write the first. Good documentation, such as actual 8 1/2 inch x 11 inch plots, kept in the project work book and a master library file will give you a quicker way to retrieve symbols now and in the future.

Scale Factors

The most common symbol found throughout an A/E's set of working drawings, changing only in size according to scale, is the annotation symbol. This type of symbol represents approximately 80% of all symbol use. You could create separate versions of the symbol for each different drawing scale, but this would consume a lot of disk storage space and be difficult to manage. A more effective method is to create a single symbol on a 1-unit by 1-unit basis. You can then use custom AutoLISP routines to read the required scale factors from your drawing variables established when you set up the drawing.

AutoLISP routines give us an effective way to use annotation symbols with various drawing scales, helping to reduce the number of symbols in a symbol library. Using AutoLISP routines to read the drawing scale factors, and to automatically insert the symbol at the correct size, is also useful when you are dealing with large quantities of annotations that are displayed at various sizes. We'll give you several examples of AutoLISP insertion routines later in the chapter, and re-address their use in Chapter Three when we discuss menu systems.

When you create real world symbols, you generally create them to their actual sizes (whether they're fixtures or furnishings). It stands to reason that once you create the symbol for a toilet, for example, there is no need to increase or decrease its dimensions since this fixture is usually a standard size.

Symbols and Layering

The connection between layer choice and symbol creation is simple — always create symbols on layer 0. Any symbol that you create on layer 0 will take on the characteristics of the layer on which you later insert it. If you create a symbol on a layer other than 0, for instance a lavatory symbol on layer PLU-150, the system associates that layer name with the symbol. This association complicates using the symbol in a drawing. Say, for example, using a layer called PLUMBING to create a lavatory symbol would leave you with a symbol with two layer names associated with it when you inserted the symbol. Such a double association makes any system of data extraction that deals solely with layers almost unusable.

Base or Insertion Points

The rule for specifying base or insertion points for symbols is also simple. Always specify an insertion point that relates to the symbol's type or function. For example, the most logical location for the insertion point of a symbol representing a wall-mounted sink is the centerline of the face that mounts to the surface of the wall. You can insert the symbol using the OSNAP function called nearest to insure that the lavatory is attached to the face of the wall. Keeping this rule of logical insertion points in mind can help you enhance the way your symbols function. If you are creating the symbols in this chapter, use 0,0 as your reference point.

Attributes and Symbols

Attributes are "special AutoCAD drawing entities that contain text." For example, you can use attributes in association with annotation symbols to supply section mark numbers, or with real world symbols to supply data for part and model numbers. You can collect this data later to create a bill-of-material or build tables of the annotation data and subsequently create schedules of room names or finishes.

You can use attributes to label any type of symbol or group of entities. Attributes can be either fixed, as in the part or model number for a given piece of equipment, or variable, as used with annotations for door numbers referencing a door schedule. Entities can also contain attribute data supplied by AutoLISP routines. This ability to assign nongraphic data to different types of symbols gives you a very effective way of automating the drawing process, as you will see later in this chapter.

Defining Attributes

You define attribute data for symbols by using the ATTDEF (Attribute

Definition) command. Attribute data is Invisible or Visible, Constant or Variable. You can define an attribute with a Preset variable, which is variable data that is not requested during symbol insertion. This provides a default value and you use it most often when you don't know the symbol's data at the time you insert the symbol. You can add to or edit this value later with the ATTDEF or DDATTE commands. If you need to edit more than a single attribute or symbol, an AutoLISP routine, such as the one we have listed at the end of this chapter, can give you the ability to update or modify groups of attributes all at one time.

❑ Invisible attributes are generally used for labeling real world symbols where the data entered should not be displayed in the drawing. By using the ATTDISP command, you can turn invisible attributes on or off to verify the data. If you take a plumbing fixture as an example, the fixture's model number is an attribute that you probably don't want to display on the drawing but would like to have associated with the symbol for verification or extraction purposes.

❑ Constant attributes are used with real world symbols when the symbol being inserted has a constant part or model number. The model number of the plumbing fixture we just suggested as an invisible attribute would also serve as an example of a constant attribute. This type of attributes are usually constant over a project, but may vary from project to project.

❑ Visible attributes are used for annotations where the numbers or characters need to be displayed. Door, room, and detail annotation symbols provide good examples of this kind of attribute, since they are used expressly for displaying a different attribute (a number or letter, for instance) for each occurrence of the symbol.

❑ Variable attributes are used with annotations when the person doing the drawing needs to be prompted for the annotation data or where the data is incremental, as in the door, room, and detail annotation symbols we just described.

Now that you have had a brief tour of the rationale behind symbols and their attributes, let's turn to creating them. If you have the AE DISK, just read along. If you want more background on creating and using symbols, refer to INSIDE AUTOCAD (New Riders Publishing).

Technique for Creating Symbols

Create the following annotation symbols on the same basis as those drawn 1-unit by 1-unit. The annotation symbols have been drawn to a smaller size to insure that the variable data that gets entered via the AutoLISP variable called USERI1 will be displayed at the right size for the scale specified through our AE-SETUP routine. We will explain USERI1 and the setup procedures in the next chapter.

 NOTE! If you are using other 3rd party menus and AutoLISP routines, there is the possibility that they may be using the USERI1 variable.

Some 1-unit by 1-unit annotations will appear out of scale when the scale factor for the drawing is applied. For example, the door annotation symbol would be much too large for most drawings if it were inserted at a size of 12 inches (six inches would be more like it). When you create annotations in the 1-unit by 1-unit fashion, you have to allow for the scale factor when the symbol is inserted.

Creating Door Annotation Symbol

The first symbol you will create is a door annotation symbol. When you use this symbol in a drawing, you use an associated number, labeling each door on the floor plan, relating specific doors to a schedule. This schedule will contain information on the size of each door, what each type is, the material each is made of, and what hardware is required. You will use this symbol later when you create the floor plan drawing in Chapter Six. The symbol itself is a simple circle with an attribute attached that is used to assign a different number to each door as the annotation symbols are inserted.

When you insert the door annotation symbol, you can use the INSERT command to place the symbol, and answer a prompt for the number you want to display within the symbol. However, at the end of the chapter we have provided a custom AutoLISP routine that you can use later to insert the symbol and number the doors automatically when you create the drawings for the project.

Door Annotation Symbol Drawing

The way you create this symbol in the following exercise insures that it will always appear at the correct size when you insert it with a drawing scale factor.

Door Symbol Drawing

Make sure AutoCAD is loaded and you are in the AE-ACAD directory

 Enter selection: 2   Edit an existing drawing named DRSYM.

Examine the drawing.

 Enter selection: 1   Begin a NEW drawing named DRSYM.

Verify the settings shown in the Chapter Two Setup Table.

Follow the command sequence below to create the symbol.

DRSYM Door Annotation Symbol

```
Command: CIRCLE
3P/2P/TTR/<Center point>: 0,0                    Pick any point.
Diameter/<Radius>: 1/8
```

```
Command: ATTDEF
Attribute modes  --  Invisible:N  Constant:N  Verify:N  Preset:N
Enter (ICVP) to change, RETURN when done: <RETURN>
Attribute tag: NUMBER
Attribute prompt: DOOR NUMBER?
Default attribute value: <RETURN>
Start point or Align/Center/Fit/Middle/Right/Style: M
Middle point: 0,0
Height <0'-0 1/2">: 1/8
Rotation angle <0>: <RETURN>

Command: END
```

It is a good idea to test the operation of DRSYM by using the INSERT command, answering the prompts just created for the attribute data.

Testing the Door Symbol

Even if you are using the DRSYM symbol from the AE DISK, test using the symbol.

❑ Load AutoCAD and select option 1. Begin a New drawing.

❑ Type in TEST for the drawing name.

❑ Use the default ACAD settings.

Now, try inserting the symbol.

DRYSYM Symbol Operation

```
COMMAND: INSERT
Block name (or ?): DRSYM
Insertion point:                            Pick any point.
X scale factor <1>/Corner/XYZ: <RETURN>
Y scale factor <default=X>: <RETURN>
Rotation angle <0>: <RETURN>
DOOR NUMBER?                                 Enter any number for testing.
```

Now that you've determined that your first symbol and its attributes work the way they should, try creating another one.

Creating Window Annotation Symbol

The next symbol is a window annotation symbol. This symbol is a polygon with an attribute attached that you can use to assign a different letter to each specified window. As with the door annotation symbol, this symbol's letter is used to relate the window location on the floor plan to a window schedule, showing size, type, and associated details.

When you insert the symbol, you can use INSERT to place the symbol and answer a prompt for the letter to be displayed within the symbol. A more productive method is to use the associated AutoLISP routine that we have provided at the end of the chapter to automate this procedure. You will use this window symbol and AutoLISP routine later in Chapter Six.

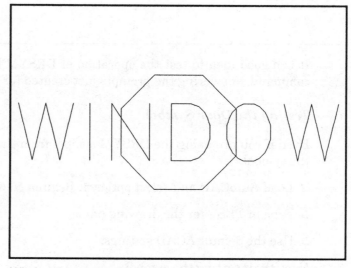

Window Annotation Symbol Drawing

Window Symbol Drawing

   Enter selection: 2   Edit an existing drawing named WINSYM.

Examine the drawing.

   Enter selection: 1   Begin a NEW drawing named WINSYM.

Verify the settings shown in the Chapter Two Setup Table.

Follow the command sequence below to create the symbol.

WINSYM Window Annotation Symbol

```
Command: POLYGON
Number of sides: 8
Edge/<Center of polygon>: 0,0                Base point for symbol creation.
```

```
Inscribed in circle/Circumscribed about circle (I/C): C
Radius of circle: 1/8


Command: ATTDEF
Attribute modes -- ·Invisible:N  Constant:N  Verify:N  Preset:N
Enter (ICVP) to change, RETURN when done: <RETURN>
Attribute tag: WINDOW
Attribute prompt: WINDOW LETTER?
Default attribute value: <RETURN>
Start point or Align/Center/Fit/Middle/Right/Style: M
Middle point: 0,0              Same base point for attribute.
Height <0'-0 1/16">: 1/8
Rotation angle <0>: <RETURN>
Command: END
```

As with the last symbol, test WINSYM's operation by using INSERT, and answering the prompts that you just created for the attribute data.

Testing the Window Symbol

Try testing the window symbol.

❑ Load AutoCAD and select option 1. Begin a New drawing.

❑ Type in TEST for the drawing name.

❑ Use the default drawing settings.

❑ Use the instructions below to insert the symbol.

WINSYM Symbol Operation

```
COMMAND: INSERT
Block name (or ?): WINSYM
Insertion point:                     Pick any point.
X scale factor <1>/Corner/XYZ: <RETURN>
Y scale factor <default=X>: <RETURN>
Rotation angle <0>: <RETURN>
WINDOW LETTER?                       Enter any letter for testing.
```

Now that you know this annotation symbol and its attributes work, move on to the next symbol.

Creating Room Annotation Symbol

The last simple annotation symbol that we will show you how to create is a room annotation symbol. You will use it to label rooms with numbers.

You can create the symbol by drawing a rectangle at the specified size, use the FILLET command to round the corners; then specify an attribute for the room number.

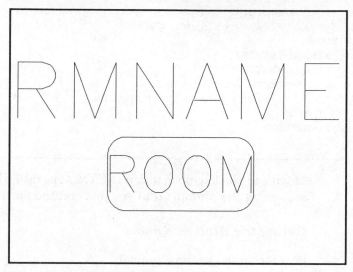

Room Annotation Symbol Drawing

Room Symbol Drawing

 Enter selection: 2   Edit an existing drawing named RMSYM.

Examine the drawing.

 Enter selection: 1   Begin a NEW drawing named RMSYM.

Verify the settings shown in the Chapter Two Setup Table.

If you are not using the AE DISK, follow the command sequence below to create the symbol.

RMSYM Room Annotation Symbol

```
Command: LINE
From point: 0,0          Starting point for symbol creation.
To point: @1/2<0
To point: @1/4<90
To point: @1/2<180
To point: C

Command: FILLET
```

```
Polyline/Radius/<Select two objects>: R
Enter fillet radius <0'-0">: 1/16
Command: FILLET
Polyline/Radius/<Select two objects>:          Select all four corners.

Command: ATTDEF
Attribute modes -- Invisible:N Constant:N Verify:N Preset:N
Enter (ICVP) to change, RETURN when done: <RETURN>
Attribute tag: ROOM
Attribute prompt: ROOM NUMBER?
Default attribute value: <RETURN>
Start point or Align/Center/Fit/Middle/Right/Style: M
Middle point: 1/4,1/8               Center of symbol and attribute.
Height <0'-0">: 1/8
Rotation angle <0>: <RETURN>

Command: ATTDEF
Attribute modes -- Invisible:N Constant:N Verify:N Preset:N
Enter (ICVP) to change, RETURN when done: <RETURN>
Attribute tag: RMNAME
Attribute prompt: ROOM NAME?
Default attribute value: <RETURN>
Start point or Align/Center/Fit/Middle/Right/Style: M
Middle point: 1/4,5/8               Center of Room Name text.
Height <0'-1/8">: 3/16
Rotation angle <0>: <RETURN>

Command: END
```

Testing the Room Symbol

Again, test the operation of RMSYM by using INSERT and answering the prompts that you just created for the attribute data.

❑ Load AutoCAD and select option 1. Begin a New drawing.

❑ Type in TEST for the drawing name.

❑ Use the default drawing settings.

❑ Insert the symbol, using the command sequences below as a guide.

RMSYM Symbol Operation

```
COMMAND: INSERT
Block name (or ?): RMSYM
Insertion point:          Pick any point.
X scale factor <1>/Corner/XYZ: <RETURN>
Yscale factor <default=X>: <RETURN>
Rotation angle <0>: <RETURN>
```

ROOM NUMBER?	Enter any number for testing.
ROOM NAME?	Enter any room name for testing.

Now, let's try something a little fancier, creating two annotation symbols that can be used together.

Creating Detail Symbol

The next annotation symbol that we want to show you is a detail symbol. You can use it alone or combine it with an arrow symbol that you will create to form a section annotation. The detail symbol will have two attributes: one for the detail or section number, and a second for the sheet number where the detail or section may be found. You will use these symbols later in Chapter Six.

Detail Symbol Drawing

Detail Symbol Drawing

Again work in your AE-ACAD directory.

Enter selection: 2   Edit an existing drawing named DETAIL.

View the drawing.

Enter selection: 1   Begin a NEW drawing named DETAIL.

Verify the settings shown in the Chapter Two Setup Table.

If you are not using the AE DISK, follow the command sequence below to create the symbol.

DETAIL Detail Symbol

```
Command: CIRCLE                              Use to create circle.
3P/2P/TTR/<Center point>: 0,0                Base point for symbol creation.
Diameter/<Radius>: 1/4


Command: LINE                                Use to create line.
From point: QUAD
of
To point: QUAD
of
To point: <RETURN>


Command: ATTDEF                              Detail or section number attribute.
Attribute modes -- Invisible:N Constant:N Verify:N Preset:N
Enter (ICVP) to change, RETURN when done: <RETURN>
Attribute tag: NUMBER
Attribute prompt: DETAIL NUMBER?
Default attribute value: <RETURN>
Start point or Align/Center/Fit/Middle/Right/Style: M
Middle point: 0,1/16                         Base point for attribute.
Height <0'-0 0">: 3/32
Rotation angle <0>: <RETURN>


Command: ATTDEF                              Sheet number attribute.
Attribute modes -- Invisible:N Constant:N Verify:N Preset:N
Enter (ICVP) to change, RETURN when done: <RETURN>
Attribute tag: SHT
Attribute prompt: SHEET NUMBER?
Default attribute value: <RETURN>
Start point or Align/Center/Fit/Middle/Right/Style: M
Middle point: 0,-1/16"                       Base point for attribute.
Height <0'-0 3/32">: <RETURN>
Rotation angle <0>: <RETURN>


Command: END
```

After creating the detail symbol, the next step is to create the section arrow.

Creating Section Arrow

Next, you are going to create a section arrow and superimpose it upon the detail symbol to form a section annotation. Create the arrow using a reference circle that is the same size as your detail symbol. Specify a base point for the arrow that relates to the center of this circle. By using this base or insertion point, you can rotate the arrow around the circle of the detail symbol when you insert the arrow using an OSNAP INSERT. That way, you show the direction of the section cut. This simple technique of defining associated base points is a great benefit — you don't have to create different arrows to point in different directions, or take time to use AutoCAD's edit commands to modify one symbol.

The following section arrow symbol is named EELEV to differentiate it from the "ARROW" symbol that many people have on their systems to indicate directions used with notes.

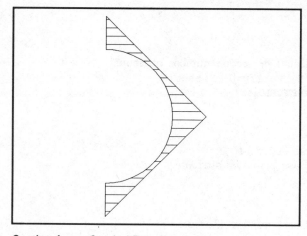

Section Arrow Symbol Drawing

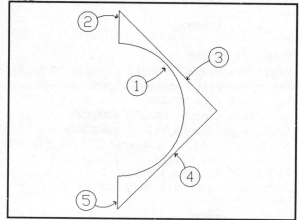

Completed Section Arrow Symbol (EELEV)

Section Arrow Symbol Drawing

    Enter selection: 2   Edit an existing drawing named EELEV.

Examine the drawing and revise if you wish.

Enter selection: 1   Begin a NEW drawing named EELEV.

Verify the settings shown in the Chapter Two Setup Table.

To help you create the symbol:

□ SNAP. Set SNAP to 1/8".

□ Follow the command sequence below to create the Section Arrow Symbol called EELEV.

EELEV Section Arrow Symbol

```
Command: ARC
Center/<Start point>: C
Center:                                              Pick any point.
Start point: @1/8"<270
Angle/Length of chord/<End point>: @1/4"<90

Command: PLINE
From point:
Current line-width is 0'-0"
Arc/Close/Halfwidth/Length/Undo/Width/<Endpoint of line>:   Select ① at top to create ②.
Arc/Close/Halfwidth/Length/Undo/Width/<Endpoint of line>:   Use to create ③.
Arc/Close/Halfwidth/Length/Undo/Width/<Endpoint of line>:   Use to create ④.
Arc/Close/Halfwidth/Length/Undo/Width/<Endpoint of line>:   Use to create ⑤.
Arc/Close/Halfwidth/Length/Undo/Width/<Endpoint of line>: <RETURN>

Command: HATCH
Pattern (? or name/U,style): LINE
Scale for pattern <1.0000>: 0.5
Angle for pattern <0>: <RETURN>
                                                     Select ①.
Select objects:
1 selected, 1 found.
                                                     Select pline ②.
Select objects:
1 selected, 1 found.
Select objects: <RETURN>
                                                     Select ①.
Command: BASE
Base point <0'-0",0'-0",0'-0">: CEN
of

Command: END
```

Testing the Component Symbols

Now that you have created the component DETAIL and EELEV symbols, test their operation and their attributes.

□ Load AutoCAD and select option 1. Begin a New drawing.

□ Type in TEST for the drawing name.

□ Use the default drawing settings.

❑ Use the sequence below to help you test the symbols.

DETAIL and EELEV Symbols Operation

```
COMMAND: INSERT
Block name (or ?): DETAIL
Insertion point:                          Pick any point.
X scale factor <1>/Corner/XYZ: <RETURN>
Y scale factor <default=X>: <RETURN>
Rotation angle <0>: <RETURN>
DETAIL NUMBER?                            Enter any number for testing.
SHEET NAME?                              Enter any number for testing.

COMMAND: INSERT
Block name (or ?): EELEV
Insertion point: INS
of                                        Select the DETAIL symbol.
X scale factor <1>/Corner/XYZ: <RETURN>
Y scale factor <default=X>: <RETURN>
Rotation angle <0>: <RETURN>
```

You tested the symbols just as you drew them. When you insert these annotations into the working drawings, use scale factors to produce the correct size. Later in this chapter. We'll show you how to use AutoLISP to set the scale and answer the attributes for you.

Real World Symbols

Real world symbols can be defined as symbols with fixed dimensions or finite characteristics. They are unique and do not require changes or modifications when you retrieve them from your symbol library for use in your drawings.

Creating Lavatory Symbol

The first The first real world symbol represents a wall-hung lavatory. Draw a rectangle to define its outside dimensions and create the inside lines defining the bowl by using OFFSET. You'll want to clean up and round the corners by using the FILLET command with a radius.

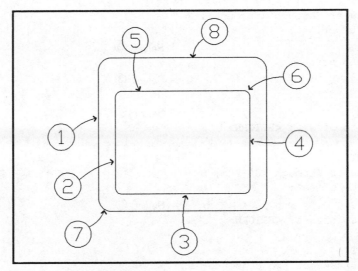

Lavatory Symbol Drawing

Lavatory Drawing

    Enter selection: 2   Edit an existing drawing named LAVRT.

View the drawing.

    Enter selection: 1   Begin a NEW drawing named LAVRT.

Verify the settings shown in the Chapter Two Setup Table.

If you are not using the AE DISK, follow the command sequence below to create the LAVRT Symbol.

LAVRT Lavatory Symbol

```
Command: LINE
From point: 0,0            Base point for ①.
To point: @1'8<0
To point: @1'6<90
To point: @1'8<180
To point: C
```

```
Command: OFFSET
Offset distance or Through <Through>: 2
Select object to offset:
Side to offset?                              Select ②.
Select object to offset:
Side to offset?                              Select ③.
Select object to offset:
Side to offset?                              Select ④.
Select object to offset: <RETURN>

Command: OFFSET
Offset distance or Through <0'-2">: 5
Select object to offset:
Side to offset?                              Select ⑤.
Select object to offset: <RETURN>

Command: FILLET
Polyline/Radius/<Select two objects>: R
Enter fillet radius <0'-0">: 3/4

Command: FILLET
Polyline/Radius/<Select two objects>:        Select ⑥.

Command: FILLET
Polyline/Radius/<Select two objects>: R
Enter fillet radius <0'-0 3/4">: 1-1/2

Command: FILLET
Polyline/Radius/<Select two objects>:        Select ⑦.

Command: BASE
Base point <0'-0",0'-0",0'-0">: MID
of                                           Select ⑧.

Command: END
```

Testing the Lavatory Symbol

You need to test the way this symbol operates.

❑ Load AutoCAD and select option 1. Begin a New drawing.

❑ Type in TEST for the drawing name.

❑ Use the default drawing settings.

❑ Use the sequence below to guide your test.

LAVRT Symbol Operation

```
COMMAND: INSERT
Block name (or ?): LARVT
Insertion point:                          Pick any point.
X scale factor <1>/Corner/XYZ: <RETURN>
Y scale factor <default=X>: <RETURN>
Rotation angle <0>: <RETURN>
```

The last symbol to create is of a flush tank type toilet.

Creating Toilet Symbol

You can create the toilet symbol by using ELLIPSE to draw the toilet bowl. Then, use OSNAP QUAD to define the tank in the correct relation to the bowl. As with all of the previous symbols, create this symbol on layer 0.

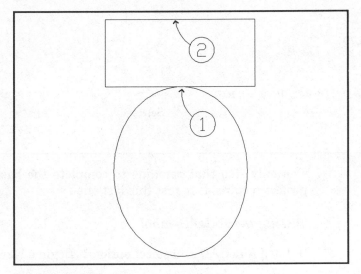

Toilet Symbol Drawing

Toilet Symbol Drawing

 Enter selection: 2   Edit an existing drawing named TOILET.

View the drawing.

Enter selection: 1   Begin a NEW drawing named TOILET.

Verify the settings shown in the Chapter Two Setup Table.

Follow the command sequence below to create the symbol.

TOILET Symbol

```
Command: ELLIPSE
<Axis endpoint 1>/Center: C
Center of ellipse:                      Pick a point.
Axis endpoint: @10<270
<Other axis distance>/Rotation: @8<0

Command: LINE
From point: QUAD                        Select ①.
of
To point: @9<0
To point: @8<90
To point: @18<180
To point: @8<270
To point: C

Command: BASE
Base point <0'-0",0'-0",0'-0">: MID
of                                      Select ②.

Command: END
```

The only step that remains to complete the basic symbols that THE project requires is to test this last one.

Testing the Toilet Symbol

❑ Load AutoCAD and select option 1. Begin a New drawing.

❑ Type in TEST for the drawing name.

❑ Use the default drawing settings.

❑ Use the test sequence below.

TOILET Symbol Operation

```
COMMAND: INSERT
Block name (or ?): TOILET
Insertion point:                        Pick any point.
X scale factor <1>/Corner/XYZ: <RETURN>
```

```
Yscale factor <default=X>: <RETURN>
Rotation angle <0>: <RETURN>
```

You have defined some useful symbols and associated attributes in this chapter. Updating or changing attributes is not difficult, especially if you only need to change a few attributes. But, as we mentioned in the defining attributes section earlier, you may find an occasion where you'd like to edit groups of attributes all at once. At the end of the next section you will find a custom AutoLISP routine that allows you to edit groups of attributes, instead of having to edit them one at a time as the standard DDATTE command requires.

AE SYSTEM Setup

Before proceeding with this section and into Chapter Three, we need to discuss setting up to use the AE MENU system and the AutoLISP routines that come with the AE DISK. We assume that your workstation already has AutoCAD and the standard tablet menu loaded. If you have not installed these items, then refer to your AutoCAD Reference Manual for the installation procedures. We will also talk about using a text editor to look at the AutoLISP routines. If you know how to create and use AutoLISP routines, you can create the book's routines using your text editor.

AE MENU System Directories

If you have already set up your directories following the instructions in Chapter One, just read along. Then, follow the instructions below to make sure that you have the correct AE-MENU and AutoLISP files that you need for the balance of this chapter. If you haven't created your directories yet, you need to do that now. Make sure that you have copied your AutoCAD overlay and configuration files from the \ACAD directory to the \AE-ACAD directory. If you haven't, go back to Chapter One and look at setting up the directories for the book.

Directories	Usage
\ (ROOT)	Used to place all Batch files
\DOS	Used to place all DOS files
\ACAD	Used to place all AutoCAD files
\AE-ACAD	Used for the book's support files
\8910	Used for the book's drawing files

AE Directory Table

If you plan on examining the AutoLISP files (or typing them in), you need to place a copy of your text editor in the \AE-ACAD directory.

Setting Up a Text Editor

Selecting a good text editor to use with AutoCAD is very important. A text editor is a valuable asset to your system's operation because it allows you to create custom menus and AutoLISP routines and to revise and update existing ones.

While we used Norton's Editor to develop the different AutoLISP and menu routines for the book, there are many other editors that will do just as well. EDLIN, PC Write or TED (both *shareware* text editors), and the text editors included with utility type programs, like Sidekick or MS Windows, will all do a good job. Just remember, you must be able to create an ASCII formatted text file with whatever editor you select. For more on text editors and their use, see CUSTOMIZING AUTOCAD and INSIDE AUTOLISP (New Riders Publishing).

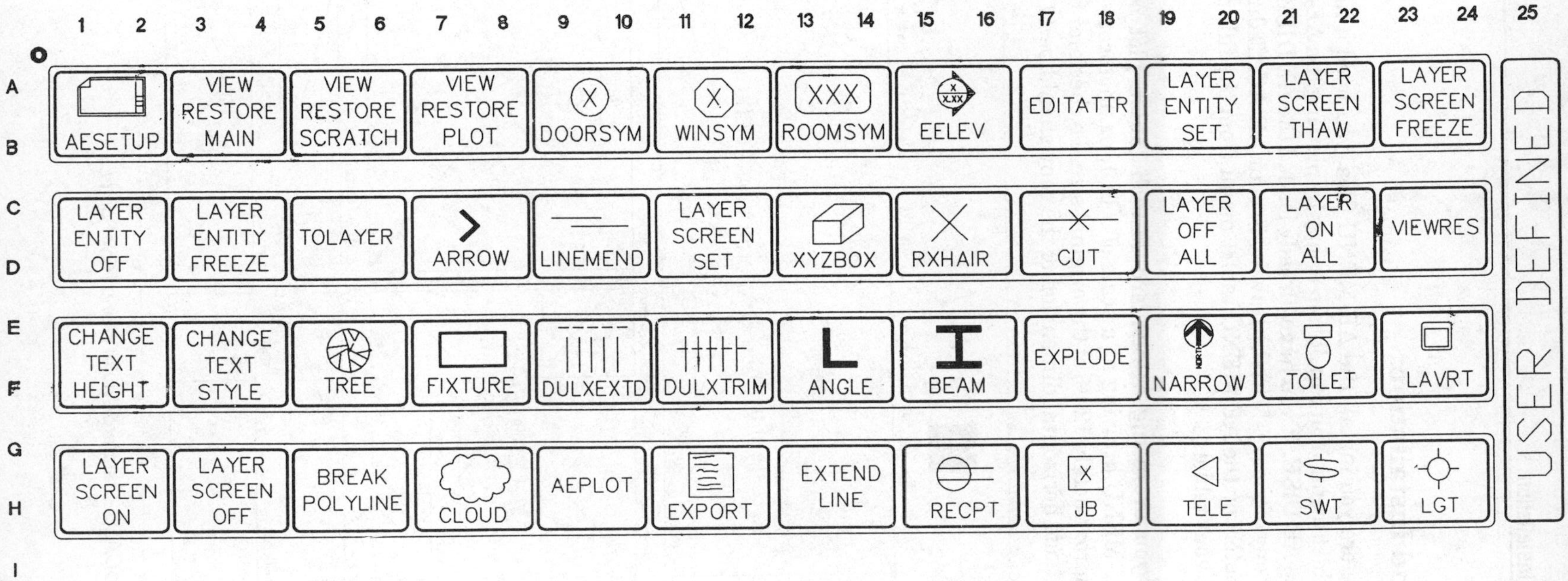

USER DEFINED

AESETUP | VIEW RESTORE MAIN | VIEW RESTORE SCRATCH | VIEW RESTORE PLOT | DOORSYM | WINSYM | ROOMSYM | EELEV | EDITATTR | LAYER ENTITY SET | LAYER SCREEN THAW | LAYER SCREEN FREEZE

LAYER ENTITY OFF | LAYER ENTITY FREEZE | TOLAYER | ARROW | LINEMEND | LAYER SCREEN SET | XYZBOX | RXHAIR | CUT | LAYER OFF ALL | LAYER ON ALL | VIEWRES

CHANGE TEXT HEIGHT | CHANGE TEXT STYLE | TREE | FIXTURE | DULXEXTD | DULXTRIM | ANGLE | BEAM | EXPLODE | NARROW | TOILET | LAVRT

LAYER SCREEN ON | LAYER SCREEN OFF | BREAK POLYLINE | CLOUD | AEPLOT | EXPORT | EXTEND LINE | RECPT | JB | TELE | SWT | LGT

The AE-MENU Tablet Menu

AE-MENU Setup and Installation

Before you install the AE-MENU files, you need a hard copy of the AE tablet menu. You can obtain a copy by plotting the AE-MENU.DWG from the AE DISK. If you have already loaded the AE DISK files, you will find a copy of the AE-MENU drawing in your AE-ACAD directory. Or, you can photocopy the AE-MENU tablet menu from the full-page illustration to fit your tablet.

If you are using a standard 11-inch tablet, plot or printer plot the AE-MENU drawing at a scale of 1 to 1, or photocopy the illustration at approximately 120% of the original size. To check for the correct size, the outside horizontal dimension of the copy should be 11 inches.

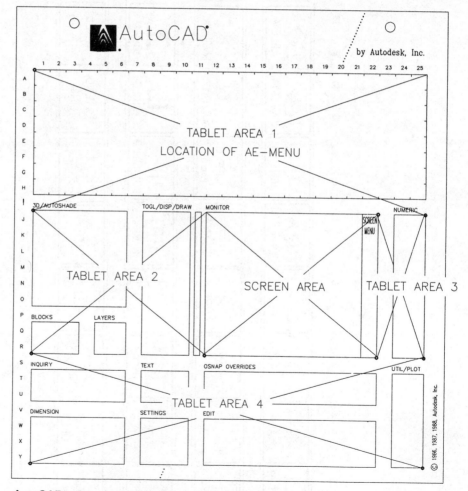

AutoCAD's Standard Tablet Menu Pick Points

Reconfiguring the AE-MENU Tablet Menu

With the drawing placed over tablet area number one, reconfigure your tablet. To reconfigure your digitizer to match the areas of the standard AutoCAD template, use the TABLET command from inside the drawing editor. AutoCAD provides a series of tablet pick points on the template to ensure the alignment of the four areas' menu areas and the screen area. These are shown in the Tablet Menu Pick Points illustration.

Select SETTINGS from the AutoCAD Screen Menu, then TABLET. Pick the three points for each area as shown in the facing illustration; and accept the default numbers, displayed by AutoCAD for the columns and rows.

Reconfiguring the Tablet Menu Areas

```
C:>CD\AE-ACAD            Change directories.
C:\AE-ACAD>
```

Load AutoCAD and from the Main Menu select the following:

```
Enter selection: 1
Enter NAME of drawing : AETEST

Command: TABLET Option (ON/OFF/CAL/CFG): CFG
Enter number of tablet menus desired (0-4) : 4
Do you want to realign tablet menu areas?   Y
```

```
Digitize upper left corner of menu area 1:        Pick point.
Digitize lower left corner of menu area 1:        Pick point.
Digitize lower right corner of menu area 1:       Pick point.
Enter the number of columns for menu area 1: 25
Enter the number of rows for menu area 1: 9

Digitize upper left corner of menu area 2:        Pick point.
Digitize lower left corner of menu area 2:        Pick point.
Digitize lower right corner of menu area 2:       Pick point.
Enter the number of columns for menu area 2: 11
Enter the number of rows for menu area 2: 9

Digitize upper left corner of menu area 3:        Pick point.
Digitize lower left corner of menu area 3:        Pick point.
Digitize lower right corner of menu area 3:       Pick point.
Enter the number of columns for menu area 3: 9
Enter the number of rows for menu area 3: 13

Digitize upper left corner of menu area 4:        Pick point.
Digitize lower left corner of menu area 4:        Pick point.
Digitize lower right corner of menu area 4:       Pick point.
Enter the number of columns for menu area 4: 25
Enter the number of rows for menu area 4: 7
```

```
Do you want to respecify the screen pointing area?   Y
Digitize lower left corner of screen pointing area:     Pick point.
Digitize upper right corner of screen pointing area:    Pick point.

Command: QUIT
Do you really want to discard all changes ? Y
```

The standard AutoCAD tablet menu is now reconfigured and all parameters are stored. Next, you want to test the AE-MENU.

AE-MENU Installation and Testing

The AE DISK contains all the files necessary for the book's menus and AutoLISP routines to work. The disk load program (AE-LOAD) copies the files into the AE-ACAD directory. (If you haven't loaded the disk files, see Chapter One.)

Verifying the AE-MENU and AutoLISP Files

Verify that you have the book's menu and support files in your AE-ACAD directory.

```
C:\AE-ACAD> DIR *.MNU
AE-MENU.MNU                      The book's menu program.
C:\AE-ACAD> DIR *.LSP
ACAD.LSP
ANGLE.LSP
BEAM.LSP
CLOUD.LSP
CUT.LSP
DOORSYM.LSP
....                            There are twenty-one files.
C:\AE-ACAD> DIR *.DWG
AEBORDER.DWG
AE-MENU.DWG
ARROW.DWG
DETAIL.DWG
DRSYM.DWG
....                            There are twenty-three symbol, prototype and drawing files.
C:\AE-ACAD> DIR *.SCR
AEPLOT.SCR                       Script file for plotting.
```

 If you don's have the AE-DISK yet. Just read along.

Now that you have verified the files are loaded, the AE-MENU.DWG is in place, and tablet reconfigured, it is time to test the AE-MENU system.

Testing the AE-MENU System

```
C:\>CD\AE-ACAD                    Change directories.
C:\AE-ACAD>
```

Load AutoCAD and from the Main Menu select the following:

```
Enter selection: 1
Enter NAME of drawing : AETEST
Command: MENU
Menu file name or . for none <ACAD>: AE-MENU
Compiling menu AE-MENU.mnu...
```

The menu system should function. Try some of the menu items.

```
Command: QUIT
Do you really want to discard all changes ? Y
```

Congratulations, you have added the AE-MENU drawing to tablet menu area one, reconfigured your tablet, loaded the AE-MENU files, and tested the system.

You are now ready to take a first look at the use of AutoLISP. For your convenience, we supply each routine's listing in the following text so that you can see what the routines look like even if you aren't using a text editor.

A Little AutoLISP

You will use four custom AutoLISP routines to automatically insert and input the attribute data of the symbols you created earlier in this chapter. These routines also serve as a good introduction to some more setup routines that we describe in the next chapter on menu systems. These short symbol insertion routines give an idea how productive it is to link AutoLISP with predrawn symbols. Instead of using multiple commands to insert and answer the attribute prompts for each symbol, you only need to use one command. In some cases, the AutoLISP routine even supplies the attribute data — all you do is answer the first prompt and then insert as many of the symbols as you want. The first three routines set the appropriate scale factors for the symbols through a user variable created by our AE-SETUP routine (detailed in Chapter Three). If you want to explore the setup routine it is the menu item, called [AESETUP], on the AE-MENU tablet menu.

Door Annotation Symbol AutoLISP Routine

The AutoLISP routine given below first prompts you for a beginning number that will be displayed within the door annotation symbol. It *then* prompts you for an insertion point. The routine then automatically steps through the numbering sequence, answering the attribute variable with the next number each time you insert the symbol.

DOORSYM.LSP AutoLISP Routine

Verify that you have the DOORSYM.LSP file in your AE-ACAD directory.

Read along, or create the AutoLISP file.

Testing the DOORSYM.LSP Routine

Once you have created the symbol and defined the starting door number using this AutoLISP routine, the only task left is designating where to place the symbol. Now that you have a copy of the routine, load AutoCAD and test its operation.

➤ *NOTE! To test the next four routines without having the AE-SETUP data to supply the user variable USERI1, you will have to change it manually. To do this, type in the setvar command and values shown after the first command prompt line. Later, when you use the AE-MENU, this menu item is called [DOORSYM] on the table menu.*

DOORSYM.LSP Operation

Load AutoCAD and start a TEST drawing.

```
Command: (setvar "USERI1" 48.0)          Set USERI1 for 1/4" scale.
Command: (load "DOORSYM")                This loads the routine.
C:DOORSYM
Command: DOORSYM                         This starts the routine.
Enter Starting Door Number:              Enter any number to test.
Insertion Point:                         Pick a point for insertion.
```

This routine will repeat the insertion of the door symbol until you hit <RETURN> at the Insertion Point prompt.

If you are curious about what the routine looks like, here is its listing.

```
(prompt "Loading DOORSYM utility...")
(defun C:DOORSYM ( / strt pt1 ntext rmtext lay)
   (setvar "cmdecho" 0)
   (setq lay (getvar "clayer"))                         ;save current layer name.
   (command "LAYER" "S" "SYM-010" "")                   ;Set layer for symbols
   (setq strt (getint "\nEnter Starting Door Number:  "))
   (if strt
     (progn
        (setq pt1 (getpoint "\nInsertion Point:  "))
        (while pt1                                       ;go until <ENTER>
           (setq ntext (itoa strt)                       ;text door number
                 rmtext (getstring "\nEnter Room Name:  :))
           (command "INSERT" "\ae-acad/RMSYM" pt1        ;draw door annotation
                 (getvar "USERI1") "" "" ntext rmtext)
           (setq strt (1+ strt)
           pt1 (getpoint "\nInsertion Point:[or ENTER to
                 cancel]"))
                 cancel]"  ))
      )                                                  ;while loop
     )                                                   ;progn
   )                                                     ;if
   (command "LAYER" "S" lay "")
(setvar "cmdecho" 1)
(gc) (princ)
)                                                        ;defun
```

AutoLISP Routine for [DOOR–ANNOTATION]

If all went well, this short DOORSYM.LSP routine prompted for the symbol's starting number; then inserted the symbol, adding one to the number each time.

Window Annotation Symbol AutoLISP Routine

You have already tested the window annotation symbol with its associated attribute. The next step is to automate its use with an AutoLISP routine, just as you did with the door symbol.

The following AutoLISP routine is similar to the DOORSYM.LSP routine, except letters are used as attributes instead of numbers. The program implements this use by converting the beginning letter into its ASCII equivalent. It then steps the number up by one each time you insert the symbol.

WINSYM.LSP AutoLISP Routine

 Verify that the WINSYM.LSP file is in your AE-ACAD directory. Examine it with your text editor.

Read along and look at the AutoLISP listing.

Testing the WINSYM.LSP Routine

Now that you have verified that you have the WINSYM.LSP routine, load AutoCAD and test the routine's operation. Later when you use the AE-MENU, the WINSYM routine is called by the menu item, [WINSYM] on the tablet menu.

WINSYM.LSP Operation

Load AutoCAD, and create a TEST drawing file.

```
Command: (setvar "USERI1" 48.0)              Set USERI1 for 1/4" scale.

Command: (load "WINSYM")                     Load the WINSYM.LSP file.
C:WINSYM
Command: WINSYM                              Start the routine.
Insertion Point:                             Pick a point for insertion.
```

This routine will repeat the insertion of the window symbol until you hit <RETURN> at the Insertion Point prompt.

The short AutoLISP routine used in this technique will automatically start at the letter "A" and step through the alphabet as each symbol is inserted.

➡ *NOTE! Be careful when you create a symbol with multiple attributes, if the those attributes are to be supplied by an AutoLISP routine. The routine must input the attributes in the reverse that you defined them when you created the symbol. The last definition is requested first.*

If you want to look at the WINSYM routine, the listing is shown below.

```
(prompt "Loading WINSYM utility...")
(defun C:WINSYM ( / strt pt1 strt1)
    (setvar "cmdecho" 0)
    (setq lay (getvar "clayer"))                   ;save current layer name.
    (command "LAYER" "S" "SYM-010" "")             ;Set layer for symbols
    (setq strt1 (strcase (getstring "\nStarting Letter:  "))
        strt (ascii strt1))                        ;convert to number
     );setq calculations
    (if (> strt 90) (setq strt (- strt 32))        ;make sure upper
    );if
    (setvar "CMDECHO" 0)
    (setq pt1 (getpoint "\nInsertion Point:  "))
    (while (and pt1 (<= strt 90))                  ;stop at Z
        (command "INSERT" "/ae-acad/WINSYM" pt1    ;draw window
                (getvar "USERI1") "" "" (chr strt))
        (setq strt (1+ strt)                       ;incr. window letter
            pt1 (getpoint "\nInsertion Point: [or ENTER to
cancel]"))
    );while loop
 (command "LAYER" "S" lay "")

(setvar "cmdecho" 1)
(gc) (princ)
);defun
```

AutoLISP Routine for [WINDOW–ANNOTATION]

Having created the window symbol with its associated attribute, you will be able to automate the process of specifying windows in Chapter Six by using this custom AutoLISP routine.

Room Annotation Symbol AutoLISP Routine

You already created the room annotation symbol with its associated attribute, so again, let's define an AutoLISP routine to automate the symbol's insertion.

The ROOMSYM.LSP routine is operationally the same as the two previous routines except that the basic symbol, a rectangle, can contain more than one number or character. As with the DOORSYM routine, you specify a beginning number; as you insert additional symbols, the routine automatically increases this number.

ROOMSYM.LSP AutoLISP Routine

 Verify that you have a copy of the ROOMSYM.LSP file in your AE-ACAD directory. Use your text editor to examine it.

 Read along and look at the listing below.

Now that you have installed, or used your text editor to create the ROOMSYM.LSP routine, load AutoCAD and test the routine's operation. When you use the routine from the AE-MENU, it is called by the menu item [ROOMSYM].

ROOMSYM.LSP Operation

Load AutoCAD and create a TEST drawing.

```
Command: (setvar "USERI1" 48.0)            Set USERI1 for 1/4" scale.

Command: (load "ROOMSYM")
C:ROOMSYM
Command: ROOMSYM
Enter Starting Room Number:                Enter a number for testing.
Insertion Point:                           Pick a point for insertion.
Enter Room Name:                           Enter a room name for testing.
```

This routine will repeat the insertion of the room symbol until you hit <RETURN> at the Insertion Point prompt.

The ROOMSYM AutoLISP routine is similar to the first two routines that you tested. Specify a starting number, and the routine adds one each time a symbol is inserted. You still have to answer the prompt for the room name by typing it in.

If you want to examine the routine, take a look at the following the listing.

```
(prompt "Loading ROOMSYM utility...")
(defun C:ROOMSYM ( / strt pt1 ntext lay)
   (setvar "cmdecho" 0)
   (setq lay (getvar "clayer"))                      ;save current layer name.
   (command "LAYER" "S" "SYM-010" "")                ;Set layer for symbols
   (setq strt (getint "\nEnter Starting Room Number:  "))
   (if strt
(progn
       (setq pt1 (getpoint "\nInsertion Point:  ")
       (while pt1
           (setq ntext (itoa strt))                  ;text room number
           (command "INSERT" "/ae-acad/DRSYM" pt1    ;draw room annotation
                   (getvar "USERI1") "" "" ntext rmtext)
           (setq strt (1+ strt))                     ;increment room no
           pt1 (getpoint "\nInsertion Point: or ENTER to
               cancel "))
       );while loop
     );progn
   );if
(command "LAYER" "S" lay "")

(setvar "cmdecho" 1)
(gc) (princ)
)                                                     ;defun
```

AutoLISP Routine for [ROOM–ANNOTATION]

You have already created the component symbols DETAIL and EELEV. The following AutoLISP routine will automatically create the section annotation using these two symbols.

Detail and Arrow AutoLISP Routine

EELEV.LSP automates the process of defining the attribute data and placing the section symbol. It is similar in function to the door, window, and room AutoLISP routines. The EELEV.LSP routine will prompt you for an insertion point, and requests the section, or detail number, and the sheet number. It will then insert the EELEV Symbol using the same insertion point as the detail symbol. It prompts you for the symbol's rotation angle.

EELEV.LSP AutoLISP Routine

 Verify that you have a copy of the EELEV.LSP file in your AE-ACAD directory. Examine it with your text editor.

 Read along and look at the AutoLISP listing below.

Now that you have the EELEV.LSP routine, load AutoCAD and test the routine's operation. When you use the AE-MENU, this routine is called by the [EELEV] menu item.

Testing the EELEV.LSP Routine

EELEV.LSP Operation

Load AutoCAD and start a TEST drawing.

```
Command: (setvar "USERI1" 48.0)          Set USERI1 for 1/4" scale.

Command: (load "EELEV")
C:EELEV
Command: EELEV
Insertion Point:                         Pick any point to insert DETAIL symbol.
Number - Above:                          Specify detail number.
Number - Below:                          Specify drawing sheet number.
Rotation angle <0>:                      Specify angle for EELEV symbol.
```

This routine will repeat the insertion of the symbols until you hit <RETURN> at the Insertion Point prompt.

This short EELEV.LSP routine combines two symbols to create one and prompts for the section number, sheet number, and direction the arrow is to point.

Here is the listing for the routine:

```
(prompt "Loading EELEV utility...")
(defun C:EELEV ( / lay numabv numblw pt)
   (setvar "cmdecho" 0)
   (setvar "attreq" 1)                              ;prompt attributes
   (setq lay (getvar "clayer"))
   (command "LAYER" "SET" "SYM-010" "")
   (setq numabv (getstring "\nNumber - Above: "))
   (setq numblw (getstring "\nNumber - Below: "))
   (setq pt (getpoint "\nSelect point: "))          ;get the point?
   (command "INSERT" "\ae-acad/DETAIL" pt           ;insert detail
         (getvar "USERI1") "" "" numabv numbw)
   (command "INSERT" "/ae-acad/EELEV" pt
         (getvar "USERI1") "" pause)
   (command "LAYER" "S" lay ""))

(setvar "cmdecho" 1)
(gc) (princ)
)                                                   ;defun
```

AutoLISP Routine for [ELEVATION–SYMBOL]

This next routine lets you select and edit multiple attributes.

Edit Attribute AutoLISP Routine

The following AutoLISP utility allows you to edit multiple attributes with the DDATTE command. The native DDATTE command only lets you edit one attribute at a time. To edit multiple attributes, you have to restart the command, resulting in numerous key strokes and selections. With the following AutoLISP utility, called EDITATTR.LSP, this operation is reduced to selecting the attributes, pressing the <RETURN> key, and editing the selected attributes one at a time — without having to restart the DDATTE command. When you use the AE-MENU, the routine is called by the [EDITATTR] menu item.

The EDITATTR.LSP routine comes in handy when a major revision to a drawing requires a major change in symbols and associated attributes that have already been defined. You will use this routine later in Chapter 10 to edit attributes in drawing title blocks.

EDITATTR.LSP AutoLISP Routine

 Verify that you have a copy of the EDITATTR.LSP file in your AE-ACAD directory. Examine it with your text editor.

 Read along and look at the AutoLISP file listing.

Testing the EDITATTR.LSP Routine

Now that you have the EDITATTR.LSP routine, load AutoCAD and test the routine to see how it works.

EDITATTR.LSP Operation

Load AutoCAD and create a TEST drawing.
```
Command: (load "EDITATTR")
C:EDITATTR
Command: EDITATTR
Select Attribute Blocks to Edit:          Select as many symbols as you wish.
```
This routine will load each attribute with the DDATTE command for editing.

Here is the listing that creates the custom command called EDITATTR.

```
(prompt "Loading EDITATTR utility...")
(defun C:EDITATTR ( / selct indx entnm)
   (setvar "cmdecho" 0)
   (command "UNDO" "GROUP")
   (prompt "\nSelect Attribute Blocks to Edit: ")        ;create selection set
    (setq selct (ssget)
         indx 0                                          ;initialize pointer
         entnm (ssname selct indx)                       ;get entity name
   )                                                     ;setq calculations


   (while entnm                                          ;until no entities
     (command "DDATTE" entnm)                            ;DDATTE command
     (setq indx (1+ indx)                                ;increment index
          entnm (ssname selct indx)                      ;get next attribute
     )                                                   ;setq calculations
   )                                                     ;while loop
   (command "UNDO" "END")
   (setvar "cmdecho" 1)
)                                                        ;defun
```

AutoLISP Routine for [EDIT–ATTRIBUTE]

The concepts, symbols, and routines that we have presented in this chapter will give you a good basis for creating or expanding your own symbol and AutoLISP libraries. You will see how these, and other symbols and insertion routines, can make creating working drawings much easier when you create the working drawings for the simulated branch bank project in later chapters.

Summary

Placing symbols and their associated data into drawings can be time-consuming. With conventional methods, you often use a template to trace the required symbols and then letter the associated data by hand or with a lettering medium. You have to keep track of this data and make sure that you don't repeat data. If changes have to be made, you have to erase and revise the affected symbols, making coordination errors a distinct possibility.

AutoCAD provides a more accurate system for inserting and tracking symbols than conventional, hand-drafting methods. Using AutoLISP as an automated input and revision tool, you can gain more consistency between drawings, especially those whose symbols have relational references (such as door symbols to door schedules). If you use symbol libraries, assign attributes, and take advantage of custom AutoLISP routines, the tasks of placing symbols, adding data, and making revisions become minor more manageable problems.

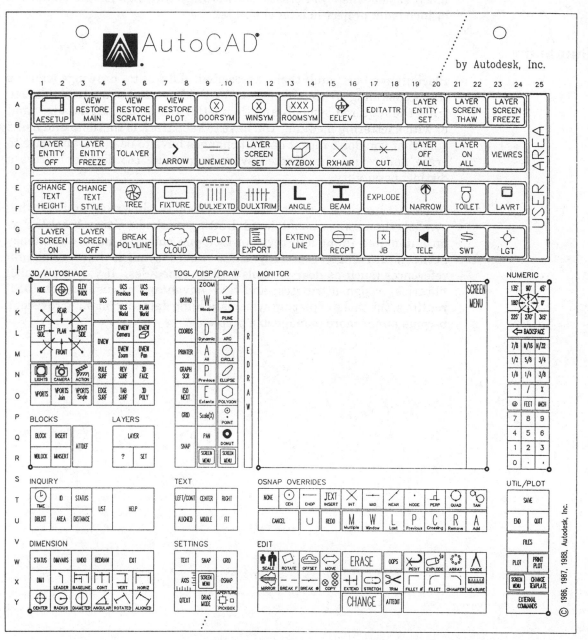

The Menu System

CHAPTER 3

Menu Routines and AutoLISP

Introduction

As an experienced AutoCAD user, you have probably come to the realization that there is no canned menu system that you can buy, from Autodesk or any one else, that will perfectly fit your firm's menu needs. Fortunately, AutoCAD was designed with customization in mind. In fact, it is one of AutoCAD's strongest selling points. Customizing and managing the customized system are two of the many on-going processes inherent in optimizing your productivity. In this chapter, we will begin to explore customizing the menu functions of AutoCAD to meet our (and hopefully your) production needs.

From a broad perspective, customization can mean anything from adding a few symbols on a tablet menu to building a complete menu system. How completely you want to customize your system usually depends on how experienced you are as an AutoCAD user, as well as on the types of projects you do. The more familiar you become with how the system functions, the more likely you are to see a faster route, or wish for more direct access to particular functions than the existing menu systems allow. To customize a menu system, you simply arrange the tools that you require to be at your finger tips. The way you go about it depends how much time you have to devote to the endeavor or how much money you are willing to spend.

Customization Choices

The first option is to stay in-house with the job of customizing the system. This can prove to be the most cost effective if there is a person or group of people already within your ranks skilled enough and willing to take on the task. The benefits of this choice are that you become intimately familiar with customizing the system's menu functions and tailoring the results for your exact needs.

The second option is to hire a third party or consultant outside of the firm who will be capable of meeting your customization requirements. The down-side to this approach is its cost, and whether you can find available consultants who can understand exactly what you want. Finally, there is the question of whether these consultants can meet your schedule. The benefits to the consultant approach are it will take less of your time, and you can learn a lot from the consultant's experiences.

The third option is to purchase a third-party menu system. This is generally the quickest method of customization. A well-designed menu package should provide you with a workable system that can meet most of your needs. Make sure the third-party menu system lets you customize their functions further. As your project progresses or your project types change, you can alter and add to the menu options. As an example of this option, our AE-MENU is a customization of an AEC menu system, called ARCH-T, from KETIV TECHNOLOGIES, Inc.

Our customized menu system, which we call the AE-MENU system, is useful in itself and is provided for you on the companion AE DISK. Even if you don't use our menu system, or you have one of your own, the examples and techniques we draw from the AE-MENU system will be useful for improving your productivity on AutoCAD in the A/E environment.

Implementation Prerequisites

The custom menus and various AutoLISP utilities presented in this chapter are available for your use on the AE DISK. Use them as they are; you'll find them useful tools in their own right. In addition, the menu topics that we discuss in this chapter will help you understand how our menu system works, and may show you some areas where you can make improvements in your own AutoCAD system.

You can also use the menus and routines from the AE-MENU system as a basis for implementing further customizations into your own AutoCAD system. In order to do this, you should have more than a casual knowledge of AutoCAD and be familiar with AutoLISP programs. We emphasize this understanding because you must be careful to test new programs, and test any changes that you make to existing programs, to be sure that they function correctly and are without errors. DO NOT alter the current working version of your menu system or a valuable drawing until you have thoroughly tested any changes or additions that you've made. Please refer to CUSTOMIZING AUTOCAD, and INSIDE AUTOLISP (New Riders Publishing) for more information on menu and AutoLISP topics where you feel you may need more help. Before we get into the AE-MENU, let's explore some basic menus system concepts.

Menu System Concepts

A good menu system is the key to rapid operation of the drafting functions developed by AutoCAD. Menus provide access to functions activated by single commands, macros, and AutoLISP programs. A simple selection from either of the two main types of menus, screen or tablet, gives you the power to create, move, edit, and change data at will. To give you all the options for customization that you need, you will probably want to use both the screen and tablet menu areas.

We will take examples from the AE-MENU System, which we use daily in our A/E firm, to show several possibilities of how menu systems can be customized. You can create and edit menu systems with any word processing program. To experiment with our menu system program, found in the file AE-MENU.MNU on the companion AE DISK, your word processing package will need to be able to handle a large document file. (The complete menu program is approximately 40,000 bytes in size. We show a partial listing in Appendix B.)

Menu syntax is important in creating and revising menus. You can create or intensify problems if you are not careful when you make changes to the programs. Several levels and paths exist within the format used to define AutoCAD menus. Follow our recommended logic; it will help you to eliminate some time consuming guesswork and calculations.

The following section gives an overview of the concepts involved in creating menus, including a brief description and example of each of the AutoCAD Main Menu sections. Following this, we will show some of the modifications that we made to the standard AutoCAD menu to create the AE-MENU.MNU program. For a more complete description of the formats and functions of AutoCAD menu areas, we again refer you to INSIDE AUTOCAD, or CUSTOMIZING AUTOCAD.

AutoCAD's Standard Menu Areas

When AutoCAD was first introduced, it had one simple screen menu area. As programming techniques improved and users indicated the desire for more power and easier ways to perform drawing tasks, AutoCAD added multiple screen menus, Button, and pull-down menus with associated icon menus, and tablet menu areas. The increase in function that followed caused an increase in complexity.

While today's AutoCAD menu system is easy for the novice or student to follow, users who must produce drawings day-in and day-out find that its user friendliness can get in the way of productivity. This, in itself, is one of the primary reasons to customize. Having foreseen the wide range in the users' skill levels and desires for direct interaction with the system, Autodesk, from the very beginning, gave us the means with which to change the standard menu system, or create an entirely new menu system.

The following descriptions and illustrations of portions of each section of the standard AutoCAD menu system, ACAD.MNU, should give you some insight into what to expect when either creating or enhancing your own menu system. If you examine the ACAD.MNU file with a text editor, you will find that each main menu section is defined by *** (three asterisks) first, followed by the section name. Submenu sections or pages are defined with ** (two asterisks) first, followed by the subsection name. If you want to follow along, you can examine a listing of ACAD.MNU with a text editor to see how these sections are organized. If you don't want to use your editor, we will provide enough sample listings so that you can get an idea of what the menu looks like, and how it is organized.

Button Menu Section

Each section of the AutoCAD menu program is delineated by what is known as a file section label. The commands or functions defined between labels are activated by the device name in the label. The *** indicates to AutoCAD that the name to follow is a main menu type. The button menu section, delineated with the ***BUTTONS file section label, is designed to add user-defined functions to the buttons on the puck of the digitizer tablet. These functions can range from a simple <RETURN> to activating elaborate custom AutoLISP routines.

The button menu section is the quickest area to find because it is the first section of the menu system program. You can gain rapid access to this section by using a good text editor, and you can add permanent drawing, or task-dependent functions. Being able to activate functions with puck buttons saves you a lot of hand and eye movement and can definitely make you more productive.

The ***BUTTONS label shown below signals AutoCAD that the functions that follow are activated by the buttons of the pointing device of the digitizer.

AutoCAD's Standard Button Menu

```
***BUTTONS                          Main Menu Label.
;                                   Used by AutoCAD to denote an Enter.
$p1=*                               Used to activate the Pull Down Menu.
^c^c                                Used to denote a Cancel.
^B                                  Used to toggle Snap mode on/off.
^O                                  Used to toggle Ortho mode on/off.
^G                                  Used to toggle Grid on/off.
^D                                  Used to toggle coordinate display.
^E                                  Used to toggle Isoplane left/top/right.
^T                                  Used to toggle Tablet mode on/off.
***AUX                              Main Menu Label for Auxiliary "function box."
```

The different sections of AutoCAD's menu system can contain as many commands as you can add to a section, with the exception of the button section. One button is always reserved for selecting or picking and cannot be reprogrammed. If you have a four button puck, you can only use three buttons for customized commands.

Pull Down and Icon Menu Sections

Autodesk added the Advanced User Interface — AUI to AutoCAD Release 9. This interface consists of a customizable menu bar, pull down and icon menus, and associated dialogue boxes. With the addition of the AUI, an entire new dimension in customization is possible. Now you can select pull down menus from the menu bar that appears when the crosshairs of your pointing device enter the drawing editor's reference area at the top of your screen. After making your selection, the pull down menu appears in the drawing area. Then, you can select additional commands or symbols from the pull-down menu.

Dialogue boxes, which can be used as part of pull down menus, appear in the main drawing area and allow quicker editing of drawing settings, layers, and attributes, among other things. Dialogue boxes can contain areas for you to input text into or are used to display icons for you to select. You cannot customize the actual structure of dialogue boxes.

The following is a partial listing of AutoCAD's standard pull down menu section named TOOLS. This area's file section label is ***POP1. Pull down menus can be labeled 1 through 10, signifying the location of the pull down menu in the menu bar area of the drawing editor.

AutoCAD's Tools Pull Down Menu

Standard Pull Down Menu

```
***POP1                                 POP1 Label.
[Tools]                                 Pull Down Menu Title.
 [OSNAP]^c^c$p1= $p1=* OSNAP \          Used to activate the OSNAP command.
 CENter                                 Used to select CENTER option.
ENDpoint                                Used to select ENDPOINT option.
INSert˙                                 Used to select INSERT option.
[INTersection]INT                       Used to select INTERSECTION option.
MIDpoint                                Used to select MIDPOINT option.
NEArest                                 Used to select NEAREST option.
NODe                                    Used to select NODE option.
```

Items that are defined in brackets, for example [Tools], are text descriptions. What comes after the brackets is the actual command performed by the selection of the item. Use the following line as an example:

```
[OSNAP]^c^c$p1= $p1=* OSNAP \
```

The bracketed OSNAP serves as a descriptor of the functions this menu selection provides. The text of the rest of the line activates the OSNAP command.

If nothing follows the [ ], then selecting that item will do nothing.

The next menu section that we can look at is the icon menu section. An icon is a graphic representation which, when selected, will initiate a macro or command. AutoCAD uses the icon menu section as a way to let you change font styles, select different 3D views, or define different UCS's. In addition, a major advantage to using this area is that after you create slide libraries of symbol drawings, you can create simple representations (icons) and place them in an icon menu. Then, you can select the symbol you want to insert after you view the icon from the screen first. Icon menus can be activated from any section of the AutoCAD menu program.

Let's take an example from the icon menu shown below in the standard AutoCAD menu system. This menu displays the available 3D Views and sets the current view, which you can list by using the VPOINT command.

A major difference exists between the icon section and other sections of ACAD.MNU. What appears in the brackets [ ] is not the command or a description of the command, but a call to a slide library (ACAD.SLB). What appears within parentheses ( ) is the name of the slide that will be displayed. What follows the parentheses is the command, or function, that is performed when the icon is selected, similar to what is done in other menu sections.

```
[acad(ul)]^C^CVPOINT R;<135;
```

The acad indicates that this is a call to the slide library named ACAD. The ul is the name of slide that makes up this portion of the icon menu. The remainder of the line consists of the command that will be executed if you select this icon for rotating the viewpoint.

The following icon menu area has a file section label of ***ICON and an associated submenu section, specified as **3DViews.

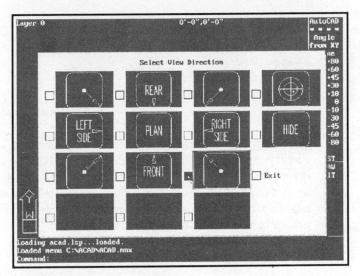

AutoCAD's 3DViews Icon Menu

```
Icon Menu

**3DViews                              Submenu Label.
[Select View Direction]               Icon Menu Title.
[acad(ul)]^C^CVPOINT R;<135;          See above for description.
[acad(l)]^C^CVPOINT R;<180;
[acad(ll)]^C^CVPOINT R;<225;
[acad(user)]
[acad(u)]^C^CVPOINT R;<90;
[acad(p)]^C^CVPOINT 0,0,1
[acad(lo)]^C^CVPOINT R;<270;
[acad(user)]
[acad(ur)]^C^CVPOINT R;<45;
[acad(r)]^C^CVPOINT R;<0;
[acad(lr)]^C^CVPOINT R;<315;
[acad(user)]
[acad(t)]VPOINT;;
[acad(h)]$S=X $S=HIDE
[ Exit]^c^c                           Used to exit the Icon.
```

Selecting symbols through icon menus lets you search a large symbol library quickly, and see the block or symbol on the screen before selecting it for insertion. A major drawback in using icon menus is that if you revise or add even one symbol, you have to recreate the entire slide library. This can become time consuming and leaves room for errors as the library grows in size and complexity.

Screen Menu Section

The screen menu section is a good example of how complex AutoCAD's menu system, with its main menu, submenus, and sub-submenus, has become through years of development. This complexity can impact systems management. If a new release of AutoCAD comes out, you may have to rewrite portions of your custom system to be able to use new features. You can avoid major rewrites under these circumstances by planning how you use your customized systems, and by developing good documentation. See CUSTOMIZING AUTOCAD for ways to design easily updatable menus.

To employ the correct syntax for menu calls, you first specify the indicator for the type of menu called — buttons, pull down, icon, screen, or tablet — then you specify an equal sign and the name of the submenu. Refer to the following table for the required menu codes.

```
SYNTAX
INDICATOR          USAGE
$B                 Button Menu
$A1                Auxiliary Menu
$P(#)              Pull Down Menu with specified
                   menu number
$I                 Icon Menu
$S                 Screen Menu
$T(#)              Tablet Menu with specified menu
                   number
$<INDICATOR>=      Will return to the last SubMenu or
                   Page
```

AutoCAD's Standard SubMenu Syntax

The following listing displays portions of the standard ***SCREEN menu section of AutoCAD. This menu appears first when a drawing is loaded. It is full of calls to submenus. Notice that you can include AutoLISP routines as part of submenu calls. This is illustrated by the [ASHADE] menu call shown in the listing below.

A Portion of AutoCAD's Screen Menu

```
***SCREEN                                  Main Menu Label.
 **S                                       SubMenu Label.
 [AutoCAD]^C^C$S=X $S=S $P1=POP1 $P3=POP3        Used to return, reset Main Menu
sections.
 [* * * *]$S=OSNAPB                        Call to OSNAP SubMenu.
 [Setup]^C^C^P(progn(prompt "Loading setup...   ")(load
"setup")) ^P$S=X $S=UNITS                  Load AutoLISP Setup routine.
```

The following items are calls to different Submenus or Pages.

```
[BLOCKS]$S=X $S=BL
[DIM:]$S=X $S=DIM ^C^CDIM
[DISPLAY]$S=X $S=DS
[DRAW]$S=X $S=DR
[EDIT]$S=X $S=ED
[INQUIRY]$S=X $S=INQ
[LAYER:]$S=X $S=LAYER ^C^CLAYER
[SETTINGS]$S=X $S=SET
[PLOT]$S=X $S=PLOT
[UCS:]$S=X $S=UCS1 ^C^CUCS
[UTILITY]$S=X $S=UT

[3D]$S=X $S=3D
```

The following is an AutoLISP routine to set up elements for AutoShade.

```
[ASHADE]^C^C^P(progn(setq m:err *error*)(prin1))(defun *error* (msg)(princ
msg)+
 (setq *error* m:err m:err nil)(princ))(cond ((null C:SCENE)(vmon)+
(if (/= nil (findfile "ashade.lsp"))(progn (terpri);+
(prompt "Please wait...  Loading ashade.  ")(load "ashade")+
(menucmd "S=X")(menucmd "S=ASHADE")(setq *error* m:err m:err nil))(progn
(terpri);+
(prompt "The file 'Ashade.lsp' was not found in your current search
directories.")+
(terpri)(prompt "Check your AutoShade Manual for installation instructions.");+
(setq *error* m:err m:err nil)(princ))))+
(T (setq *error* m:err m:err nil)(menucmd "S=X")(menucmd "S=ASHADE")(princ)))
^P
```

[SAVE:]^C^CSAVE   AutoCAD command to save the drawing to disk.

The screen menu section is the most complex menu section used for drawing input. Read INSIDE AUTOCAD or CUSTOMIZING AUTOCAD for a more in-depth discussion of this segment of the menu system.

Tablet Menu Section

When AutoCAD allowed for the use of more input devices, the digitizing tablet became a natural extension for executing custom macros and activating AutoLISP routines. AutoCAD provides one tablet menu section, tablet area one, specifically for adding your own commands. This offers an easy and quick way to enhance or customize AutoCAD without changing any of the standard AutoCAD tablet menu code found in the three other standard tablet menu sections. Code for the tablet menu section is embedded in the last half of the AutoCAD menu system and is hard to find unless you use a good text editor.

The customizable top portion of the AutoCAD template is defined by the code contained in the section designated ***TABLET1 (Tablet area one). The following is a partial listing of this section. (Please note that AutoCAD has changed the format for this section from Release 9. Instead of numbering items [T1-1] thru [T1-200], they are labeled [A-1] to [A-25] for the first row through [H-1] to [H-25] for the last row. There are still 200 individual areas that you can define).

One difference between the tablet and other sections is that what appears in the [ ] is an address on the tablet, not a command or descriptor. What follows each label or address is the section in which to define the command or macro to be activated by picking that address.

```
A Portion of AutoCAD's Tablet Menu

***TABLET1
[A-1]
[A-2]
[A-3]
[A-4]
[A-5]
[A-6]
[A-7]
[A-8]
[A-9]
[A-10]
```

You create code for tablet menu sections using macros. Now that we have finished our overview of the standard AutoCAD menu system sections, we'll briefly review AutoCAD's macro command structure and syntax before moving on to the AE-MENU.

Macro Command Structure

The following is a quick refresher on the procedures of macro creation. Macros are a language which communicates commands to AutoCAD. Any command in AutoCAD may be used in writing custom macros. Creating macros is the first step in learning and understanding the flow of commands and responses between you and AutoCAD. Macros let you move through a series of commands and responses with very little interaction with the system. Here is an example. Say you always begin your drawings with the following command sequence:

Layer Set

```
Command: LAYER
?/Make/Set/New/ON/OFF/Color/Ltype/Freeze/Thaw: S
New current layer : EXT-010
?/Make/Set/New/ON/OFF/Color/Ltype/Freeze/Thaw: <RETURN>
```

By observing how this command operates while you use the keyboard for input, you can easily determine how to reproduce the function in a custom menu or macro. You can add the command sequence just listed to the tablet menu section by using some of the simple characters listed in the table below.

CHARACTER	USAGE
;	A semicolon is used to specify a <RETURN>.
\	A backslash is used to pause for user input.
" "	A space is used to represent pressing the space bar on the keyboard.
^c	A control C is used as a Cancel.
+	A plus is used as an indicator to go to the next line of a macro sequence.

AutoCAD's Macro Codes

The keyboard input when using the LAYER command and our previous layer set sequence looks like this:

LAYER <RETURN> S <RETURN> EXT-010 <RETURN>< RETURN>

Note the two <RETURN>s at the end of the sequence. The layer command requires two <RETURN>s to complete the command. To convert this into a macro, the table tells us that the <RETURN>s will have to be replaced with semicolons. The macro would look like this:

LAYER;S;EXT-010;;

And if you wanted to use the first address location, [A-1], of tablet area one to activate this macro, you would add the following to the listing of that tablet menu:

[A-1]^c^c^cLAYER;S;EXT-010;;

The addition of the three ^c characters will cancel any other commands when the macro is selected so they don't interfere with the macro's operation.

If you follow these simple principles, you can use any command available with AutoCAD in a macro. You can link commands together in the same manner. In fact, you can make a series of commands as long as you want if you add a plus + at the end of each line. AutoCAD uses this character as an indication to proceed to the next line, not to stop.

The AE-MENU System

The AE-MENU system is our own customization of a reduced set of a menu system from KETIV TECHNOLOGIES, called ARCH-T, adapted to the requirements of this book. (To learn more about the full KETIV menu, see the information in the back of the book.) The process we chose in customizing KETIV's menu system was one of the three options we mentioned of producing a customized menu system.

We chose this menu system as one that fit our needs fairly closely. You can use our techniques and methods of customization to modify another menu system or create your own. The AE-MENU system, includes the symbols and AutoLISP routines that you copied (or created) in Chapters One and Two as well as additional routines that you will encounter as you progress through the rest of the book.

The AE-MENU.MNU program is designed for Release 10 tablet section one, provided by the standard AutoCAD menu. If you are using the AE DISK, we assume that you have loaded the AE-MENU, and installed the AE-MENU tablet menu. If you have not installed the AE-MENU, refer back to Chapter Two for instructions on how to install the menu.

➡ *NOTE! The AE-MENU system contains a special ACAD.LSP file which can overwrite the existing ACAD.LSP file when the AE DISK files are installed. If you have customized this file, or are using another menu system that may have changed it, it is best to rename your existing ACAD.LSP file to ACAD.OLD before you install the AE-MENU system DISK files in your AE-ACAD directory.*

Use the following sequence to verify that your AE-MENU system is installed.

Testing the AE-MENU System

C:> **CD\AE-ACAD**   Change directories.

 Verify that you have a copy of the AE-MENU.MNU file and the ACAD.LSP files from the AE DISK in your directory.

C:\AE-ACAD>

Load AutoCAD and from the Main Menu select the following:

Enter selection: **1**
Enter NAME of drawing : **AETEST**

Command: **MENU**
Menu file name or . for none <ACAD>: **AE-MENU**
Compiling menu AE-MENU.mnu...

The menu system should now function.

```
Command: QUIT
Do you really want to discard all changes ? Y
```

 Read along.

If you do not have the AE DISK, but you do want to use the AutoLISP routines that we have listed throughout the book, you will need to type them in with a text editor. As you create the programs, load the AutoLISP routines and test each as it is described in the text.

AE-MENU Tablet Area One Diagram

USER DEFINED

	1–2	3–4	5–6	7–8	9–10	11–12	13–14	15–16	17–18	19–20	21–22	23–24
B	AESETUP	VIEW RESTORE MAIN	VIEW RESTORE SCRATCH	VIEW RESTORE PLOT	DOORSYM	WINSYM	ROOMSYM	EELEV	EDITATTR	LAYER ENTITY SET	LAYER SCREEN THAW	LAYER SCREEN FREEZE
C/D	LAYER ENTITY OFF	LAYER ENTITY FREEZE	TOLAYER	ARROW	LINEMEND	LAYER SCREEN SET	XYZBOX	RXHAIR	CUT	LAYER OFF ALL	LAYER ON ALL	VIEWRES
E/F	CHANGE TEXT HEIGHT	CHANGE TEXT STYLE	TREE	FIXTURE	DULXEXTD	DULXTRIM	ANGLE	BEAM	EXPLODE	NARROW	TOILET	LAVRT
G/H	LAYER SCREEN ON	LAYER SCREEN OFF	BREAK POLYLINE	CLOUD	AEPLOT	EXPORT	EXTEND LINE	RECPT	JB	TELE	SWT	LGT

AE-MENU Tablet Area One Diagram

AE-MENU Tablet Area One

As we mentioned in our description of the tablet menu sections earlier in the chapter, tablet area one comes from AutoCAD ready for customization. Areas two, three, and four are filled in with standard AutoCAD commands, but can also be customized. Area one is where the AE-MENU.MNU program and all the symbols, programs, and routines developed for this book are located.

Below we show a partial listing of the AE-MENU item called [AE-SETUP], which we are going to use as an example of how menu routines operate and how they are organized. You can select this routine from the tablet; it contains a main screen and submenus, user-defined variables, prototype drawings, a border drawing, units settings, and view specifications.

Using the AE MENU System Setup

To show you how the AE-SETUP routine works, we will walk you through it to set up a test site plan drawing.

AE-MENU Main Screen Menu

When AE-SETUP is selected from the tablet overlay, the screen menu, called the AE-SETUP (shown below) appears. This is the menu code which is in brackets in the menu file. While the bracketed text may be the name of a command, it can also be a description of the function performed by commands when that option is picked.

```
Layer 0                              0'-0",0'-0"              AE-SETUP
                                                               MENU
                                                              DRAW'G
                                                               TASK

                                                              CIVIL
                                                              -PLAN-
                                                             -LANDSC-

                                                             ARCHTECT
                                                              -PLAN-
                                                              -ELEV-
                                                              -SECT-
                                                              -DETL-

                                                             ENGINEER
                                                              -STRU-
                                                              -MECH-
                                                              -PLUM-
                                                              -ELEC-

Loading acad.lsp...loaded.
Loaded menu C:\ACAD\AESYS.mnx
Command:
```

AE-SETUP Screen

To use the menu to set up the site plan drawing, follow the next instruction sequence.

❏ Load AutoCAD and begin a NEW drawing called TEST-SP.

❏ Load the AE-MENU with the MENU command.

❏ Start by initializing the setup routine from the AE-MENU tablet menu by selecting the menu item called [AESETUP].

❏ Select [-PLAN-] from the CIVIL section of the AESETUP1 menu that appears in the righthand screen menu area.

❏ The prototype drawing SITE-PT (which we defined in Chapter One) will then be automatically inserted into the drawing area of the screen, providing all the required layers for the site plan drawing.

❏ A submenu, AESETUP2, will then be called up and the screen defined for it will appear on your display.

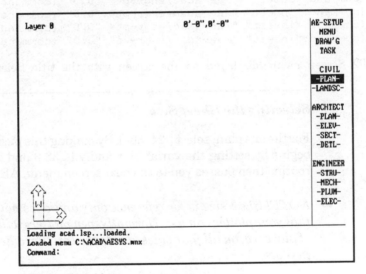

Drawing Task Menu

The following listing shows a part of the AE-MENU screen menu section.

```
Main AE-SETUP Menu

**AESETUP1
[AE-SETUP]
[MENU    ]
[DRAW'G  ]

[TASK    ]

[   CIVIL ]
[ -PLAN- ](command "insert" "/ae-acad/site-pt" "0,0" (command));$S=AESETUP2
[-LANDSC-](command "insert" "/ae-acad/site-pt" "0,0" (command));$S=AESETUP2

[ARCHTECT]
[ -PLAN- ](command "insert" "/ae-acad/plan-pt" "0,0" (command));$S=AESETUP2
[ -ELEV- ](command "insert" "/ae-acad/plan-pt" "0,0" (command));$S=AESETUP2
[ -SECT- ](command "insert" "/ae-acad/sect-pt" "0,0" (command));$S=AESETUP2
[ -DETL- ](command "insert" "/ae-acad/sect-pt" "0,0" (command));$S=AESETUP2

[ENGINEER]
[ -STRU- ](command "insert" "/ae-acad/engr-pt" "0,0" (command));$S=AESETUP2
[ -MECH- ](command "insert" "/ae-acad/engr-pt" "0,0" (command));$S=AESETUP2
[ -PLUM- ](command "insert" "/ae-acad/engr-pt" "0,0" (command));$S=AESETUP2
[ -ELEC- ](command "insert" "/ae-acad/engr-pt" "0,0" (command));$S=AESETUP2
```

The AESETUP2 Screen menu, displayed on the screen with the title Select Sheet Size, gives you five options.

Selecting the Sheet Size

For the site plan, select [-24x36-]. By making this pick, an AutoLISP routine begins by setting the variables x and y to 36.0 and 24.0, respectively. The routine then passes you to the next screen menu, AESETUP3.

➠ *NOTE! Sheet size is for reference only and does not indicate the true size that your plotter can use. Normally, pen plotters will lose 1 to 1 1/2 inches of plot area, on all four sides, due to the traction wheels that hold the sheet in place.*

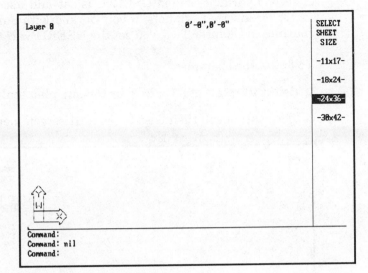

Sheet Size Selection Menu

Here is the menu listing.

```
Sheet Size
**AESETUP2
[SELECT  ]
[SHEET   ]
[SIZE    ]

[-9x11-  ](setq x 11.0)(setq y 9.0);$S=AESETUP3

[-11x17- ](setq x 17.0)(setq y 11.0);$S=AESETUP3

[-18x24- ](setq x 24.0)(setq y 18.0);$S=AESETUP3

[-24x36- ](setq x 36.0)(setq y 24.0);$S=AESETUP3

[-30x42- ](setq x 42.0)(setq y 30.0);$S=AESETUP3
```

Selecting Drawing Scale

The following drawing scale screen menu, AESETUP3, is a little more complicated than the two you've already reviewed. Two different variables are defined when you select a scale factor for a drawing from the screen. A variable, called scale, is set and used in the AutoLISP routine that does the setup operation.

A second variable, called USERI1, is set and used to supply the scale factor for inserted symbols. If you worked with our AutoLISP insertion routines in Chapter Two, you set the USERI1 variable manually.

To follow the example:

❑ Select the scale of [-1"=20'-] for the site plan scale.

❑ The routine will then pass to the next screen menu, AESETUP.

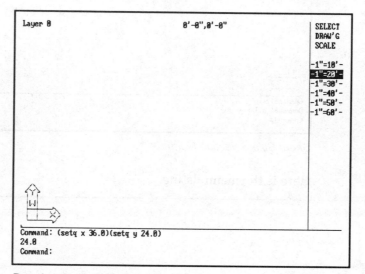

Drawing Scale Selection Menu

Here is the menu listing.

```
Drawing Scale
**AESETUP3
[SELECT  ]
[DRAW'G  ]
[SCALE   ]

[ -1/16"-](setq scale 192.0)(setvar "USERI1" 192.0);$S=AESETUP
[ -1/8"- ](setq scale 96.0)(setvar "USERI1" 96.0);$S=AESETUP
[ -1/4"- ](setq scale 48.0)(setvar "USERI1" 48.0);$S=AESETUP
[ -1/2"- ](setq scale 24.0)(setvar "USERI1" 24.0);$S=AESETUP
[ -3/4"- ](setq scale 16.0)(setvar "USERI1" 16.0);$S=AESETUP
[-1 1/2"-](setq scale 8.0)(setvar "USERI1" 8.0);$S=AESETUP
[  -3"-  ](setq scale 4.0)(setvar "USERI1" 4.0);$S=AESETUP
[ -FULL- ](setq scale 1.0)(setvar "USERI1" 1.0);$S=AESETUP


[-1"=10'-](setq scale 120.0)(setvar "USERI1" 120.0);$S=AESETUP
[-1"=20'-](setq scale 240.0)(setvar "USERI1" 240.0);$S=AESETUP
[-1"=30'-](setq scale 360.0)(setvar "USERI1" 360.0);$S=AESETUP
[-1"=40'-](setq scale 480.0)(setvar "USERI1" 480.0);$S=AESETUP
[-1"=50'-](setq scale 600.0)(setvar "USERI1" 600.0);$S=AESETUP
[-1"=60'-](setq scale 720.0)(setvar "USERI1" 720.0);$S=AESETUP
```

Selecting the Border Drawing

The last screen menu, AESETUP, prompts you with [SELECT EITHER
Y or N BELOW TO BEGIN SET-UP ROUTINE] and [DO YOU WANT
SCRATCH AREA -YES- -NO-] to determine how to begin the drawing
area setup.

□ Select [-YES-] for the Site Plan.

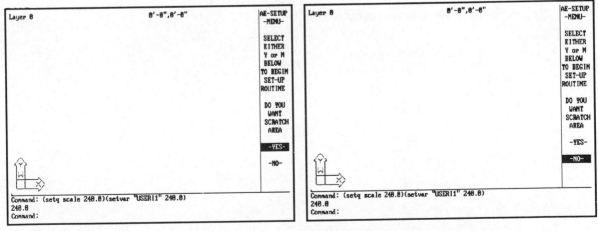

AEBORDER Selection   AEBORDER Drawing

After making this selection, our AutoLISP setup routine begins using the
variables defined by your choices in the previous menus. Its first

operation is to set the drawing limits for the AEBORDER drawing that was created in Chapter One. Then it configures the required VIEWs and sets the default to MAIN. The last two functions set the UNITS and display all of the layers, by pages, in the screen area so that you can select the layer on which to begin drawing.

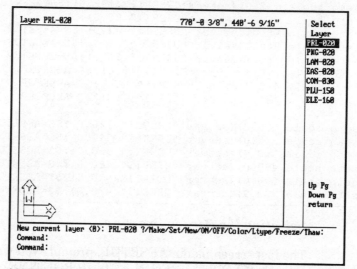

MAIN Drawing Area

If you had answered [-NO-] above, the standard AutoCAD border drawing would be inserted and only a PLOT view would be set up.

Here is the menu listing.

```
Border Drawing & Setup
**AESETUP
[AE-SETUP]
[ -MENU- ]
[        ]
[ SELECT ]
[ EITHER ]
[ Y OR N ]
[ BELOW  ]
[TO BEGIN]
[ SET-UP ]
[ROUTINE ]
[        ]
[DO YOU  ]
[ WANT   ]
[SCRATCH ]
[ AREA   ]
[        ]
[ -YES- ]^C^C(setq x1 (* x scale)) (setq y1 (* y scale))+
(setq x2 (* x1 2)) (setq y2 (* y1 2))+

(command "limits" "0,0" (list x2 y2) "insert" "/ae-acad/AEBORDER" "0,0" "c"+
(list x2 y2) "0" "view" "w" "main" "0,0" (list x1 y1)+
"view" "w" "plot" "0,0" (list x1 y1) "view" "w" "scratch" (list x1 y1)+
(list x2 y2) "view" "r" "main"))) (setvar "LUNITS" 4)^D;+
(command "ucs" "o" (list x1 y1) "ucs" "s" "scratch" "ucs" "")+
(setvar "LUNITS" 4)^D;LAYLOAD;LAY-UTIL;LAYERSET;
[        ]
[ -NO-  ]^C^C(setq x1 (* x scale)) (setq y1 (* y scale))+
(command "limits" "0,0" (list x1 y1) "insert"+
"border" "0,0" "c" (list x1 y1) "0" "zoom" "a"+
"view" "w" "plot" "0,0" (list x1 y1))
(setvar "LUNITS" 4)^D;LAYLOAD;LAY-UTIL;LAYERSET
```

In the example of setting up the site plan drawing, you saw how, by selecting from a tablet overlay, you can execute a very streamlined and useful series of commands and responses through associated screen menus and AutoLISP routines.

Ending Your Test Drawing

When you are done examining the AE-MENU's setup routine, end your drawing. You will use this same procedure for setting up the site plan drawing in Chapter Four when you begin the simulated branch bank drawings.

This brief look at menu's setup routines illustrate the basics involved in customizing AutoCAD, particularly how screen menu areas, tablet areas, and AutoLISP routines can work together. Using customized screens and functions, you can cut down on the amount of time you have to spend getting ready to draw. The AutoLISP routines shown in these first few examples were small, and don't reflect the power that AutoLISP can provide in customizing and using AutoCAD. Using AutoLISP effectively can add a big boost to your productivity, as you will see in the next section.

AutoLISP and Productivity

AutoLISP is a subset of the LISP programming language and was developed by Autodesk to enhance the ability to customize AutoCAD. This programming language gives you and AutoCAD the flexibility to control the operation of commands, create new commands, and enhance the operation of applications. You can export the variables used by AutoLISP programs to help in other functions in or out of the drawing environment. For example, AutoLISP can give you access to data useful for generating specification guides, estimates, and reports of many types. You can merge AutoLISP routines with macros, and you can initiate routines through any of the command input areas we've described.

As you proceed through the next section, you will see how custom AutoLISP routines can simplify and enhance the operation of some everyday drawing tasks. For example, one of the most time-consuming chores that face AutoCAD operators in today's AE offices is the complexity of dealing with multiple layers. Trying to page through the large numbers of layers in the floor plans of major projects is frustrating. Finding the right layer to SET, turning layers ON and OFF, along with all the FREEZE and THAW operations that are required, can consume large amounts of your time.

At the end of our AE-SETUP routine, we included commands to invoke two AutoLISP routines, LAYLOAD and LAY-UTIL, that will help alleviate some of this complexity and drudgery. The first routine, called LAYLOAD, creates a list within AutoCAD of all of the layers that are present at the time and loads those layers to the utility, called LAY-UTIL, which then displays them in the screen menu area. Because we added these and some other simple, short routines, you'll be able to manipulate layers with only a page through and simple screen menu selections to SET, turn ON/OFF, FREEZE, and THAW. You can even modify the LAYLOAD routine to create a list of blocks that will be displayed for insertion, again, by simply selecting them from the screen menu area.

Invoking AutoLISP Routines

AutoLISP comes to you as a program with a standard set of variables and is pre-loaded by AutoCAD when the system is initiated. If you want more information about how AutoLISP works, look at INSIDE AUTOLISP (New Riders Publishing). For now, we will give you a brief overview of the options that AutoCAD provides to you in invoking AutoLISP functions.

❑ First, you can create stand alone programs and activate them by loading and running each as it is needed.

❑ Second, you can incorporate AutoLISP routines with macros and activate these routines from a menu area.

❑ Third, you can input one line routines directly at AutoCAD's command line.

❑ Finally, you can add short routines to the ACAD.LSP file.

All four methods are effective ways to use AutoLISP. We have already illustrated incorporating routines into macros and menu sections earlier in this chapter. In the following section, we will show you how we added routines, including our LAYLOAD and—LAY-UTIL routines, to the ACAD.LSP file. This enables you to initiate the routines via short commands at the command line. We will also show you routines that you can access directly by loading them, or picking them from the AE-MENU.

Memory

Memory problems may occur when you run large (or numerous) AutoLISP programs. An error message stating "out of node space" appears (a node being defined as a memory structure capable of representing all AutoLISP data types). Any further use of new variables is disallowed until older variables are removed. Extended memory and the addition of Extended AutoLISP have opened some additional node space for variables, but you still must be careful not to load your programs with too many variables. If you encounter "out of node space" errors, see INSIDE AUTOLISP, or your AutoCAD Reference Guide for help.

AutoLISP Utilities

We have found the following AutoLISP utilities and modifications to the ACAD.LSP file extremely useful. We use them in our AE-MENU. These routines present unique functions that basic AutoCAD does not offer at this time. These routines, as well as others that we will present in later chapters, are on the companion AE DISK.

LARBYENT, TOLAYER, LINEMELD.LSP and ACAD.LSP AutoLISP Routines

 Verify you have the following files in your AE-ACAD directory: LARBYENT.LSP, TOLAYER.LSP, LINEMEND.LSP, and a copy of the book's ACAD.LSP file.

 Just read along, or create the AutoLISP files with your text editor.

The first routine, titled LARBYENT, is a custom utility that allows you to either turn OFF, FREEZE, or SET a layer by selecting an entity on the desired layer. This utility supports the AE-MENU menu items, called [LAYER–ENTITY–SET], [LAYER–ENTITY–OFF], and [LAYER–ENTITY–FREEZE].

Now that you have verified that you have a copy of LARBYENT.LSP, load AutoCAD and test the operation.

LARBYENT.LSP Operation

Load AutoCAD and begin a new drawing called TEST. Create some entities on different layers.

```
Command: (load "LARBYENT")
C:LARBYENT
Command: LARBYENT
Loading LARBYENT utility...            Loading indicator.
Off/Freeze/Set :                       Enter O, F, or S for operation.
Select entity for layer to            Select entity on desired layer.
```

Next is a custom utility, TOLAYER, that lets you CHANGE an entity's layer to that of an entity selected on another layer. This routine supports the AE-MENU item called [TOLAYER].

Test TOLAYER.LSP by loading AutoCAD, and using a test drawing to test the routine's operation. Create some entities on different layers.

TOLAYER.LSP Operation

```
Command: (load "TOLAYER")
C:TOLAYER
Command: TOLAYER
Loading TOLAYER utility...            Loading indicator.
Select objects to be changed:        Pick object to change layer.
Pick an object on the desired layer: Pick any entity on desired layer.
```

This next custom routine, LINEMEND, prompts for two lines then replaces the two lines with one, maintaining the proper end points. It supports the AE-MENU item called [LINEMELD].

Test its operation the same way you tested the two previous routines.

LINEMEND.LSP Operation

```
Command: (load "LINEMEND")
C:LINEMEND
Command: LINEMEND
Loading LINEMEND utility...
Select first line:
Select second line:
```
Loading indicator.
Select first line to mend.
Select second line to mend with first.

Routine will automatically repeat itself until an ENTER.

Using the Modified ACAD.LSP File

The custom routines (that we list below), modify the ACAD.LSP file. They are made up of ten short AutoLISP routines that let you call standard AutoCAD commands with the keyboard entry of only a few characters; and several AutoLISP utilities that provide more flexibility in controlling layers. These layer routines support the other layer menu items on the AE-MENU, letting you easily turn on or off, set, freeze, or thaw multiple layers by selecting the layers from the screen menu area.

Testing the ACAD.LSP File Routines

If you have installed the book's ACAD.LSP file from the AE DISK in your AE-ACAD directory, these utility routines are loaded automatically by AutoCAD each time you enter the drawing editor.

To test the operation of these routines, load AutoCAD and enter the drawing editor. AutoCAD will tell you that it is loading ACAD.LSP. Examine the listing below to select routines to test.

The following listings show all the AutoLISP routines that you have examined so far, starting with the LAYBYENT.LSP routine, and ending with the our modified ACAD.LSP file.

```
(prompt 2"Loading LARBYENT utility...")
(defun C:LARBYENT ( / typ ent params entlayer curlayer)
(setvar "cmdecho" 0)
(setq typ (strcase (getstring "\nOff/Freeze/<Set>:  ")))
(if (= typ "")  (setq typ "S"))
(if (= typ "O") (setq typ "OFF"))
(prompt "\nDefine entity for layer to ")
(cond ((= (substr typ 1 1) "S") (princ "SET."))
      ((= (substr typ 1 1) "O") (princ "turn OFF."))
      (t (princ "FREEZE."))
)                                                     ;cond
(setq ent (entsel))
(while (null ent)
   (prompt "\nNo entity found.")
   (prompt "\nDefine entity for layer to ")
   (cond ((= (substr typ 1 1) "S")  (princ "SET."))
         ((= (substr typ 1 1) "O") (princ "turn OFF."))
         (t (princ "FREEZE.))
   )                                                  ;cond
   (setq ent (entsel))
)                                                     ;while
 (setq ent (car ent)
 (setq params (entget ent)
 (setq entlayer (cdr (assoc 8 params)))
 (setq curlayer (getvar "clayer"))
 (if (= entlayer curlayer)
   (if (= typ "S") t
     (prompt "\nThe entity choosen is on the current layer.")
   )
   (command "layer" typ entlayer "")
)

 (setvar "cmdecho" 1)
 (gc) (princ)
)                                                     ;defun
```

AutoLISP Routine for [LAYER-BY ENTITY-OFF, FREEZE, or SET]

```
(prompt "Loading TOLAYER utility...")
(defun C:TOLAYER ( / objselct newobj lyrname)
   (prompt "\nSelect objects to be changed:  ")
   (setq objselct (ssget))                   ;user selects
    (if objselct
      (progn
         (setq newobj
             (car (entsel "Pick an object on the desired layer:  ")))
         (if newobj
            (progn
               (setq newobj (entget newobj));get the entity
                (setq lyrname (cdr (assoc 8 newobj))) ;get layer name
 vp
               (command "CHANGE" objselct "" "PROP" "LAYER" lyrname "") ;perform
                                            CHANGE

               )                            ;progn
            )                               ;if
         )                                  ;progn
      )                                     ;if
   )                                        ;defun
```

AutoLISP Routine for [TO-LAYER]

```
(prompt "Loading LINEMEND utility...")
(defun C:LINEMEND ( / ent1 ent2 params1 1pt10 2pt10 old10
                      1pt11 2pt11 old11 dist1 dist2 dist3 dist4
                      distlist maxdist pt1 pt10 pt11)
   (setvar entsel "\nSelect first line: "))
   (setvar "cmdecho" 0)
   (setvar "orthomode" 1)
   (while  ent1)
      (setq ent1 (car ent1))
      (redraw ent1 3)
      (setq ent2 (car (entsel "\nSelect second line:  ")))
      (while (null ent2)
         (prompt "\nNo entity found.")
         (setq ent1 (car (entsel "\nSelect second line:  )))
         (if (/= (cdr (assoc 0 (entget ent2))) "LINE")
            (progn
               (prompt "\nLines Only!")
               (setq ent2 nil)
            )                                              ;progn
         )                                                 ;if
      )                                                    ;while
(redraw ent2 3)
(setq params1 (entget ent1)
         1pt10 (cdr (setq old10 (assoc 10 params1)))
         1pt11 (cdr (setq old11 (assoc 11 params1)))
         params2 (entget ent2)
         2pt10 (cdr (assoc 10 params2))
         2pt11 (cdr (assoc 11 params2))
         dist1 (distance 1pt10 2pt10)
         dist2 (distance 1pt10 2pt11)
         dist3 (distance 1pt11 2pt10)
         dist4 (distance 1pt11 2pt11)
         distlist
           (list
              (list dist1 1pt10 2pt10)
              (list dist2 1pt10 2pt11)
              (list dist3 1pt11 2pt10)
              (list dist4 1pt11 2pt11)
           );list
         maxdist (max dist1 dist2 dist3 dist4)
         pts (cdr (assoc maxdist distlist))
         pt10 (car pts)
         pt11 (cadr pts)
```

```
            params1 (subst (cons 10 pt10) old10 params1)
            params1 (subst (cons 11 pt11) old11 params1)
      )                                                  ;setq calculations
   (entdel ent2)
   (entmod params1)
   (entupd ent1)
   (setq ent1 (entsel "\nSelect first line:  "))
 )                                                       ;while loop
 (setvar "cmdecho" 1)
 (gc) (princ)
)                                                        ;defun
```

AutoLISP Routine for [LINE-MEND]

The following listing shows the modified ACAD.LSP file.

```
(vmon)
(setvar "cmdecho" 0)

(DEFUN C:BR () (COMMAND "BREAK"))
(DEFUN C:ZE () (COMMAND "ZOOM" "E"))
(DEFUN C:EN () (COMMAND "END"))
(DEFUN C:ZW () (COMMAND "ZOOM" "W"))
(DEFUN C:ZP () (COMMAND "ZOOM" "P"))
(DEFUN C:ER () (COMMAND "ERASE"))
(DEFUN C:OF () (COMMAND "OFFSET"))
(DEFUN C:MO () (COMMAND "MOVE"))
(DEFUN C:CO () (COMMAND "COPY"))
(DEFUN C:FI () (COMMAND "FILLET"))
(DEFUN C:CH () (COMMAND "CHANGE"))
(DEFUN C:EL () (COMMAND "ERASE" "L" ""))
(DEFUN C:CR () (COMMAND "CIRCLE" "2P"))
(DEFUN C:LI () (COMMAND "LINE"))
(DEFUN C:RE () (COMMAND "REDRAW"))

;This routine creates the layer list
(defun C:LAYLOAD (/ tmp)
  (setq layers (cdr (assoc 2 (tblnext "LAYER" T))))
  (setq layers (list (cdr (assoc 2 (tblnext "LAYER"))) layers))
  (while (setq tmp (tblnext "LAYER"))
        (setq layers (cons (cdr (assoc 2 tmp)) layers))
  )
)

;This program loads the layer list into the screen area
(defun C:LAY-UTIL (/ mo bw lay num lim valid)
  (if (not (boundp layers)) (setq layers (C:LAYLOAD)))
  (setq mo 0 lay nil)
```

```
  (setq len (1- (length layers)))
  (setq num (fix (/ len 18)))
  (setq lim (rem (1+ len) 18))
  (while (not lay)
    (if (< mo num) (setq tt1 18) (setq tt1 lim))
    (setq bw 0)
    (while (< bw tt1)
         (grtext bw (nth (+ bw (* mo 18)) layers))
        (setq bw (1+ bw))
    )
    (while (< bw 20)
        (grtext bw "          ")
        (setq bw (1+ bw))
    )
     (cond ((= mo 0) (grtext 18 "next"))
           (( mo num) (progn (grtext 18 "next") (grtext 19 "previous")))
           (T (grtext 19 "previous"))
    )
    (setq valid nil)
    (while (not valid)
      (setq temp (nth 1 (grread)))
      (cond
          ((and (/= mo num) (= temp 18))
            (progn
              (setq mo (1+ mo))
              (setq valid T)
            )
         )
          ((and (= temp 19) (/= mo 0))
            (progn
              (setq mo (1- mo))
              (setq valid T)
            )
         )
          ((< temp tt1)
            (progn
              (setq lay (+ temp (* mo 18)))
              (setq valid T)
            )
         )
      )
    )
  )
  (grtext)
  (setq tt1 (nth lay layers))
)
;This routine sets the layer from a screen menu pick
(defun C:LAYERSET()
  (command "LAYER" "SET" tt1 "")
)
;This routine turns the layer off from a screen menu pick
(defun C:LAYEROFF()
  (command "LAYER" "OFF" tt1 "")
```

```
)
;This routine turns the layer on from a screen menu pick
(defun C:LAYERON()
   (command "LAYER" "ON" tt1 "")
)
;This routine thaws the layer from a screen menu pick
(defun C:LAYERTHW()
   (command "LAYER" "THAW" tt1 "")
)
;This routine freezes the layer from a screen menu pick
(defun C:LAYERFRZ()
   (command "LAYER" "FREEZE" tt1 "")
)
```

ACAD.LSP Listing

Summary

We have presented portions of the AE-MENU.MNU menu program to show one way you can enhance AutoCAD. The improvements made via a customized tablet menu, related screen menus and AutoLISP routines, all contribute to speeding up drawing setup tasks. They cut down on repetitious tasks, commands, and keystrokes. This illustrates only a fraction of what you can accomplish with your own system. Use what we have shown as a base for customizing your system to your own requirements. The time you spend integrating your procedures and customizing AutoCAD's functions is always worthwhile.

Careful management, planning, and documentation assure that the tasks of collecting, sorting, and maintaining your system data will be more efficient, and that the long term gains in ease of use and streamlined function will far outweigh the effort you put in. If you manage and maintain your modifications, they will continue to contribute to your productivity for as long as you care to use them.

Now let's see how to put the productivity tools that we've assembled in this first section of the book to work. In the next two sections, you'll create the major working drawings of the simulated branch bank project — using the custom symbols, AutoLISP routines, and menus routines that we have discussed so far, plus a lot of special techniques and approaches that we will show you as we go along.

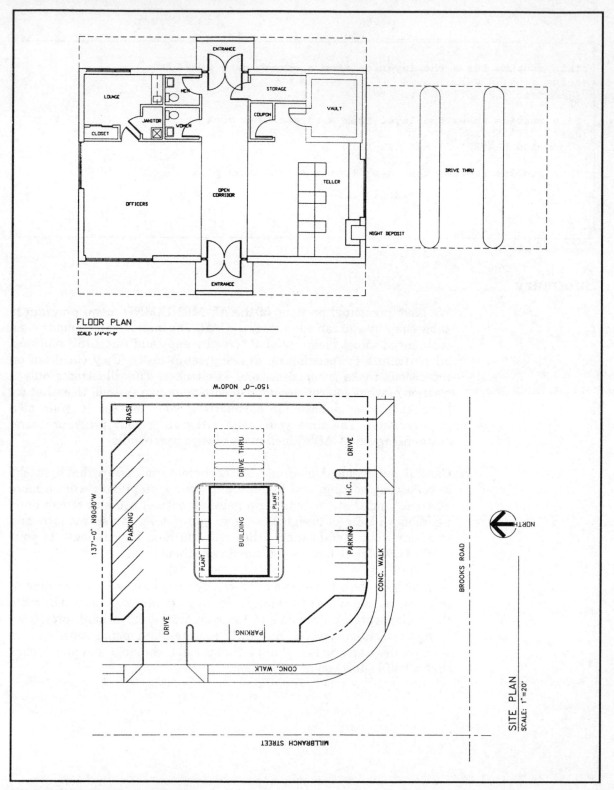

Site and Floor Plan Design Drawings

CHAPTER 4

Design Drawings with AutoCAD

Design Drawings

You begin the design drawing process by obtaining extensive information about the actual site. Your research normally includes determinations of utilities availability, soils investigation reports, and a topographic survey. Also, you must make a thorough assessment of the client's additional requirements and ideas for finishes, costs, furnishings, and so on. After the basic concept for the project has been approved, you apply your experience and skill as an A/E professional to produce the more definitive drawings for the project, commonly called design drawings.

Design Drawing Procedures and AutoCAD

The procedures that you will follow in making the design drawings for the simulated branch bank project are those that would actually occur in an A/E firm's development of drawings for buildings like this, and almost any other type.

Whether you use AutoCAD to develop your designs, or you draft them by hand, you will have to invest the same amount of time in determining the client's requirements, making your assessment of the site and organizing the work to follow. Either way, you will probably have to make revisions to your drawings at some point in the process. These revisions might be based on the client's response to the design you have proposed, input from review boards and government agencies, or on the new requirements and ideas which always seem to crop up.

If you use a manual drafting technique, including the popular overlay drafting system, your design drawing documents may undergo so many changes and revisions that you may ultimately require additional time and resource management to handle them. Almost all of us have experienced having to scrap much of our completed work and have found ourselves going back to the drawing board.

Although AutoCAD will not cure the ever-present reasons for revisions and changes, it does make it much easier to directly apply data from a previous solution to the current solution. CAD systems in general have always excelled in the revision process compared to conventional manual techniques. The reasons are simple. The computer's eraser is faster than a *pink pearl* in even the most capable hands, and the CAD system can also draw faster than the most proficient and experienced draftsmen.

The purpose of this book is not to show you how to expedite revisions or changes to the branch bank (or any other project, for that matter), but to give you a road map and short cuts for producing working drawings more efficiently. You can apply the techniques we illustrate to any project, and you can shorten the delays you experience due to revisions.

Advantages of the AE-MENU System in Design Drawings

The advantages you gain by using AutoCAD for design drawings will be augmented when you use the AE-MENU System and the routines presented in the exercises. We see three major advantages.

Increased Speed with AutoCAD

With AutoCAD's basic editing commands, you can create a design and make revisions to evaluate alternative schemes. One of the primary benefits of using AutoCAD for design drawing is that it gives you the ability to move blocks of data and make corrections without the redrawing effort common to manual systems. The custom menus and AutoLISP routines of the AE-MENU speed up the design process even further by automating more design tasks.

Flexibility with AutoCAD

Flexibility is another fundamental benefit in using AutoCAD for design drawings. You can explore more ideas and make changes without having to redraw alternatives from scratch. You can make multiple copies of a base drawing within the system and develop variations of the design to present to your client.

With AutoCAD Release 10, a full three-dimensional representation of the design can be achieved. Being able to see how a design will actually look, and being able to generate perspectives of *what if* solutions quickly gives you more time to be creative. Viewing three-dimensional solutions can also help you determine what critical areas must be studied and detailed further. The exercises in the next chapter will show techniques for developing a 3D design.

Transferability of Data

By following the techniques detailed in the previous chapters for system set up, standardization, and customization, you can be sure that your drawings will be consistent through all phases of development. Organizing the project around the use of AutoCAD also helps you make design decisions earlier and therefore insures that they are included in each phase of development and are represented in the final production process.

With very little effort, you can reuse data developed for one phase in producing the drawings of another phase. You can reuse and enhance the data from design drawings with technical information to produce working drawings. You can also transfer the data in the form of background drawings to engineers or other consultants for their portion of the work in completing the final drawing requirements.

Project Introduction

The project design, branch bank, is started in this chapter. The project consists of approximately 1700 square feet. The spaces required by the client are a public area, an officer's area, booths for the tellers, a vault, a lounge, toilets, a safety deposit box review area and drive through tellers.

Project Design Drawings

In the first series of command sequences and techniques, we will show you how to use AutoCAD with the AE-MENU to produce the key elements of a site plan design drawing for the branch bank. This plan will contain information about the building site, the building itself, as well as the locations of all other site improvements.

In a second series of command sequences and techniques, we will show you how to produce significant portions of a floor plan design drawing for the project. This plan will provide information on the general layout of the building and reflect the design criteria. It will include enough detail about particular design features, such as the teller's area, and special construction, such as the vault, to convey the appearance and function of the building.

The materials for the project include load bearing masonry, metal studs, gypsum board, steel beams and joists, aluminum storefront and glazing, and a built-up roof and metal canopy system.

Site Plan Design Drawing

The site plan design drawing depicts the layout of the building on the site with associated parking and walk areas.

To create the site plan design drawing, you can follow the project sequence shown in the following table. If you are looking for specific techniques, these are flagged in the table with a diamond symbol, and an icon in the chapter's text.

```
Project Sequence for the Site Plan Design Drawing

Drawing setup.
AutoLISP drawing utilities.
◆ Technique for Property Line Input.
◆ Technique for Creating Setback Lines.
◆ Technique for Creating Setback Block.
◆ Technique for Calculating Areas.
Place the base plan block from Floor Plan Sequence.
Add streets and walks.
Add pavement, drives, parking areas, curbs, and curb cuts.
Define pavement edges.
Add South drive, islands, and parking area.
Add West drive, islands, and parking area.
Add trash and North parking areas.
Add the South and West parking striping.
Define curbs.
Define curb cut slopes.
◆ Technique for Adding the North Parking Area Striping.
Enhance building lines.
Insert North arrow.
Add text.
```

Floor Plan Design Drawing

The floor plan design drawing for the project will indicate all exterior walls, interior partitioning, doors, window, fixtures, casework and special equipment such as the remote teller units. We will use the following sequence for the development of the floor plan design drawing:

```
Project Sequence for the Floor Plan Design Drawing

Drawing setup.
◆ Technique and AutoLISP Routine for Floor Plan Area
Layout.
◆ Technique for Adjusting the Areas.
◆ Technique for Defining Wall Thicknesses.
Define sub-areas.
Create the base plan block for use in the Site Plan.
Correct layering.
Move design layout to the Main area.
Insert doors.
Insert plumbing fixtures and lounge cabinets.
Add teller work positions.
Define drive through islands.
Define roof and changing the canopy overhangs.
Add windows.
◆ Technique for Defining Mullions.
Add text.
```

As we indicated in the Introduction, we have set up two ways for you to practice the concepts and techniques that we present in each chapter. First, you can follow each project sequence and complete the drawings as if they were part of a real project. Second, you can read along with the project sequences and use AE-SCRATCH drawings to experiment with the techniques. Either method will provide you with the same benefits — increased productivity and the understanding of how to apply AutoCAD to your work environment.

Method One

If you have the AE DISK, you have the completed site plan drawing, called 8910SP-D, and the completed floor plan 8910FP-D. These drawings were copied to your 8910 directory when you loaded the AE DISK drawing files. (See Chapter One, if you haven't copied these files to your hard disk yet.) You can load the drawing, and just read along with the sequences. To practice a drawing technique, use the [VIEW–RESTORE–SCRATCH] command from the AE-MENU, and practice the technique in the SCRATCH area of the drawing.

If you want to create the two design drawings from start to finish, change to your drawing directory, called 8910. Use the following sequence to start the site plan drawing.

❑ Load AutoCAD and select Option 1, Begin a New drawing.

❑ Name the site plan design drawing 8910SP-D.

❑ Use the [AESETUP] routine from the AE-MENU to setup the drawing using the SITE-PT.DWG as the prototype drawing. (This is the same menu sequence that we used in Chapter Two.)

❑ Move to the SCRATCH area to experiment with any techniques before you apply them to the drawing.

Method Two

If you want to practice the techniques in this chapter, but not complete the drawings, experiment using an AE-SCRATCH drawing, calledAE-SCR4. This drawing is defined in the table below. Change to your drawing directory called 8910. Use the following sequence as a guide to get started.

❑ Load AutoCAD and select Option 1, Begin a NEW Drawing.

❑ Name the drawing AE-SCR4.

❑ Set up for the techniques by using the AutoCAD Setup Routine, selecting the Architectural unit type, 1/4"=1" scale, D–24"x36" sheet size, and setting the other options defined in the following table.

COORDS	GRID	SNAP	ORTHO	UCSICON
ON	10'	6"	ON	WORLD

UNITS	Set UNITS to 4 Architectural.
	Default the other UNITS settings.
LIMITS	0,0 to 720',480'
ZOOM	Zoom All.

Layer name	State	Color	Linetype
0	On	7 (white)	CONTINUOUS
MIS-010	On	1 (red)	CONTINUOUS
BLD-010	On	1 (red)	CONTINUOUS
PRL-020	On	5 (blue)	CONTINUOUS
PRK-020	On	2 (yellow)	CONTINUOUS
STE-020	On	2 (yellow)	CONTINUOUS
WIN-080	On	5 (blue)	CONTINUOUS
SPL-100	On	1 (red)	CONTINUOUS

AE-SCR4 Setup Table

Now that you have set up an 8910SP-D (or AE-SCR4) drawing, you can start the site plan sequence.

Site Plan Design Drawing Sequence

To begin the design drawing phase with the site plan, first, define the property lines and the required setbacks. The book's project sequence and techniques will step you through the tasks to complete the major features proposed to meet the requirements for this site plan.

Laying Out the Property Lines

The first step is to lay out the property lines. Since we usually take data for these lines from the surveyor's field notes, we are going to show you how to simplify the steps involved in using this data by changing the UNITS "System of angle measure:" to five — Surveyor's Units.

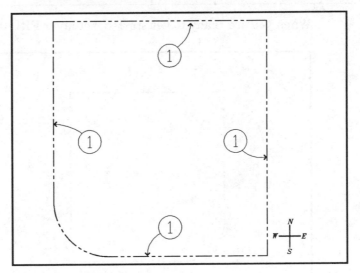

Site Plan Property Lines Illustration

① CREATING THE PROPERTY LINES. One way you could create the property lines would be to set layer PRL-020 as current, and use the LINE command to draw the property lines; then, use the FILLET command to add the curve required at the street intersection. But we show a better way to input the property lines below.

Technique for Property Line Input

Our technique shows you how to use absolute and relative coordinates for placing property lines, using the "Surveyor" method for angle input. This lets you use data directly from surveyor's notes, and you do not have to transpose the data to feet and inches. Move to the SCRATCH area to experiment with the technique before using it in the MAIN drawing area. Select a random XY coordinate to begin laying out the property lines to practice the technique.

Setting Up Property Line Technique Drawing

    Enter selection: 2   Edit an existing drawing named 8910SP-D.

Select the [VIEW—RESTORE—SCRATCH] command from AE-MENU.

Enter selection: 1   Begin a NEW drawing named AE-SCR4.

Verify the settings shown in this chapter's AE-SCRATCH Setup Table.

When you are ready to get started, set layer PRL-020 as current.

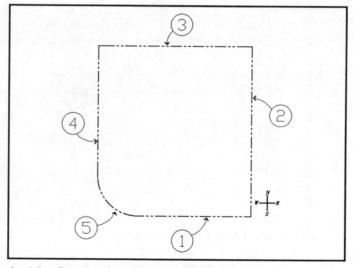

Applying Property Line Technique

Property Lines

```
Command: LINE
From point:                          Select a point.
To point: @150'<N0d0'W               Property line ①.
To point: @137'<N90d0'W              Property line ②.
To point: @150'<S0d0'W               Property line ③.
```

```
To point: C                              Property line ④.

Command: FILLET                          Set radius of 35'.

Command: FILLET
Polyline/Radius/<Select two objects>:    Select property lines ① and ④.
```

This technique uses the relative coordinate system of data input, and LINE and FILLET to create the property lines.

Benefits

By using the absolute and relative coordinates from the surveyor's field notes, you can place the site's property lines accurately and quickly.

Technique for Creating Setback Lines

Local zoning regulations usually require that you establish setback lines within prescribed distances from the property lines.

One way you can add the setback lines is to use the LINE command. You can draw reference lines extending the distance of the offset perpendicularly from each property line. Then, use these lines to draw the setback lines. But a simpler approach, as you'll see in the first of the three techniques we're about to show you, is to use OFFSET with the required distances, building upon existing data (the property lines) to create new data (the setback lines). Then, you can easily edit and convert the new lines to their correct layer.

After you create the setbacks, we will show you how to create a block from the new setback lines so you can use them later as a guide for laying out the design floor plan.

In a final setback technique we'll show you how to use the property lines and setback lines to calculate the area of each, and how to subtract the setback area from the site area to determine square foot usage of the property.

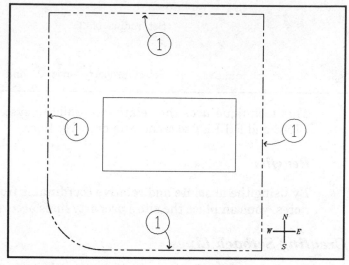

Setback Area Illustration

① ADDING SETBACK LINES. Use the technique we show below for adding the setback lines to the site plan. If you are creating the drawing, practice the technique in the SCRATCH area.

Setting Up Property Line Technique Drawing

  Enter selection: 2   Edit an existing drawing named 8910SP-D.

Select the [VIEW–RESTORE–SCRATCH] command from AE-MENU.

Enter selection: 1   Begin a NEW drawing named AE-SCR4.

Verify the settings shown in this chapter's AE-SCRATCH Setup Table.

When you are ready to get started, set layer MIS-010 current. If you are using the AE-MENU, use the tablet menu by selecting [LAYERS–BY SCREEN–SET].

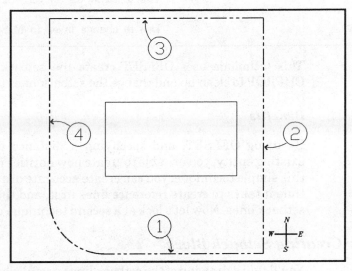

Applying Setback Technique

Defining Setback Area

```
Command: OFFSET
Offset distance or Through <Through>: 50'6
Select object to offset:
Side to offset?
Select object to offset: <RETURN>
```
           Select property line ①.
           Select a point toward the North.

```
Command: OFFSET
Offset distance or Through <50'-6">: 17'6
Select object to offset:
Side to offset?
Select object to offset: <RETURN>
```
           Select property line ②.
           Select a point toward the West.

```
Command: OFFSET
Offset distance or Through <17'-6">: 56'
Select object to offset:
Side to offset?
Select object to offset: <RETURN>
```
           Select property line ③.
           Select a point toward the South.

```
Command: OFFSET
Offset distance or Through <56'-0">: 36'
Select object to offset:
Side to offset?
Select object to offset: <RETURN>
```
           Select property line ④.
           Select a point toward the East.

Command: **TRIM**   Use to clean up all corners.

Command: **CHPROP**   Use to change layer to MIS-010.

This technique uses OFFSET create the setback lines, and TRIM and CHPROP to clean up and change the setback lines to the appropriate layer.

Benefits

By using OFFSET, and specifying a distance and direction from an existing entity, you are able to create new entities from old entities. Using this simple technique, you can create accurate drawings in about half the time it takes to create reference lines first, and then use them to add the setback lines. Now let's look at a second technique using the setback lines.

Technique for Creating Setback Block

You'll find a drawing of the setback lines useful when you begin the layout of the floor plan design drawing. It will serve as a guide showing the area of the site that the plan may cover.

Name the block SETBACK, and use WBLOCK to create the entity so that you can insert it into the floor plan design drawing later.

Creating SETBACK Block

```
Command: WBLOCK
File name: SETBACK
Block name: <RETURN>
Insertion base point: INT
of                                          Select the lower left corner.
Select objects: W
First corner: Other corner: 4 found.
Select objects: <RETURN>

Command: OOPS                               To bring back the drawing.
```

By adding the WBLOCK to the setback technique, you can use the data later in other drawings.

Benefits

This setback technique shows how to transplant data by inserting a drawing or a portion of a drawing into another drawing. Don't be fooled by its simplicity — a lot of productivity is gained by being able to use data over and over in different drawings.

Now that you have completed the property and setback lines, let's look at a third technique to convert these lines into polylines and calculate the different areas of the site.

Technique for Calculating Areas

To calculate the sizes of the areas defined so far on the site plan drawing, use the PEDIT command. First, convert the property lines into polylines, and second, convert the setback lines into polylines. AutoCAD will treat them as separate entities. Then, using the AREA command, you can calculate the square inches, square feet and perimeter of each area. If you want to calculate the difference between the area within the setback and the total area of the site, AREA will also do this subtraction.

Calculating Areas

```
Command: PEDIT                                  Convert property lines into polylines.

Command: PEDIT                                  Convert setback lines into polylines.

Command: AREA                                   Area of Site.
<First point>/Entity/Add/Subtract: E
Select circle or polyline:                      Select property line.
Area = 2921344.24 square in. (20287.1128 square ft.), Perimeter = 558'-11 3/4"

Command: AREA                                   Area of Setback.
<First point>/Entity/Add/Subtract: E
Select circle or polyline:                      Select setback line.
Area = 523044.00 square in. (3632.2500 square ft.), Perimeter = 254'-0"

Command: AREA
<First point>/Entity/Add/Subtract: A
<First point>/Entity/Subtract: E
(ADD mode) Select circle or polyline:                Select property line.
Area = 2921344.24 square in. (20287.1128 square ft.), Perimeter = 558'-11 3/4"
Total area = 2921344.24 square in. (20287.1128 square ft.)
(ADD mode) Select circle or polyline: <RETURN>
<First point>/Entity/Subtract: S
<First point>/Entity/Add: E
(SUBTRACT mode) Select circle or polyline:           Select setback line.
Area = 523044.00 square in. (3632.2500 square ft.), Perimeter = 254'-0"
Total area = 2398300.24 square in. (16654.8628 square ft.)
(SUBTRACT mode) Select circle or polyline: <RETURN>
<First point>/Entity/Add: <RETURN>

Command: END
```

By using PEDIT, you were able to convert the property lines (five entities), and the setback lines (four entities) into single polylines. This let you use the AREA command, selecting these lines as single entities.

Benefits

The key to this technique is converting the various lines and arcs into single polyline entities so the AREA command give you an accurately calculation. You'll see the advantage, if you have ever tried to use AREA to select points along an arc. You know how time-consuming and inaccurate these area calculation can be.

Setting Aside the Site Plan

You have now created the basic design data for the site plan. We'll come back to this plan later in the chapter. Right now, we want to move on to the floor plan. We want to create the layout of the floor plan design drawing; then, use the SETBACK block from the site plan sequence to verify the fit of the building layout on the site plan.

Floor Plan Design Drawing Project Sequence

You can approach the sequences for the floor plan design drawing the same way you did the site plan drawing. You can either complete the entire drawing as if it were a real project (with a drawing name of 8910FP-D), or just practice the techniques. If you are using the AE DISK, you have the completed drawing.

To just practice the techniques, you can use the same AE-SCR4 drawing defined at the start of the chapter.

Major Steps

Here are the three major steps that you will work through:

- Begin the sequence by defining the various areas of the floor plan.

- Move these areas into position, add the wall thickness and create a base plan drawing that can be used later.

- Then, add the doors, plumbing fixtures, cabinets, and furnishings to the plan. The last additions will be the drive through islands, windows, and text.

Setting Up Floor Plan Design Drawing

Change to 8910 directory. Load AutoCAD.

 Enter selection:2   Edit an existing drawing named 8910FP-D.

Select [VIEW–RESTORE–SCRATCH] from AE-MENU to work in SCRATCH area.

 Enter selection:1   Begin a NEW drawing named AE-SCR4.

Verify the settings shown in this chapter's AE-SCRATCH Setup Table.

Defining Floor Plan Areas

Let's start by defining the floor plan areas. There are eight major areas in the floor plan shown in the Main Area Layout Table below. You can use the LINE command to draw each area independently and adjust them later. However, we think the AutoLISP technique following the table shows a more productive way of laying out the major areas.

AREA NAME	NUMBER	AREA DIMENSIONS	
		X	Y
PUBLIC	1	48' x	19'
LOUNGE	2	14' x	11'
TOILETS	3	6' x	11'
COUPON/STORAGE	4	11' x	11'
VAULT	5	8' x	10'
DRIVE THRU AREA	6	32' x	28'
ENTRANCE (REAR)	7	10' x	6'
ENTRANCE (FRONT)	8	10' x	6'

Main Area Layout Table

The illustration following shows these layout areas.

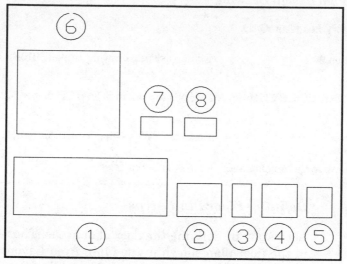

Area Layout Illustration

① CREATING THE MAJOR AREAS OF THE DESIGN FLOOR PLAN.
You could simply take the dimensions given in the table, refer to the area
layout illustration, and create the eight areas with the LINE command.
But, take a look at what the following AutoLISP routine can do in
speeding up the layout.

Technique for Floor Plan Area Layout

There is an AutoLISP routine called XYZBOX.LSP on the AE DISK. if you
are using the AE DISK, you need to verify that you have the routine in your
AE-ACAD directory. Then, you can use the XYZ BOX command of the
AE-MENU to lay out the areas, using the interior requirements listed in the
Main Area Layout table above.

The menu's XYZ BOX command invokes the AutoLISP routine to create the
areas with polylines.

XYZBOX.LSP AutoLISP Routine

 Copy or install the XYZBOX.LSP file in AE-ACAD and examine.

 Just read along, or create the AutoLISP file.

➥ *NOTE! The routine will add a Z distance or thickness of these areas. Adding a value for this data will help you create the 3D presentation drawing in Chapter Five.*

If you want to examine th XYZ box utility, here is the listing.

```
(prompt "Loading XYZBOX utility...")
(defun C:XYZBOX ( / oldblip mark wid lg str ang el th var1)
    (setvar "cmdecho" 0)
    (setq wid 0.0
          lg 0.0
          str ""
          ang (getvar "SNAPANG")
          el (getdist "\nEnter Elevation:  ")          ;get elevation
          th (getdist "\nEnter Thickness:  ")          ;get thickness
)                                                       ;setq calculations
(command "ELEV" el th)                                  ;set the elevation
(setq str (getpoint "\nLower left corner point:  "))
(while str                                              ;only if entered
    (setq var1 (getdist str (strcat "\nX-length of rectangle:  ")))
    (if (/= var1 nil)                                   ;if not empty
    (setq lg var1)
    )
    (setq var1 (getdist str (strcat "\nY-length of rectangle:  ")))
    (if (/=var1 nil)                                    ;if not empty
        (setq wid var1)
    )
    (command (if (= PLINE 0) "LINE" "PLINE")            ;draw the box
            str (polar str ang lg)
            (polar (getvar "LASTPOINT") (+ ang (/ pi 2)) wid)
            (polar (getvar "LASTPOINT") (+ ang pi) lg)
            "C"
     )
     (setq str (getpoint "\nLower left corner point:  "))
   )
   (setvar "cmdecho" 1)
   (gc) (princ)
)
```

XYZBOX.LSP AutoLISP Routine

Testing the XYZBOX.LSP Routine

Now that you have a copy off XYZBOX.LSP, load AutoCAD and test its operation. Set your layer BLD-010 current. If you are using the

AE-MENU, use the tablet [LAYER–BY SCREEN–SET]. To use the command from the AE-MENU, select the menu item [XYZBOX]. Use the sequence below in a scratch area, or use a scratch drawing to test its operation.

Elevation and Thickness

XYZBOX prompts first for an elevation and a thickness. For the areas you want to define, use zero for an elevation and 12 feet for the thickness. (You will use the layout data later in Chapter Five to create the 3D presentation drawings.) The XYZBOX routine prompts you for the X, or horizontal, dimension and the Y, or vertical, dimension. Select an arbitrary point at the bottom of the drawing editor to begin.

XYZBOX.LSP Operation

```
Command: (load "XYZBOX")
C:XYZBOX
Command: XYZBOX
Enter Elevation:              Enter the elevation of the area.
Enter Thickness:              Enter the thickness of the area.
Lower left corner point:      Pick the lower left corner point.
X—length of rectangle:        Enter the X dimension.
Y—length of rectangle:        Enter the Y dimension.
```

This routine will repeat the corner point, X and Y dimension until you terminate it with an <RETURN>.

Applying the Layout Technique

If you are just testing the technique, continue with your scratch drawing.

If you are completing the entire floor plan drawing, continue with your drawing, called 8910FP-D. You need to insert the SETBACK block (which you created in the first part of the chapter) into the MAIN area of your floor plan drawing.

Once you have inserted the SETBACK block, move to the SCRATCH area to do the initial layout (using XYZ BOX) so that you can experiment. When you are done, move your completed floor plan design drawing from the SCRATCH area to the MAIN area to check its size against the SETBACK block.

Follow the sequence below to create the first area for the floor plan drawing.

Setting Up Area Layout Technique Drawing

    Enter selection: 2   Edit an existing drawing named 8910FP-D.

Select the [VIEW–RESTORE–SCRATCH] command from AE-MENU.

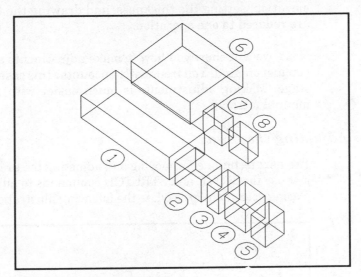

Enter selection: 1   Begin a NEW drawing named AE-SCR4.

Verify the settings shown in this chapter's AE-SCRATCH Setup Table.

When you are ready to get started, set layer BLD-010 as current.

❑ Create the first area (Public Area) using the sequence below as a guide.

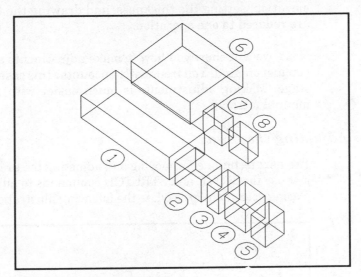

Applying Area Layout Technique

Applying Main Area Layout

```
Command: (LOAD "XYZBOX")
C:XYZBOX

Command: XYZBOX
Enter Elevation: 0
Enter Thickness: 12'
Lower left corner point:          Select any point in lower left corner.
X-length of rectangle: 48'        Create area ①.
Y-length of rectangle: 19'
```

This technique shows how you can use an AutoLISP routine to define shapes or areas, minimizing operations.

Creating the Other Main Areas

② through ⑧ Create the rest of the main areas using XYZ BOX. Get your X and Y values from the Main Area Layout Table. Use 0 elevation, and a 12 foot thickness value for each area.

Benefits

Using AutoLISP to define areas is a productive tool when you can define the X, Y, and Z dimensions in one operation, as XYZBOX does. In this case, drawing tasks, composed of six different operations (setting the elevation, setting the thickness, and drawing the four sides of an area), are reduced to one operation.

Next, we will show you how to make adjustments and relocate the areas you just created. You frequently encounter this case at this layout process stage. Making adjustments is much easier with AutoCAD than using manual methods.

Technique for Adjusting the Areas

The next technique for moving and adjusting the areas is quite simple: you just use the MOVE and STRETCH commands to adjust the area layout by placing your main areas. Use the following illustration as your guide.

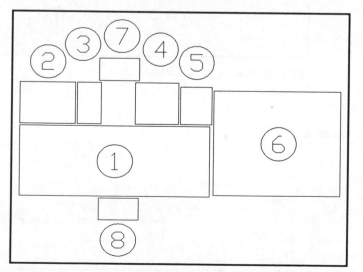

Applying Design Layout Technique

Adjusting Layout

Command: **MOVE**   Repeat this command until all areas are placed properly.

Command: **STRETCH**   Repeat this command until all areas are adjusted to their proper size.

By using STRETCH and MOVE, you can adjust the plan to fit together and function as the design dictates.

Benefits

This drawing sequence shows how to automate the design process by combining different techniques. You used AutoLISP to combine AutoCAD drawing functions and AutoCAD's innate ability to edit and modify drawings to simplify and speed up the area layout process.

Adding Exterior and Interior Wall Thickness

Now that you have completed the areas of the design floor plan, the next task is to add wall thickness. You can use the OFFSET command to do this. But, you created these areas with polylines using the XYZBOX routine. To get differing wall thicknesses, you will need to explode the polylines to work with individual entities.

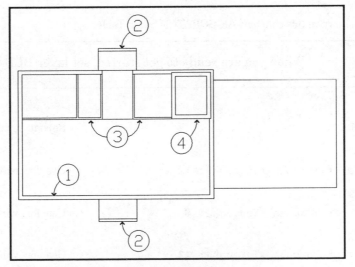

Exterior and Interior Walls Illustration

The exterior and vault walls should be 12 inches thick, the entrance and windbreak walls should be eight inches thick, and the interior walls should be five inches thick.

Technique for Defining Wall Thicknesses

In the following technique, use OFFSET to modify the area layout by adding wall thicknesses. Explode all of the areas with the EXPLODE command before you use OFFSET.

➡ *NOTE! You may be wondering why we used polylines instead of lines to create these areas. It is easier to move and adjust areas created with polylines, since they are single entities, than to adjust areas made up of groups of lines. You can always convert the areas back to single entities for additional editing.*

Setting Up Wall Thickness Technique Drawing

Enter selection: 2   Edit an existing drawing named 8910FP-D.

Select the [VIEW–RESTORE–SCRATCH] command from AE-MENU.

Enter selection: 1   Begin a NEW drawing named AE-SCR4.

Verify the settings shown in this chapter's AE-SCRATCH Setup Table.

When you are ready to get started, set layer BLD-010 as current.

Defining Walls

Command: **EXPLODE**   Repeat until all areas are exploded.

Command: **OFFSET**
Offset distance or Through <Through>: **12**   Use for exterior walls ①.

Command: **OFFSET**
Offset distance or Through <Through>: **8**   Use for wind break walls ②.

Command: **OFFSET**
Offset distance or Through <Through>: **12**   Use for vault walls ④.

Command: **OFFSET**
Offset distance or Through <1'-0">: **5**   Use for interior walls ③.

Command: **FILLET**   Use for all corners.

Laying out the initial floor plan design was easy using XYZBOX. Adding wall thickness to these layout areas is also easy, using the EXPLODE, OFFSET and FILLET sequence.

Defining SubAreas

The last task to perform in laying out the floor plan design, is to add the required subareas to the interior and exterior. These areas consist of closets, toilets, utility and other interior rooms, plus the drive through and islands of the exterior canopy area. The dimensions and locations for these are defined both by name and number in the SubArea Layout Table and the SubArea Layout illustration below.

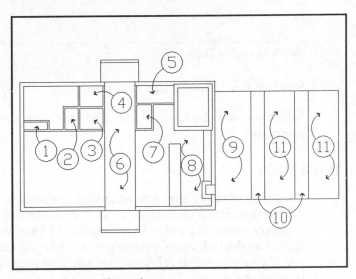

SubArea Layout Illustration

① CREATING THE SUBAREAS OF THE DESIGN FLOOR PLAN. Use the dimensions from the following table and the illustration above to create these eleven areas with the OFFSET command.

❑ Set layer BLD-010 current — AE-MENU Tablet [LAYER–BY SCREEN–SET].

❑ EDITING THE SUBAREAS. You may want to use TRIM, FILLET and EXTEND to clean up any of the subareas that you feel need cleaning up.

AREA NAME	NUMBER	AREA DIMENSIONS	
		X	Y
CLOSET	1	6'-6" x	2'-0"
JANITOR	2	3'-6" x	5'-3"
WOMEN	3	6'-0" x	5'-6"
MEN	4	6'-0" x	5'-6"
STORAGE	5	10'-6" x	4'-2"
CORRIDOR	6	8'-0" x	32'-4"
COUPON	7	4'-0" x	6'-0"
TELLER	8	12'-0" x	19'-6"
DRIVE THROUGH	9	9'-6" x	28'-0"
ISLAND	10	3'-6" x	28'-0"
DRIVE THROUGH	11	7'-9" x	28'-0"

SubArea Layout Table

Saving Your Floor Plan Drawing

When you are done, save your floor plan drawing by using the SAVE command.

Creating the Base Plan

Let's stop to think ahead to what tasks remain to complete the site plan design drawing. You need to be sure that the outline of the building fits properly within the setbacks of the site. An easy way to do this is to create a block of the building's outline now so that you can use it to verify the fit when you resume development of the site plan design drawing a little later in the chapter.

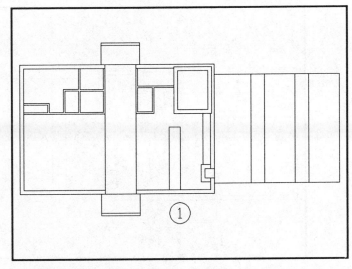

BASEPLAN Block Illustration

① CREATING THE BASE PLAN BLOCK. Use the WBLOCK command and select the entire plan to create a block named BASEPLAN.

➡ *NOTE! Even though a specific BLOCK may not be present in the drawing, the WBLOCK command lets you select entities to write out to create a separate drawing file.*

Correcting Layering

One drawback to using existing entities to create or define new entities (as you did with OFFSET in creating the layout subareas) is that the new entities may be on the wrong layer. To correct this, use the CHPROP (change properties) command to move the entities to their correct layers.

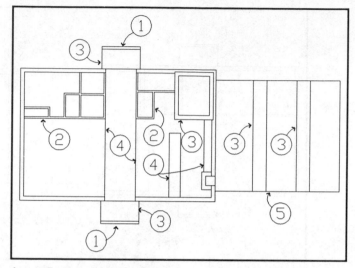

Layer Designations Illustration

① through ⑤ CORRECTING LAYER DESIGNATIONS OF DESIGN LAYOUT DRAWING. Use the CHPROP command to change all entities to their correct layers. Refer to the Correcting Layering Table below and the Layer Designations illustration above as guides.

ILLUSTRATION NUMBER	NEW LAYER
1	MAS-040
2	STD-050
3	CON-030
4	SPL-100
5	ROF-070

Correcting Layering Table

Now is a good time to use the SETBACK block created earlier to see if the floor plan design layout will fit on the site as required.

Moving Design Layout to the Main Area

If you are completing the entire drawing, and you have the areas of the floor plan set in the approximate size and required relationship to each other, you can use the SETBACK block created from the site plan to verify the fit.

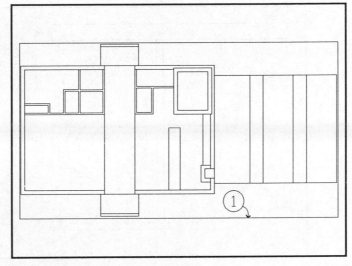

Checking Design Layout Illustration

① CHECKING DESIGN LAYOUT. You can use the MOVE command to transfer the design layout from the SCRATCH area to the MAIN area. Or, if you are just practicing the techniques, INSERT the SETBACK block into the scratch area of your drawing, or into AE-SCR4 scratch drawing.

Check to see how well the design fits according to the SETBACK block. Use the illustration above as a guide.

Inserting Doors

If you have been using AutoCAD for a while, you have probably created a door symbol. If you don't have one of your own, use the LINE and ARC commands to create one. (See the illustration below.) The doors for the building are all three feet wide.

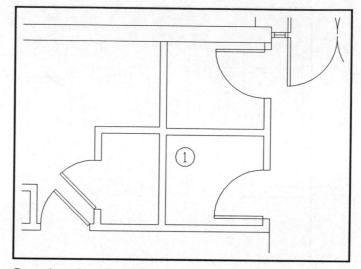

Doors Locations Illustration

① PLACING DOOR SYMBOL. You can use INSERT to place the door symbol, and use TRIM to remove the excess wall lines to define the door opening.

❑ Set layer DRS-080 current — AE-MENU Tablet [LAYER–BY SCREEN–SET].

Inserting Plumbing Fixtures and Lounge Cabinets

The next thing to do is to add the plumbing fixtures to your drawing. You can either insert the TOILET or LAVRT symbols that were created in Chapter Two, or, if you are using the AE-MENU, select them from the tablet area.

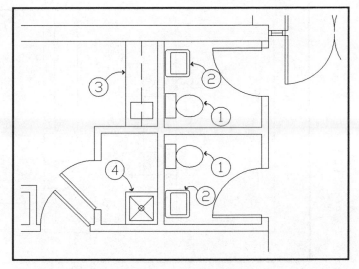

Plumbing Fixtures and Lounge Cabinets Illustration

❑ Set layer PLU-150 current — AE-MENU Tablet [LAYER–BY SCREEN–SET].

① PLACING TOILETS. Insert the toilet symbol, [TOILET], into your drawing in the locations illustrated above, by using INSERT command or by using the AE-MENU.

② PLACING LAVATORIES. Select the symbol, [LAVRT], from the AE-MENU or use the INSERT command to place the lavatories into the drawing as illustrated above.

③ CREATING LOUNGE CABINETS. Quickly draw in the base cabinet in the lounge area, using OFFSET with a distance of 24 inches. To represent the wall cabinet, offset the line, you just created, 12 inches back toward the wall. These entities should go on layer SPL-100, and you need to change the wall cabinet linetype to dashed. Use CHPROP for both of these tasks.

④ ADDING JANITOR SINK. Use LINE to create the janitor sink.

Adding Teller Work Positions

Next define the work positions at the teller counter. You can use OFFSET to create the required lines, then use FILLET and TRIM for cleanup. Use the following illustration to guide you.

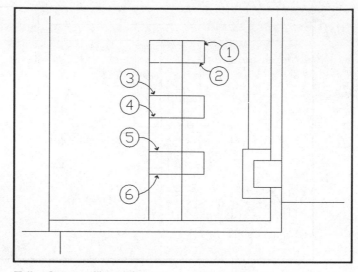

Teller Counter Illustration

❑ Set layer SPL-100 current — AE-MENU Tablet [LAYER–BY SCREEN–SET].

① through ⑥ ADDING THE TELLER COUNTER WORK POSITIONS. Add work positions to the original teller counter, created during the layout design phase, by using OFFSET with the dimensions specified in the Teller Counter Dimension Table below (and shown in the illustration above).

ILLUSTRATION NUMBER	OFFSET DISTANCE
1	2'
2	2'
3	3'
4	2'
5	3'
6	2'

Teller Counter Dimension Table

❑ CLEANING UP WORKSTATION ADDITIONS. Use FILLET and TRIM to remove any unnecessary lines.

Defining Drive Through Islands

The XYZBOX routine defined the drive through islands by laying them out as rectangles. The front and rear edges of these islands are actually rounded. So, the next task is to correct these areas.

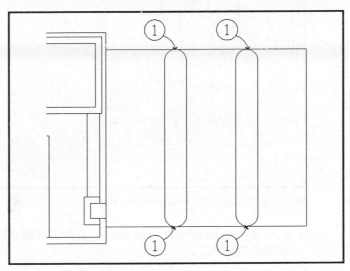

Drive Through Islands Illustration

❑ Set layer CON-030 current — AE-MENU Tablet [LAYER–BY SCREEN–SET].

① CORRECTING THE FRONT AND REAR EDGES. First use the CIRCLE command with the 2-Point option and set OSNAP to NEAR to draw a circle at each end of the islands.

❑ MOVE TO CORRECT LOCATION AND CLEAN UP THE ISLANDS. Try using the MOVE command with OSNAP set to QUAD for the base point on circle. Use OSNAP PERP for the To point: and select the canopy line. Now, you can use TRIM to remove the inner halves of the circles by selecting the vertical lines.

Defining Roof and Changing the Canopy Overhangs

Let's look at defining the roof and canopy overhangs.

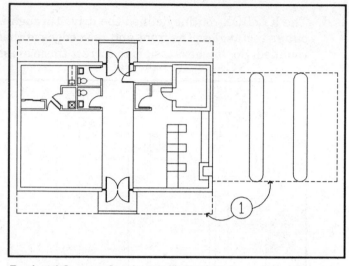

Roof and Canopy Overhangs Illustration

❑ Set layer ROF-070 current — AE-MENU Tablet [LAYER–BY SCREEN–SET].

① DEFINE ROOF OVERHANGS. This is pretty simple — just draw a line with an OSNAP of END on each side of the front and rear entrances, extending from each side of the wind break walls to the corners of the building.

❑ CHANGE ROOF AND CANOPY LINETYPE. Use CHPROP, selecting the roof and canopy overhangs, and change the linetype to DASHED.

Adding Windows

This is probably a good time to add the windows to the building. You can use a series of commands — OFFSET, TRIM, BLOCK and DIVIDE — to accomplish this task.

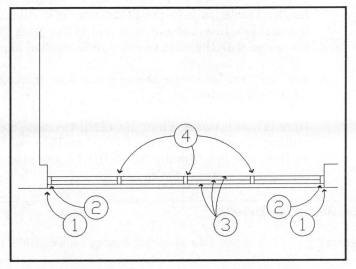

Windows Illustration

❑ Set layer WIN-080 current — AE-MENU Tablet [LAYER–BY SCREEN–SET].

① DEFINING THE WINDOW OPENING. Use OFFSET and select the left vertical exterior wall line and specify the first distance as 1'-4" to define the left jamb. Then repeat the command, selecting the first offset line, specifying a distance of 12 feet to define the right jamb.

② ADDING JAMB WINDOW FRAMES. Again, use OFFSET command. This time specify a distance of two inches, select the window opening lines, and offset toward the center of the window to create the frame members.

③ ADDING THE FRAME AND GLASS LINES. Now, use OFFSET yet again, with the default distance of two inches, select the horizontal exterior wall line and repeat the offset operation until you have defined both sides of the frame and glass line.

❑ CLEAN UP WINDOW DRAWING. Simply TRIM the excess lines.

④ ADDING THE MULLIONS. You have some vertical mullions to create for the windows. We think this calls for a technique.

Technique for Defining Mullions

The objective of this technique is to add equally spaced rectangles (defining mullions) centered on a line representing the glass of the window.

You *could* use the LIST command, select the glass line, and with the length reported, divide it by the number of mullions. Then, you can define the mullion. Insert one at each end of the glass line and use the COPY command with the relative coordinate method to place the others.

But our next technique shows you a way to accomplish this task in a fraction of the time.

Here is how it works: First, use LINE to create a rectangle to define the mullion. Second, create a block of the mullion and use DIVIDE to place it on the glass line. Finally, EXPLODE these blocks and trim the excess lines that pass through the mullions.

Setting Up Mullion Technique Drawing

 Enter selection: 2   Edit an existing drawing named 8910FP-D.

Select the [VIEW–RESTORE–SCRATCH] command from AE–MENU.

Enter selection: 1   Begin a NEW drawing named AE-SCR4.

Verify the settings shown in this chapter's AE-SCRATCH Setup Table.

When you are ready to get started, set layer WIN-080 as current.

➡ *NOTE! In order to get equal spacing, you will need to change or stretch the glass line, on each end, the width of one half of the mullion.*

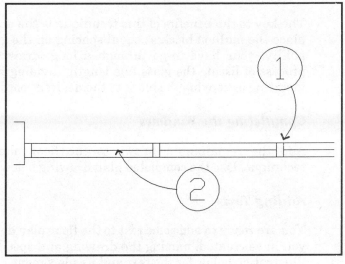

Applying Mullion Technique

Defining Windows

```
Command: LINE                                Draw a 21"x4" rectangle to represent the mullion ①.


Command: BLOCK
Block name (or ?): MUL
Insertion base point:                        Select center of MUL.
Select objects: W
First corner: Other corner: 4 found.
Select objects: <RETURN>


Command: DIVIDE
Select object to divide:                     Select glass line ②.
<Number of segments>/Block: B
Block name to insert: MUL
Align block with object? <Y> <RETURN>
Number of segments: 4


Command: EXPLODE                             Repeat for all MUL block.


Command: TRIM
```

Use trim to remove the portion of the glass line that passes through the mullions and at both ends.

By using DIVIDE with an associated block, you can create an equally spaced window, or storefront section quickly and easily.

Benefits

The key to the benefits of this technique is the ability to use DIVIDE to place the mullion block at equal spacing on the glass line. Without this ability, your have to go through a long, drawn out, and error-prone process of listing the glass line length, dividing it mathematically, and inserting or copying the block to the desired coordinates.

Completing the Windows

Add the remaining windows to the floor plan, using the mullion technique. Use the completed plan drawing below as your guide.

Adding Text

You are ready to add some text to the floor plan describing the areas that you have created, naming the drawing and specifying its scale. See the illustration below for the text and its placement.

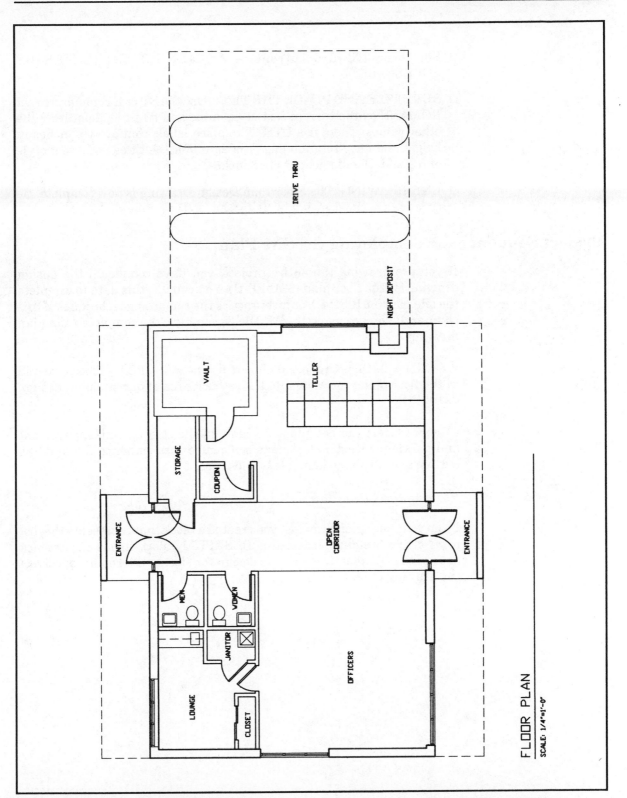

FLOOR PLAN

SCALE: 1/4"=1'-0"

Completed Floor Plan Design Drawing

❑ Set layer TXT-010 current — AE-MENU Tablet [LAYER–BY SCREEN–SET].

❑ SET STYLE AND PLACE THE TEXT. Use the STYLE command to set ROMANS as the current text style name and font file, defaulting the other values. Then, use DTEXT to place labels that are six inches in height and have rotation angles of zero. For the titles use a text style of ROMAND and a height of 18 inches.

❑ END DRAWING. The Floor Plan Design Drawing is now complete. Be sure to save the drawing by using the END command.

Project Sequence for Completing the Site Plan

If you are drawing the entire project, you have completed the design drawing for the floor plan (8910FP-D). You can use this data to complete the site plan (8910SP-D), or just practice the remaining techniques of this chapter. If you are using the AE DISK, you have the completed site plan drawing.

We will begin this sequence by inserting the BASEPLAN block (created in the floor plan sequence) into the setback area of our site plan design, 8910SP-D.

After we insert the baseplan, we can add the streets, walks, pavement and parking with striping, make some curb cuts, enhance the building outline and finish by adding text and titles.

Placing the Base Plan

From your floor plan drawing, you created a block that represents the *foot print* of the building. Insert this BASEPLAN block inside the setback area of the site plan to check the design for size, and if required, correct the drawing to match the area.

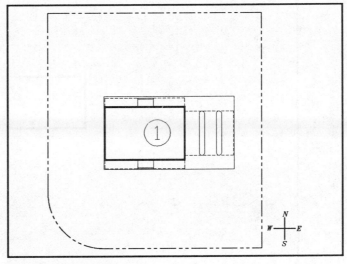

Baseplan Illustration

□ Set layer MIS-010 current — AE-MENU Tablet [LAYER–BY SCREEN–SET].

① PLACING THE BASEPLAN BLOCK WITHIN THE SETBACK AREA. Use INSERT, specify the block named BASEPLAN, and place it within the setback area.

Now that the building is in place, the next areas that need defining are the streets and walks that border the site.

Adding Streets and Walks

Just as you used OFFSET to define the setback area, you can use the command again to define the streets and walks. Specify exact dimensions for the locations of these entities by using the property lines as the selected objects to be offset. Specifying these entities at a dimension from the property lines will ensure that the site plan drawing is accurate and takes into account all allowances for rights-of-way and traffic flow patterns.

➡ *NOTE! Use the North arrow in the illustration below to help you follow the directions of entities selected in the following sequence.*

The following series of commands may seem redundant at times, but so is real life! Practice makes perfect.

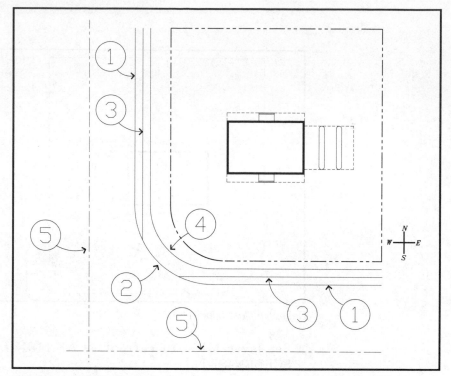

Streets and Walks Illustration

❑ LAYER. Set layer STR-020 current — AE-MENU Tablet [LAYER–BY SCREEN–SET].

① ADDING STREET CURB LINES. Add the street curb line that borders the site on the West by using OFFSET, a distance of 23 feet, selecting the West property line, and offsetting toward the West. You can add the street curb line that borders the site on the South by again using OFFSET, a distance of 15 feet, selecting the South property line, and offsetting toward the South. Also you need to use the CHPROP command to change these lines from the PRL-020 layer to STE-020.

② ADDING CURVE AT INTERSECTION OF STREETS. To add this curve or arc to the drawing, use FILLET, specify a radius of 40 feet, and select the two curb lines you just created.

③ ADDING CONCRETE WALK LINES. Now add lines to represent the concrete walk on the curb side. Use OFFSET, specify a distance of five feet, select the curb line that borders on the West side of the site, and offset toward the East. To finish the walk, repeat the offset toward the East again by selecting the new line.

To create the walks on the South side, just repeat OFFSET with the same distance, select the South curb line, and offset toward the North twice.

④ ADDING CURVE TO CONCRETE WALK. Use FILLET with a radius of 30 feet, selecting the West and South walk lines that are adjacent to the property lines, to place the curve. Use EXTEND to extend the walk lines that are adjacent to the curb lines to the curve at the intersection.

⑤ ADDING THE STREET CENTER LINES. For the West street center line, use OFFSET, specify a distance of 30 feet, select the West curb line, and offset toward the West. For the center line of the South street, repeat OFFSET; specify a distance of 43 feet, select the South curb line, and offset toward the South. To complete the sequence, use CHPROP to change the linetype that represents the street center lines to CENTER, and change the length of each line so that they cross at the intersection.

Now that you have the streets and walks defined, go ahead and define the paved area for parking and drives.

Adding Pavement, Drives, Parking Areas, Curbs and Curb Cuts

This next section is a rather lengthy section. We have included it in its entirety. First, because these additions would be necessary to complete a site plan drawing; and second, because a little practice with this type of exercise can improve your drawing productivity.

If you are simply studying the techniques for this chapter, please read this section through, it may give you some insight on how to use different AutoCAD commands in a sequence to improve your productivity, as well.

Major Steps

The first step in defining the paved areas of the site is to draw in the pavement edges as they relate to the property lines. Then you can define the drives and parking areas as illustrated below. Finally, you can add the curbs and place the striping in the South and West areas.

➡ *NOTE! The Site Plan Areas illustration below is for reference only. The areas are labeled to help you in this portion of the site plan sequence, if you are working through the sequence.*

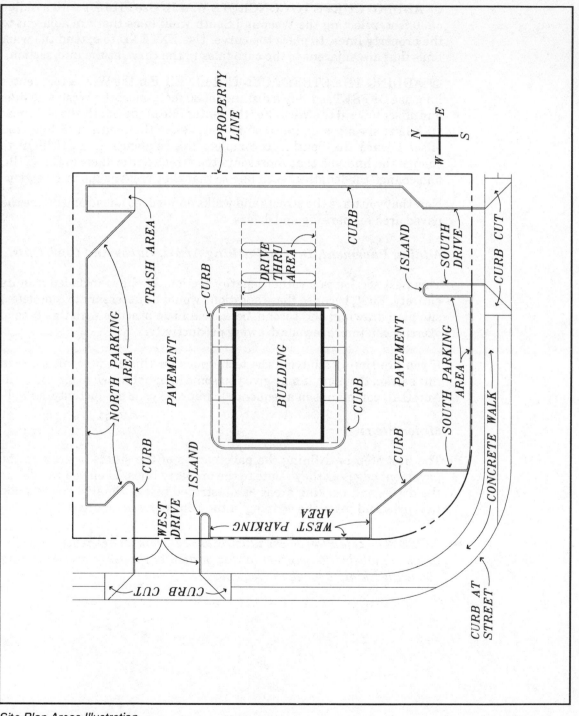

Site Plan Areas Illustration

Defining Pavement Edges

Start by adding some of the edges of the pavement to the site plan. To define the North, East, and West edges, use OFFSET. You'll do the South and the remaining West edges when you create the parking areas.

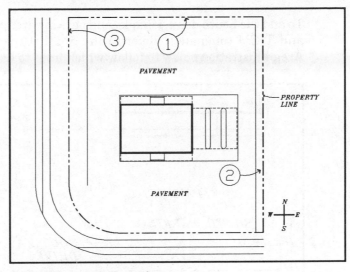

Adding Pavement Illustration

❑ Set layer PKG-020 current — AE-MENU Tablet [LAYER–BY SCREEN–SET].

① DEFINING NORTH PAVEMENT EDGE. Make the first OFFSET by specifying a distance of 5'-6", selecting the northern property line, and offsetting toward the South. This gives you the North edge of the pavement.

② DEFINING EAST PAVEMENT EDGE. Now use OFFSET again. This time, specify a distance of five feet, select the East property line, and offset toward the West to create the East pavement edge.

③ DEFINING WEST PAVEMENT EDGE. Use OFFSET once again. Specify a distance of 12'-8". Select the West property line, and offset toward the East. That will give us the West edge of the pavement.

❑ CLEAN UP CORNERS AND CORRECT LAYERS. You'll need to correct the overlapping lines at the corners with FILLET. Use CHPROP to change pavement edge lines to the correct layer, PKG-020.

Adding South Drive, Islands and Parking Area

To add the South drive, islands, and parking area, use the OFFSET, ARC, and TRIM commands. Refer to the South Drive, Islands and Parking Area illustration below to follow which lines to use for each operation.

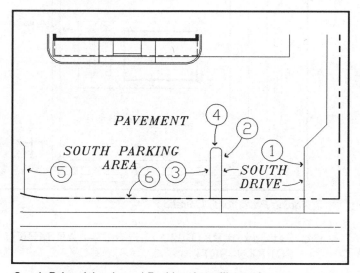

South Drive, Islands and Parking Area Illustration

① DEFINING SOUTH DRIVE EASTERN EDGE. To create the East drive edge, use OFFSET, specify a distance of eight feet, select the edge of the pavement line, and offset toward the West.

② DEFINING SOUTH DRIVE WESTERN EDGE. To create the West edge of the drive, OFFSET again, only this time specify a distance of 28 feet, select the East edge, and offset toward the West.

③ DEFINING THE ISLAND. Again, OFFSET is used to define the island by offsetting the drive line a distance of four feet toward the West.

④ DEFINING THE ISLAND CURVE. Now, create the curve of the island by using the ARC command with the 2-point option, and specifying a radius of two feet. Use MOVE and TRIM to locate and complete this curved section of the island.

⑤ DEFINING SOUTH PARKING AREA. The South parking area is 65 feet long. You can use the OFFSET command with this distance, selecting the West island line, and offset toward the West to define this area.

⑦ DEFINING PAVEMENT EDGE OF SOUTH PARKING AREA. With OFFSET, specify a distance of six inches, select the South property line, and offset toward the North to add the southern edge of the South parking area. You'll need to use CHPROP to change these lines to layer PKG-020.

Adding West Drive, Islands and Parking Area

You can add the West drive, islands and parking area b following the same sequence used with the South areas, using the OFFSET, ARC and TRIM commands. Refer to the West Drive, Islands and Parking Area illustration below to follow which lines to use to do what operation.

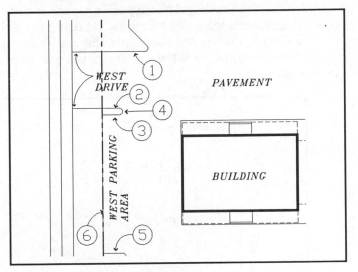

West Drive, Islands and Parking Area Illustration

① DEFINING WEST DRIVE EASTERN EDGE. Using OFFSET, specify a distance of 18 feet, select the edge of pavement line, and offset toward the South to create the West drive edge.

② DEFINING WEST DRIVE SOUTHERN EDGE. OFFSET is used to create the southern edge of the drive specifying a distance of 24 feet. Select the southern edge, and offset toward the South.

③ DEFINING THE ISLAND. Now use OFFSET again, this time with a distance of four feet to the South, to define this island.

④ DEFINING THE ISLAND CURVE. Use the ARC command with the 2-point option and a radius of two feet to create the curve. If needed, use the MOVE and TRIM commands to locate and complete this curved section of the island.

⑤ DEFINING WEST PARKING AREA. The West parking area is 60 feet long. Select the South island line, and offset toward the South to define this area.

⑥ DEFINING PAVEMENT EDGE OF WEST PARKING AREA. Specify a distance of six inches, select the West property line and offset toward the East to add the western edge of the West parking area. You'll need to use CHPROP to change these lines to layer PKG-020.

Adding Trash and North Parking Areas

Adding the Trash and North parking areas is a little different because the islands are angled at 45 degrees. By using the SNAP command with the ROTATE option, you can rotate the crosshairs to this angle, and draw these lines just as if the crosshairs were at right angles to the drawing area.

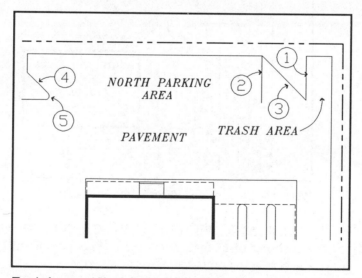

Trash Area and North Parking Area Illustration

① DEFINING THE TRASH AREA. To start with, let's use OFFSET, specifying a distance of 10 feet, select the eastern edge of the pavement, and offset toward the West. Then you can use OFFSET again, with a distance of 18 feet, to define the southern edge of this area.

② ADDING A REFERENCE LINE. To help define the diagonal East edge of the North parking area, create a reference line by offsetting the West edge of the trash area a distance of 18 feet.

③ DEFINING EASTERN EDGE OF THE NORTH PARKING AREA. First, use SNAP with the option to ROTATE the crosshairs and specify 45 degrees. Then you can draw a line (with the aid of OSNAP END) from the trash area to the pavement line, as illustrated above. Use TRIM to remove the North edge of pavement line and the ERASE command to remove the reference line.

④ ADDING THE WESTERN EDGE OF THE NORTH PARKING AREA. The western edge line of the North parking area is 99'0 from the eastern edge line. Use OFFSET with this distance and offset the diagonal East edge line toward the West. You may want to use the LINE command to square off the northern corner.

⑤ ADDING THE CURVE TO THE WESTERN EDGE OF THE NORTH PARKING AREA. To add the curve, use FILLET with a radius of three feet and select the northern edge of the West drive and the diagonal West edge of the North parking area. This is also a good time to correct the layering of these lines with CHPROP.

Adding the South and West Parking Striping

This next sequence involves adding the parking stripes to the South and West parking areas. Again, you can make use of OFFSET.

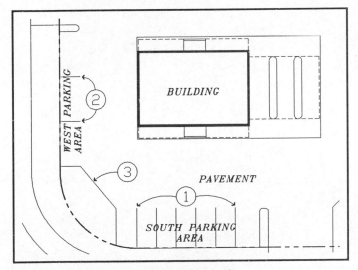

South and West Parking Striping Illustration

① PARKING STRIPING FOR THE SOUTH AREA. Use OFFSET with a distance of nine feet to the East, select the West edge of the parking area, and repeatedly offset this line until you add all the stripes. Be sure to leave the last parking space wider than the rest so that it can serve the handicapped.

② PARKING STRIPING FOR THE WEST AREA. Again, use OFFSET, this time with a distance of 20 feet, select either the northern edge (offsetting toward the South) or the southern edge (offsetting toward the North), to add the striping for the West parking area. Either direction produces the same results in the same number of operations.

③ COMPLETING THE PAVEMENT. The last required pavement line extends at an angle from the western edge of South parking area to the southern edge of the West parking area. Use the LINE command with OSNAP END to draw this line in.

Defining Curbs

If you can still tell North from South after the previous series of steps, congratulations! Just stick with us a little longer and you will be finished with the site plan.

Ready? Let's define the curbs that border the pavement, parking areas, and building.

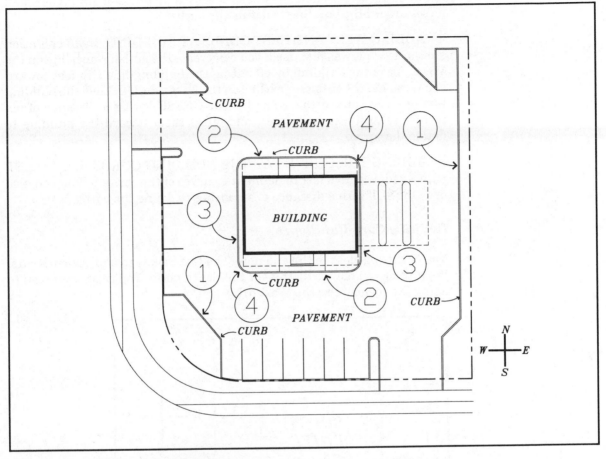

Pavement and Building Curbs Illustration

❑ Set layer CON-030 current — AE-MENU Tablet [LAYER-BY SCREEN-SET].

① ADDING CURBS TO THE PAVEMENT AND PARKING AREAS. All the curbs are the standard six inches. You can simply OFFSET all the outside pavement edges that border the site toward the outside to create these curbs. Use the FILLET command with a 0 radius to close the corners.

② ADDING NORTH AND SOUTH CURBS AT BUILDING. The first thing to do to create these curbs is to EXPLODE the BASEPLAN block. Then you will be able to use OFFSET, select the exploded lines of the North and South walls of the building, and offset them a distance of 8'6" to create our North and South outside curb lines. Then, reuse OFFSET again on these outside curb lines. Specify a distance of six inches, and offset toward the building to define the curbs.

③ ADDING WEST AND EAST CURBS AT BUILDING. In an operation similar to what was just used, you can create the outside curb line on the West side of the building by offsetting the building line two feet toward the West. For the outside curb line on the East side, OFFSET the building line one foot toward the East. Repeat OFFSET with a distance of six inches, select these curb lines, and offset them toward the building to define the curbs.

④ ADDING THE CURVES AT THE BUILDING CURBS. Use FILLET with a radius of five feet to define the curves of the corners. You can also use OFFSET with a distance of six inches to define the curbs.

Defining Curb Cut Slopes

You're getting close to the end now. The last task is to add the curb cuts. We'll show you how to do this for the South drive. You're on your own to apply the sequence to the West drive.

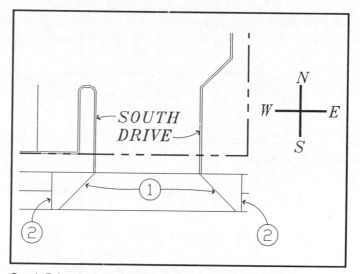

South Drive Curb Cut Illustration

① DEFINING CURB CUT SLOPE. First, use SNAP command to rotate the cross hairs to 45d. Draw a line from the intersection of the drive curb lines, at the walk, to the line that represents the street. Repeat this for the other side of the drive.

② DEFINING CURB CUT EXTENT. Now, rotate the cross hairs back to 0 and draw a line one foot from the intersection of the last line and the street line to the back side of the concrete walk. Do the other side as well.

③ COMPLETING THE CURB CUT. To trim the walk line that passes through the curb cuts, we suggest you use TRIM and select the last two vertical lines.

You have defined all the major features of the site plan drawing. To complete the drawing, all you have left to do is to add the North Arrow, text and titles. No more offsetting, we promise.

Adding the North Parking Area Striping

This phase of defining the parking lot striping consists of adding 45-degree diagonal lines to represent the striping for the North area.

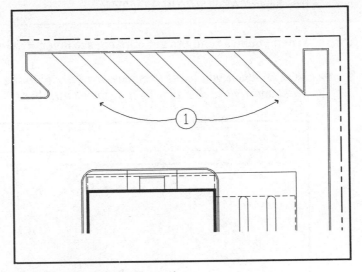

North Parking Striping Illustration

❏ Set layer PKG-020 as current — AE-MENU Tablet [LAYERS–BY SCREEN–SET].

① ADDING THE NORTH PARKING AREA DIAGONAL STRIPING. You could define the North parking area diagonal striping by using SNAP, rotating the crosshairs 45 degrees, and using the LINE command to draw each stripe. This would be tedious. Take a look at the following technique, using the ARRAY command, which accomplishes the same thing much more easily.

Technique for Adding the North Parking Area Striping

 In the following simple technique, use ARRAY, select the existing West curb line, and specify a distance of 10'-4" to add diagonal parking stripes. You can also use ARRAY with the same distance (see the note below) and select the East inside curb line to create the diagonal striping.

Setting Up Diagonal Striping Technique Drawing

 Enter selection: 2   Edit an existing drawing named 8910SP-D.

Select the [VIEW—RESTORE—SCRATCH] command from AE-MENU.

Enter selection: 1   Begin a NEW drawing named AE-SCR4.

Verify the settings shown in this chapter's AE-SCRATCH Setup Table.

When you are ready to get started, set layer PKG-020 as current.

➥ *NOTE! When you want to generate an array to the left and/or down, you must specify the distance as a negative number.*

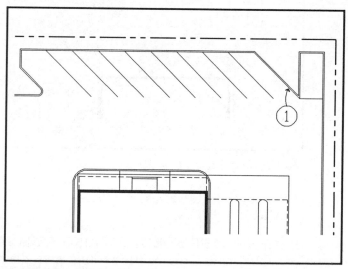

Applying the Striping Technique

Defining Parking Striping

Command: **ARRAY**

Select objects: Select the West curb line ①.

```
1 selected, 1 found.
Select objects: <RETURN>
Rectangular or Polar array (R/P): R
Number of rows (——) <1>: <RETURN>
Number of columns (||||) <1>: 9        Specify the number of stripes.
Distance between columns (||||): -10'4  Specify a negative distance.
```

By using ARRAY and selecting an existing entity, we can space other like-natured entities automatically.

Benefits

The benefit of this technique is that ARRAY reduces a task that would normally take a several operations into a single, more manageable step.

Now that you've finished the parking features of the site, you need to enhance the building lines.

Enhancing Building Lines

The following sequence, while simple in nature, produces worthwhile results when the drawing is plotted. By using PEDIT, you can add a width to the building lines so that they stand out more.

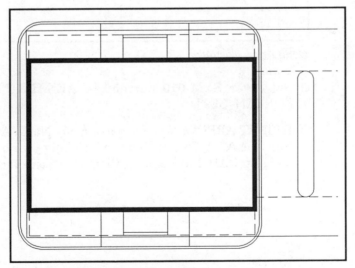

Enhanced Building Lines Illustration

① CHANGING THE BUILDING FOOT PRINT TO A BOLD LINE. Simply use the PEDIT command to convert the existing lines representing the exterior wall of the building into Plines and specify a width of 12 inches to add a bold appearance to the building outline.

Inserting North Arrow

Most A/E firms have their own version of a North Arrow. You can use yours, or, if you are using the AE-MENU System, select ours from the tablet. See the Completed Site Plan Design Drawing illustration below for placement.

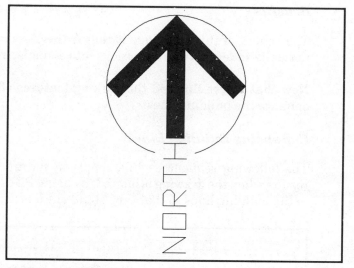

North Arrow Illustration

❑ Set layer SYM-010 current — AE-MENU Tablet [LAYER–BY SCREEN–SET].

❑ INSERT ARROW. If you are using our North Arrow, select [NORTH] from the AE-MENU. If you are inserting your own, and it is drawn to one AutoCAD drawing unit, then use a scale factor of 240.

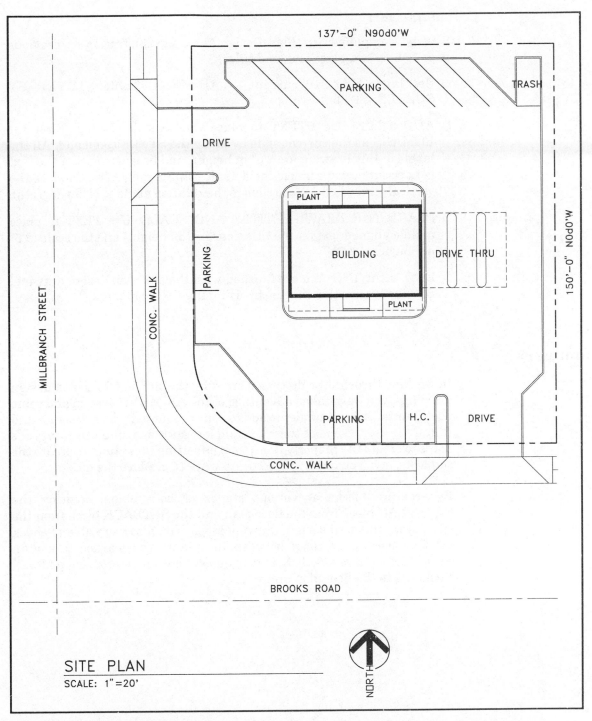

SITE PLAN
SCALE: 1"=20'

Completed Site Plan Design Drawing

Adding Text

See the Completed Site Plan Design Drawing illustration on previous page to help you place text, and titles.

❑ Set layer TXT-010 current — AE-MENU Tablet [LAYER–BY SCREEN–SET].

❑ ADD TEXT. Use DTEXT to place the text illustrated. First set ROMANS as the current style by using the STYLE command. All the labels in this drawing are 2'6" in height, and because most of the text is horizontal, you can use the default rotation angle (E). To place the vertical text, you'll have to change the rotation angle to N (90 degrees).

❑ PLACE THE DRAWING TITLE AND SCALE. Use TEXT to place middle aligned text for the title (four feet in height) and the scale (2'6" in height).

❑ END DRAWING. You are finished with the site plan design drawing. Be sure to save the drawing by using the END command.

Summary

In working through the process of creating the site and floor plan design drawings, you have taken advantage of the AE-MENU system and some useful AutoCAD techniques to add to your productivity as a designer and AutoCAD user. Some of these techniques, such as using the surveyor's notes to create the property lines; then offsetting those lines to create the setbacks, give you quick and accurate ways to produce the design.

By creating blocks at various stages of both plans, creating the BASEPLAN block from the floor plan and the SETBACK block from the site plan, you saved some redrawing steps. You also were able to check for dimension consistency between the plans. You learned to use an AutoLISP routine, XYZBOX, to help quickly lay out, adjust, and add wall thickness to the floor plan areas.

Even if you only followed a few of our drawing sequences, you certainly became acquainted with the capability of the OFFSET command to help create components for both plans. By following this chapter's sequences, you have laid a foundation for the 3D presentation drawings and more detailed working drawings that you will produce in subsequent chapters.

Now that we have a well-defined set of design drawings. Let's see how the building looks in 3D. And let's make a *movie* of it.

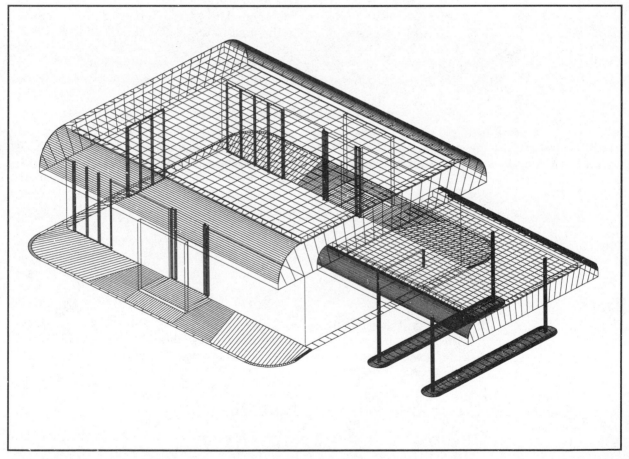

Three Dimensional AutoCAD Drawing

CHAPTER 5

3D Presentations and Animation

Release 10, AutoShade and AutoFlix

Introduction

There are as many methods of *presentation* as there are architects. Each individual has a preferred method. The objective in any presentation, however, is to communicate the *idea* to the client. How well the idea is understood and accepted depends on how thoroughly the design solution is portrayed, regardless of what technique is used to present it. We, as architects and engineers, take great pride in our artistic prowess in this phase of the practice. Sometimes how well we communicate the idea takes a back seat to how slick we can make the presentation look.

However, we can go well beyond the usual presentation of hand-drawn perspective drawings by using AutoShade, a 3D rendering program, and AutoFlix, a 3D animation program. Any presentation can be made more impressive by including the multi-colored views and animations from AutoCAD generated designs. Beyond how good these programs can make our presentations look, they can also make a contribution towards providing the client with more thorough and easy-to-understand explanations of our design ideas. They help provide the kind of quality design services, in artistic caliber and substance, that we all strive to give our clients.

Through the sequences and techniques presented in this chapter, we will show you how to generate AutoShade renderings and an AutoFlix *movie* of the design solution developed in Chapter Four for the simulated branch bank project.

Advantages to Using AutoShade and AutoFlix

Using AutoShade and AutoFlix to create presentations of AutoCAD generated design drawings represent an advanced integration of computer assisted design efforts. You can benefit from using AutoShade to generate, color, and shade perspective drawings of your designs by the added flexibility, accuracy, and time savings the program makes possible. AutoFlix offers you even more sophistication by letting you animate your drawings. Simulating a *stroll* around the building or through an area of particular interest not only gives the client a better feel for your design but also makes for much more exciting presentations.

Using Release 10's 3D Features for Presentation

AutoCAD Release 10 offers true 3D capabilities. In the preceding chapters we have briefly touched on two elements that are fundamental to these 3D capabilities — the World Coordinate and User Coordinate Systems. These coordinate systems control how entities are viewed in 3D space. Let's take a closer look at these coordinate systems.

Coordinate Systems

Release 10 of AutoCAD uses two coordinate systems which enable you (and AutoCAD) to keep track of 3D drawings:

- The World Coordinate System (WCS) — The WCS is a fixed cartesian coordinate system that employs the standard X,Y, and Z axes to indicate horizontal and vertical direction and 3D planes. The origin for this system is a fixed point, (0,0,0). This is the *standard* coordinate system we are all used to using.

- The User Coordinate System (UCS) — The UCS is a coordinate system that you can use to tilt and turn entities and the angles from which you view them to simplify locating 3D points. The origin for this system is not fixed and can be specified anywhere within the World Coordinate System to fit the entity you are working on. You can define a UCS orientation along any construction plane you want to work within, making complicated 3D drawings much easier to complete. You can define an unlimited number of User Coordinate Systems in any drawing.

- Coordinate System Icon — AutoCAD shows you how the current UCS is oriented through a Coordinate System Icon.

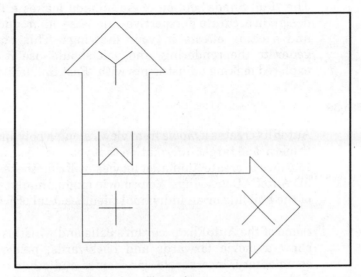

Coordinate System Icon Illustration

This icon indicates the directions of the positive X and Y axes of the current UCS and from what direction you are viewing it. It helps you keep your bearings as you change the orientation of a 3D drawing.

■ Plan View according to UCS — A *plan view* of a specified User Coordinate System is a view from directly above, with the construction plane parallel to the screen. You can dynamically specify plan views and other 3D view points.

AutoShade Basics

AutoCAD uses information on the heights of walls, windows, sills, and doors to create wireframe layouts from any angle you specify. This can save you hours of time, especially in laying out complicated projects. AutoShade takes AutoCAD's data from these wireframe layouts. It colors and applies shading to the surfaces.

You graphically place AutoShade *camera* and light icons while you are within the AutoCAD drawing to determine the perspective from which the rendering is to be seen and the qualities of the shading. You can create different *scenes* using combinations of cameras, light sources, and targets within the drawing.

The "Fast Shade" option of AutoShade lets you make color renderings of your scenes fairly quickly. You can adjust options in this rough draft until you get the results you want.

The "Full Shade" option of AutoShade makes a full rendering of your design in accurate perspective, and lets you produce exceptional realism and exciting effects in your drawings. This option takes longer to generate the rendering, and you should use it only after you have explored making adjustments with "Fast Shade."

AutoFlix Basics

AutoFlix creates a *movie* from views along a polyline *path* using the light, camera, and target information that you specify. AutoFlix uses AutoLISP routines to create animated movies utilizing these paths drawn through 3D AutoCAD drawings. Your movie might simulate a walk into the lobby of the building or an individual client's actual office.

Some of the AutoFlix program's bells and whistles let you do things like show a movie forwards and backwards, pause or stop on cue for commentary, incorporate wireframes and other renderings, and merge slides, overlays, music, text, and other applications. Using AutoFlix as part of your presentation technique can add a lot of interest as well as give your client more information about your design solution.

➡ *NOTE! If you want to experiment with AutoShade and AutoFlix, you can use the completed presentation drawing, called 89120FP-P that comes on the AE DISK, to experiment with AutoShade (and AutoFlix). If you don't have the disk, you'll have to follow our sequence of the commands through the chapter to create the drawing to use with AutoShade. We assume that you have a copy of AutoShade and AutoFlix in your AE-ACAD directory.*

Presentation Drawings and Movie Introduction

To illustrate the basic capabilities of AutoShade and AutoFlix, we chose to create our presentation drawings for this project by showing only the features of the exterior of the building and the site. Because we are showing only the exterior features, you can erase all the interior drawing data to save processing time. This will also allow you to make several changes to the exterior walls.

Before you create the AutoShade files, we will ask you to add some 3D enhancements to portions of the exterior. Then, you can perform the steps to generate an AutoShade rendering and an AutoFlix movie of the drawing. If you do not have the time to work with AutoShade and AutoFlix, our AutoCAD sequences will still help you build a 3D wireframe presentation of the building in AutoCAD. Of course, if you are using the AE DISK, you have the finished 3D drawing.

Project Presentation Drawings

We will use the following project sequence and techniques to show you how to combine your AutoCAD produced design drawings from Chapter Four with AutoShade and AutoFlix, to generate presentation drawings and a movie of the branch bank:

```
Project Sequence for the Presentation Drawings and Movies

AutoLISP drawing utility.
◆ Routine for Cutting a Line.
Prepare the Presentation Drawing Base Plan.
◆ Technique for Enhancing and Correcting Exterior
Entities.
Correct windows and frames.
Make color enhancements.
◆ Technique for Adding Columns for Canopy.
◆ Technique for Adding 3D curbs.
Define 3D entrance walks.
◆ Technique for Adding 3D Recessed Areas to Roof Base
Plan.
Add 3D roof curves.
Define 3D curved roof system.
◆ Technique for Outlining 3D Roof Curves.
◆ Technique for Defining 3D Vertical Ruled Surfaces.
◆ Technique for Adding Remaining 3D Curved Surfaces to
Canopies.
◆ Technique for Defining 3D Horizontal Surfaces.
◆ Technique for Adding 3D Ceiling Surfaces.
Define 3D entrance surfaces.
Define 3D curb surfaces.
Define 3D island surfaces.
Define 3D planting area surfaces.
Procedure for setting up AutoShade.
Procedure for loading and operating AutoShade.
Procedure for setting up and operating AutoFlix.
Generate a movie.
```

Drawing Setup for Chapter Five

As we did for the last chapter, we have set up two ways for you to approach practicing the concepts and techniques presented in this chapter. First, you can follow each project sequence and complete the drawings as if they were part of a real project. Second, you can read along with the project sequences and use AE-SCRATCH drawings to experiment with the techniques. By following either method, you will benefit by learning new ways to increase your productivity and gain a better understanding of how to apply AutoCAD, particularly its 3D capabilities, to your work environment. But to really generate the AutoShade rendering and AutoFlix movie of the design, you will have to follow Method One.

Method One

If you are using the AE DISK, you have the completed presentation drawing, called 8910FP-P. You can load this drawing and follow along with the exercises. Select the [VIEW–RESTORE–SCRATCH] command from the AE-MENU and move to the SCRATCH area to experiment with any of the techniques that catch your fancy.

If you plan to complete the presentation drawings from start to finish, load AutoCAD. Select Option 1, Begin a NEW drawing. Since you will be enhancing design drawing 8910FP-D, enter the drawing name 8910FP-P=8910FP-D. This will create the working copy of 8910FP-D and name it 8910FP-P.

Method Two

If you want to practice the techniques in this chapter, but not complete the drawings, experiment using an AE-SCRATCH drawing, called AE-SCR5, defined in the table below. With the exercise sequences that follow, you can set up the technique drawing by using the AutoCAD Setup routine and selecting the Architectural unit type, 1/4"=1" scale, D–24"x36" sheet size, and setting all other options defined in the table below.

COORDS	GRID	SNAP	ORTHO	UCSICON
ON	10′	6"	ON	WORLD

UNITS   Set UNITS to 4 Architectural.
           Default the other UNITS settings.
LIMITS   0,0 to 144′,96′
ZOOM   Zoom All.

Layer name	State	Color	Linetype
0	On	7 (white)	CONTINUOUS
CON-030	On	3 (green)	CONTINUOUS
MAS-040	On	2 (yellow)	CONTINUOUS
ROF-070	On	5 (blue)	CONTINUOUS
WIN-080	On	5 (blue)	CONTINUOUS
DRS-080	On	5 (blue)	CONTINUOUS

AE-SCR5 Setup Table

AutoLISP Drawing Utility

We will start out by showing you a simple AutoLISP routine called CUT.LSP. This routine lets you break or cut a line at a selected point with one operation. You will find this routine very useful when you are creating presentation drawings. It lets you cut a predrawn line into different parts and change them to different colors or layers.

CUT.LSP AutoLISP Routine

Verify that you have the CUT.LSP file in your AE-ACAD directory. Examine it if you wish.

Just read along, or create the AutoLISP file.

Testing the CUT.LSP Routine

To use the CUT.LSP routine, first select the line you want to cut, then select the point where you want it cut. You can use all OSNAP options in defining the cut point.

Here is what the listing looks like:

```
(prompt "Loading CUT utility...")
(defun C:CUT (/ pnta pntb)
     (setvar "cmdecho" 0)
     (prompt "\nSelect line to break:")
     (setq pnta (car (entsel)))                        ;line to cut
     (setq pntb (getpoint "\nEnter Cut Point:  "))     ;where to cut
     (command "BREAK" pnta pntb pntb)                  ;cut the line

     (setvar "cmdecho" 1)
)                                                       ;defun
```

AutoLISP Routine for [CUT–LINE]

Presentation Drawings Project Sequence

In the last chapter, you created a layout for the branch bank by using the XYZBOX AutoLISP routine. Because you specified not only the X and Y dimensions, but also a Z dimension, you have already created a 3D wireframe drawing on which to base a presentation drawing.

This drawing needs additional work to make it a true presentation drawing. In using XYZBOX, all of the entities have the same elevation and thickness. Also, some of the lines, and all of the text, that you created are not really necessary for a presentation. It is important to remember that AutoShade can process data faster with fewer lines. If certain lines are not going to display, or enhance the final drawing, then why include them? They will only slow AutoShade down.

We've chosen to illustrate the branch bank project based upon how the exterior of the building would be viewed. Therefore, the first steps in using the design drawing data are to erase all the lines and text that are unnecessary for this exterior presentation drawing, and to correct the elevations and thicknesses of the remaining lines. Then, we can enhance the drawing by adding 3D surfaces. After you've made your final 3D enhancements, you can use AutoShade and AutoFlix to create a presentation drawing and an animated walk around the exterior of the building.

Preparing the Presentation Drawing Base Plan

If you are using the AE DISK, you have the completed 3D drawing. Just read along with the exercises, practicing any techniques in the SCRATCH area.

If you are going to complete the drawing, starting from the existing floor plan drawing, erase most of the interior lines of the building. Use the following illustration as a guide.

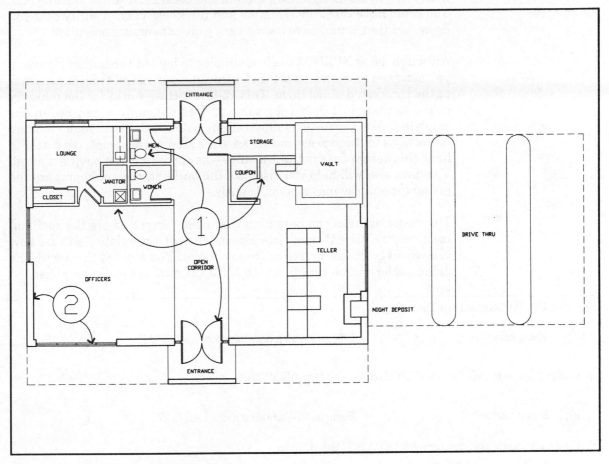

Presentation Drawing Base Plan Illustration

① ERASING INTERIOR LINES. Begin preparing the base plan by erasing all the interior entities from the floor plan drawing, leaving only the exterior walls, roof lines, windows, window frames, and drive through islands.

② ENHANCING AND CORRECTING EXTERIOR WALLS. Now that you've simplified things a bit, it will be easier to examine the entities and lines that remain and to correct any flaws that you find. The following techniques will make it easy.

Technique for Enhancing and Correcting Exterior Entities

During the design phase for the floor plan, you were probably more interested in the design itself than in the detailed settings of AutoCAD. You could have reset the elevation and thickness of each entity that you drew, but that would have been a very time-consuming operation.

Although using XYZBOX made it simpler to lay out the building's areas, you assigned an elevation of 0 and a thickness of 12 feet to all the entities at the time you defined them. This 12-foot setting works for the majority of the entities in 3D, but for some it does not, as you can see in the Applying 3D Correction illustration below. Even though making corrections to these areas means an extra step at this point, it's better to have this design data in the drawing — as opposed to not having it at all. The data also will help you visualize the building mass and prepare the presentation drawings more efficiently.

The major areas of the base plan that need correction are the roof and canopies, the drive through islands, and the entrance slabs. Let's see how you would go about correcting these entities. You created them with the default information specified with XYZBOX during the design phase.

Setting Up 3D Correction Drawing

Enter selection: 2   Edit an existing drawing named 8910FP-P.

Select the [VIEW–RESTORE–SCRATCH] command from AE-MENU.

Enter selection: 1   Begin a NEW drawing named AE-SCR5.

Verify the settings in this chapter's AE-SCRATCH Setup Table.

When you are ready to get started, select the Pull Down menu DISPLAY and VPOINT 3D.... Select the icon for the FRONT RIGHT view at +30.

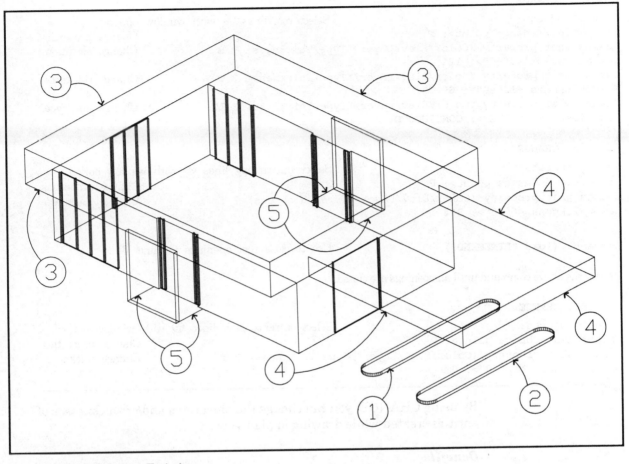

Applying 3D Corrections Technique

Drawing Corrections

```
Command: CHANGE
Select objects:                                          Select island lines ① with window option.
Properties/<Change point>: P                             Change properties.
Change what property (Color/Elev/LAyer/LType/Thickness) ? T    Change thickness.
New thickness <12'-0">: -6


Command: CHANGE
Select objects:                                          Select island lines ② with window option.
Properties/<Change point>: P                             Change properties.
Change what property (Color/Elev/LAyer/LType/Thickness) ? E    Change elevation.
New elevation <0'-0">: -6
Change what property (Color/Elev/LAyer/LType/Thickness) ? T    Change thickness.
New thickness <-0'-6">: -6
```

```
Command: CHANGE
Select objects:                                     Select roof lines ③ with window option.
Properties/<Change point>: P                                        Change properties.
Change what property (Color/Elev/LAyer/LType/Thickness) ? E         Change elevation.
New elevation <0'-0">: 12'
Change what property (Color/Elev/LAyer/LType/Thickness) ? T         Change thickness.
New thickness <12'-0">: 6'
Change what property (Color/Elev/LAyer/LType/Thickness) ? LT        Change line type.
New linetype <HIDDEN>: CONTINUOUS


Command: CHANGE
Select objects:                                     Select the canopy lines ④ with window option.
Properties/<Change point>: P                                        Change properties.
Change what property (Color/Elev/LAyer/LType/Thickness) ? T         Change thickness.
New thickness <6'-0">: 3'4


Command: (load "LINEMEND")                           Use to close roof openings for line ③.
```

Repeat the above command until all openings are closed.

```
Command: CHANGE
Select objects:                                     Select entrance slab lines ⑤ with window option.
Properties/<Change point>: P                                        Change properties.
Change what property (Color/Elev/LAyer/LType/Thickness) ? T         Change thickness.
New thickness <12'-0">: 0
```

By using CHANGE, you can change the elevations and/or thicknesses of entities created while drawing in plan view.

Benefits

This technique displays the flexibility of AutoCAD. You could have reset the elevation and thickness for each entity while creating them with XYZBOX, but having to keep up with all those individual specifications would have prevented concentration on the design. By using CHANGE, you can quickly correct the data at a time when it is more appropriate.

With these areas corrected, you can edit the windows and frames.

Correcting Windows and Frames

Using TRIM and BREAK, you can correct and clean up some unnecessary window and frame lines that were created during the design phase. You also need to change the elevation and thickness of the drive through window. It will be easier to do this if you first rotate the building into a plan view.

❑ 3D VIEW. Using VPOINT, rotate the building into plan view by specifying a VIEW POINT of 0,0,1.

③ EDITING WINDOWS. Use the TRIM and BREAK commands to remove the excess lines that represent the exterior wall and interior frame lines. OSNAP INTersection will help do this.

④ CHANGING ELEVATION AND THICKNESS OF TELLER DRIVE THROUGH WINDOW. The elevation and thickness of this window both need to be three feet. You can use options of the CHANGE command to make the corrections.

Making Color Enhancements

If your system has the ability to display the added colors shown below, changing the color you originally assigned to the layers of your drawing will enhance the way your final computer presentation looks under AutoShade. We have found (by trial and error) that the following color assignments, give the drawings more realistic shading effects in AutoShade. This can be done from a menu option. If your system does not have the ability to control color brightness and hue, then AutoShade and AutoFlix will use black and white dots, called dithering, to simulate shading and color enhancement for you.

❑ CHANGE LAYER COLORS. To change layer colors, select SETTING and then MODIFY LAYER from the Pull Down menu. If you select the Color option, you can change the colors of the layers shown in the table to the colors listed beside them.

LAYER NAME	COLOR NUMBER
ROF-070	30
CON-030	253
MAS-040	45
WIN-080 and DRS-080	173

Color Enhancements for AutoShade

By adding an even or an odd number to the new colors given above, you can either increase or decrease the layer color's hue, saturation, or brightness.

➡ *TIP! To determine the best color for your project (and also which colors are available with your system), use the SLIDE command and look at the CHROMA slide.*

Adding Columns for Canopy

Now add the columns that support the canopy at the drive through area.
The columns are six inch diameter steel pipes, one of which is located at
the center of each of the arcs that define the curved ends of the islands.
Use the CIRCLE command to add these columns.

❑ LAYER. Set layer STL-050 current — AE-MENU Tablet [LAYER–BY
SCREEN–SET].

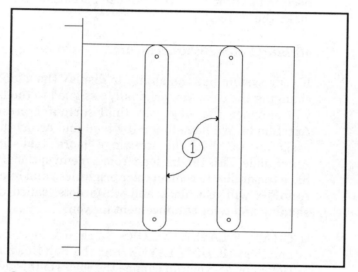

Columns at Canopy Illustration

① ADDING CANOPY COLUMNS. Use the following technique to add
your pipe columns.

Technique For Adding Columns

You can add the columns needed to support the canopy over the drive
through area, extending from the islands up to the ceiling of the canopy,
by defining a circle diameter of six inches. There is a straightforward way
of adding these columns by first setting the appropriate layer as current,
and then by specifying the elevation and thickness for each set.

Setting Up Column Technique Drawing

  Enter selection: 2   Edit an existing drawing named 8910FP-P.

Select the [VIEW–RESTORE–SCRATCH] command from AE-MENU.

Enter selection: 1   Begin a NEW drawing named AE-SCR5.

Verify the settings in this chapter's AE-SCRATCH Setup Table.

When you are ready to begin, set layer STL-050 as current and select the Pull Down menu DISPLAY and VPOINT 3D.... Select the icon for the FRONT RIGHT view at +30.

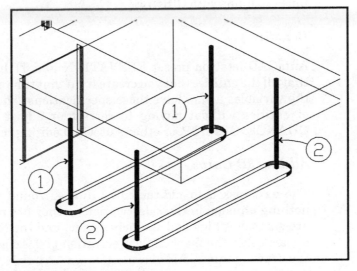

Applying Column Technique

Adding Columns

```
Command: OSNAP
Object snap modes: CEN

Command: SETVAR
Variable name or ?: ELEVATION
New value for ELEVATION <0'-0">: 0          Check current elevation.

Command: SETVAR
Variable name or ? <ELEVATION>: THICKNESS
New value for THICKNESS <0'-0">: 12'        Thickness of columns ①.

Command: CIRCLE              Create pipe columns ① with a radius of 3.

Command: SETVAR
Variable name or ? <THICKNESS>: ELEVATION
New value for ELEVATION <0'-0">: -6         Elevation top of island ②.
```

```
Command: SETVAR
Variable name or ? <ELEVATION>: THICKNESS
New value for THICKNESS <0'-0">: 12'6        Thickness of columns ②.

Command: CIRCLE                Create pipe columns ② with a radius of 3.
```

By first setting the elevation and thickness variables, you can add the
pipes columns with CIRCLE.

Benefits

AutoCAD lets you preset ELEVATION and THICKNESS variables so
that all the entities that you create from that time on (or until you change
the variables again) have those specifications. Thus, you create a group
of entities without having to correct or adjust them later with the
CHANGE, CHPROP, or other AutoCAD editing commands.

Adding 3D Curbs

The next task is to add the curbs that surround the building. There is
nothing unusual in their description — they are made of concrete, they
are six inches high and six inches wide, and the top of each is set at six
inches below the finished floor elevation of 0. To get started, first rotate
the drawing into PLAN view.

❑ LAYER. Set layer CON-030 current — AE-MENU Tablet [LAYER–BY
ENTITY–SET].

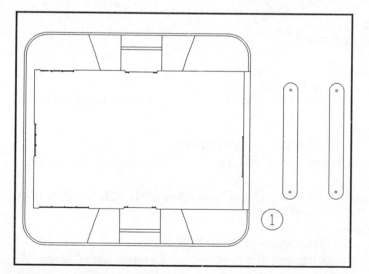

3D Curbs at Building Illustration

① ADDING BUILDING CURBS. Take a look at the following technique for creating all the 3D curbs for the building and for placing them at the right elevation.

Technique for Adding 3D Curbs

Before you can create the curbs that border the building , you have to make some fundamental design decisions. How will the curbs slope, how much will you slope the grade and paved areas around the building, and will this slope affect the function of the drive through window?

We decided to slope the immediate area between the building and curb by six inches, and set the height of the curb at six inches to allow for storm drainage away from the building. This represents a change in elevation from 0 inches for the finished floor to -12 inches for the top of the pavement. This -12-inch elevation presents a problem at the teller window. The design calls for an exterior dimension of no more than 3'-6" from the top of the pavement to the bottom of the window and a interior dimension of no more than 3'-0" from the finished floor to the bottom of the window. Because of these specifications, you will have to raise the paved area that passes next to the window by six inches.

The following two-part technique uses the OFFSET, LINE, FILLET, and 3DLINE commands to create the curbs and compensate for the required change in pavement elevation.

Setting Up Curb Techniques Drawing

    Enter selection: 2   Edit an existing drawing named 8910FP-P.

Select the [VIEW–RESTORE–SCRATCH] command from AE-MENU.

Enter selection: 1   Begin a NEW drawing named AE-SCR5.

Verify the settings in this chapter's AE-SCRATCH Setup Table.

When you are ready to begin, set layer CON-030 as current and select the Pull Down menu DISPLAY and VPOINT 3D.... Select the icon for the PLAN view.

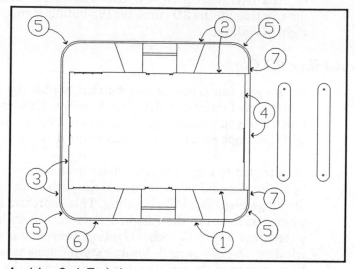

Applying Curb Technique

Adding Curbs

```
Command: OFFSET
Offset distance or Through <0'-0">: 8' 6

Select object to offset:
Side to offset?

Select object to offset:
Side to offset?
Select object to offset: <RETURN>
```
Use to create curb lines ① and ②.

Select exterior wall ①.
Any point toward the South.

Select exterior wall line ②.
Any point toward the North.

```
Command: OFFSET
Offset distance or Through <8'-6">: 2'

Select object to offset:
Side to offset?
Select object to offset: <RETURN>
```
Use to create curb line ③.

Select exterior wall line ③.
Any point toward the West.

```
Command: OFFSET
Offset distance or Through <2'-0">: 1'

Select object to offset:
Side to offset?
Select object to offset: <RETURN>
```
Use to create curb line ④.

Select exterior wall line ④.
Any point toward the East.

```
Command: FILLET
Polyline/Radius/<Select two objects>: R
Enter fillet radius <0'-0">: 5'
```
Use to set radius of corners ⑤.

```
Command: FILLET
Polyline/Radius/<Select two objects>:
```
Repeat for all corners ⑤.

```
Command: OFFSET
Offset distance or Through <1'-0">: 6
Select object to offset:
Side to offset?
Select object to offset: <RETURN>
```
Repeat for all lines and fillets that represent curbs.

Select all offset lines that represent curbs.
Any point away from building.

```
Command: ERASE
```
Erase curb lines ⑦.

```
Command: CHPROP
```
Change all curb lines to layer CON-030.

Now the curbs are defined, but at the wrong elevation. The next part of this technique shows you how to place the curbs at the correct elevations and to tie the drive through window curbs to the others. Since the curbs on the front, rear, and West sides of the building are the simplest, you can start making the elevation changes there. Then, with the building in a 3D view, ZOOM in on the drive-through side of the building and add 3D lines to represent the sloping section of the curbs that tie the front and rear curbs to the drive through curb.

3D Curbs

```
Command: CHANGE
```
Select curbs lines ① ,② ,③ ,⑤ ,⑥ set elevation to -12 and thickness to 6.

```
Command: CHANGE
```
Select curb line ④ , set thickness to -6.

Now you are ready to change the view of the building to the [FRONT RIGHT] view at +30, and add the sloped curb below the teller window.

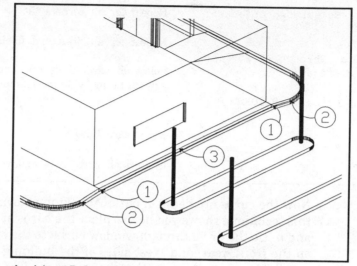

Applying 3D Curb Technique

```
Command: ZOOM
```
Define a window size large enough to display both curb corners at the teller window.

```
Command: SETVAR
Variable name or ?: ELEVATION
New value for ELEVATION <0'-0">: 0
```
                                                            Elevation of top of curb.

```
Command: SETVAR
Variable name or ? <ELEVATION>: THICKNESS
New value for THICKNESS <0'-0">: -6
```
                                                            Thickness of curb.

```
Command: 3DLINE
From point: INT
of
To point: END
```
                                          Select intersections of curb lines at ①.

```
of
```
                                          Select end of fillet ②.
Repeat 3DLINE command for other side of drive through.

```
Command: ERASE
```
                                          Select curb line ③.

You were able to add the curbs to your presentation drawing with little difficulty by offsetting the building walls. The technique is similar to the method used for creating the curbs in the site plan design drawing. You used a combination of offsets and fillets to define the outside curb lines and curves at the corners. After you changed the elevation and thickness of the North, South, and West curbs, you used the 3DLINE command to add the curb lines connecting the drive through curb with the North and South curbs.

Benefits

It was easier to create the curbs and filleted corners of three sides of the building at one elevation and thickness; editing them later (leaving the teller window curb as a separate entity to create with its own specifications), than to create each curb line at its individual specification.

Defining 3D Entrance Walks

The entrance walks are designed to be flat at the entrance and slope down six inches to the top of the curb, giving a compound curve to the walk surface. You will need to outline this area with 3DLINEs and 3DPOLYs so that you can add a 3D surface to it later in the chapter.

❏ LAYER. Set layer CON-030 current — AE-MENU Tablet [LAYER–BY SCREEN–SET].

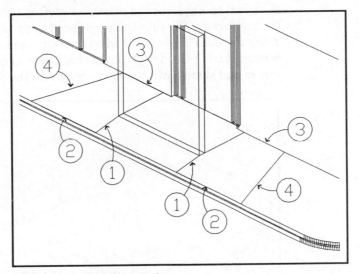

3D Entrance Walks Illustration

① DRAWING THE WALK LINE AT THE ENTRANCE. You can start by using the 3DPOLY command with OSNAP END to create a line from the building to the outside edge of the wind break wall, continuing the line with OSNAP PERP to join the top of the curb.

② DRAWING THE WALK LINE AT THE CURB. Now you can use the 3DLINE command with OSNAP END to create an eight-foot line starting at the end of the 3DPOLY you just created and ending with OSNAP NEAR so that the line is parallel to the curb.

③ DRAWING THE WALK LINE AT THE BUILDING. Now use 3DLINE and OSNAP END again, this time to create a six-foot line starting at the other end of your 3DPOLY and ending with OSNAP NEAR so that the line is parallel with the building.

④ DRAWING THE WALK LINE FROM THE BUILDING TO THE CURB. To join the building walk line to the curb walk line, just continue the 3DLINE and use OSNAP END.

❑ DRAW THE OTHER SIDE. If you use VPOINT to define a similar 3D View of the other side of the building, you can repeat this sequence to define its 3D entrance walks.

Adding 3D Recessed Areas to Roof Base Plan

Let's move on to the roof now (step lightly). As you recall, the building is designed to have curved mansard roofs covering the building and drive through areas. The first thing you need to do to get this part of the drawing ready for AutoShade is to add a recessed area to the roof base plan that you created during the design phase. Each recessed area will be built-up in the center and slope down to roof drains set in the four corners.

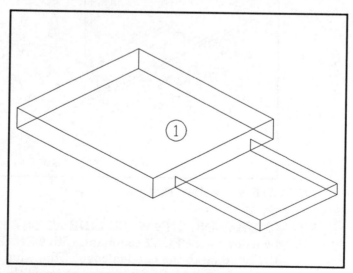

Roof Base Plan Illustration

① ADDING RECESSED AREAS OF ROOFS. We have a technique that you can use to place the 3D elements of these recessed areas of the roofs into the base plan.

Technique for Adding 3D Recessed Areas

Since the roof line of the base plan has already been defined with an elevation and a thickness, your task is to use and enhance this data to create the recessed area for the built-up roof elements.

You could spend a lot of time in 2D setting the variables for elevation and thickness and adding lines to represent each recessed area. Or, you can use OFFSET and obtain the same results with half the work. In this technique, we are going to show you how to do just that, first with the recessed area of the main roof, then with recessed area of the drive through roof.

Setting Up 3D Recessed Areas Drawing

Enter selection: 2   Edit an existing drawing named 8910FP-P.

Select the [VIEW–RESTORE–SCRATCH] command from AE-MENU.

Enter selection: 1   Begin a NEW drawing named AE-SCR5.

Verify the settings in this chapter's AE-SCRATCH Setup Table.

Start with the Roof Base Plan by setting layer ROF-070 as current and turning off all other layers.

Now you can select the Pull Down menu DISPLAY and VPOINT 3D..., and select the icon for the FRONT RIGHT view at +30.

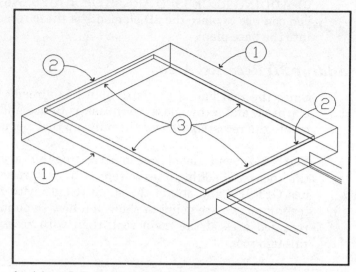

Applying 3D Recessed Area Technique to Main Roof

Adding Main Roof Recessed Area

Command: **OFFSET**   Select ①, specify distance of 6', pick a point toward the center of roof.

Command: **OFFSET**   Select ②, specify distance of 1', pick point toward the center of roof.

Command: **FILLET**   Select offset lines and correct all corners.

Command: **CHANGE**   Select the offset lines. Change elevation to 17', the thickness to 1'.

Now that you have added the recessed area to the main roof, add one to the drive through roof.

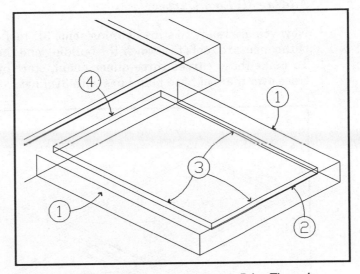

Applying 3D Recessed Area Technique to Drive Through

Adding Drive Through Roof Recessed Area

Command: **OFFSET**	Select ①, specify distance of 3'4, pick a point toward the center.
Command: **OFFSET**	Select ②, specify distance of 1', pick a point toward the center.
Command: **FILLET**	Select offset lines and correct all corners.
Command: **CHANGE**	Select the offset lines and change the elevation to 14'4, the thickness to 1'.
Command: **COPY**	Select recessed roof line and create ④.

Using this technique, you created the recessed areas of the roofs by correcting and enhancing the existing roof data from the roof base plan.

Benefits

The major benefit of this technique is it lets you add an entire new area to each roof without having to draw a single line. Instead of using LINE, you took existing 3D roof data and added the recessed area of each roof section with OFFSET.

Adding 3D Roof Curves

Now you are ready to start defining some 3D roof curves, the 3D curves of the mansard roofs that cover the building and the drive through areas. To make these curves three-dimensional, start by setting a UCS, and trace over the roof base plan lines with 3D lines.

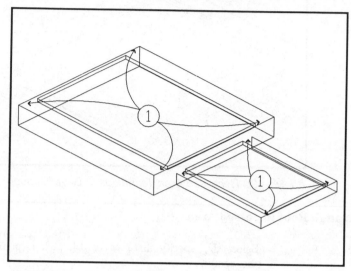

3D Roof Base Plan Illustration

① ADDING CURVES TO ROOF. To add the curves to the roofs, you can follow along with this next technique. The first step is to trace over the lines of the roof base plan with 3DLINES.

Technique for Outlining 3D Roof Curves

Defining 3D curves, or curves that have Z radii, is much easier if you define a UCS to rotate the drawing. You created the base plan for the roof using lines with elevations and thicknesses. Now you can use the 3DLINE command with a running OSNAP to trace the lines from the base plan that you want to curve. You will be working on layer 000-000.

After you complete your trace, define 3DLINES with an elevation and thickness to tie the ends of the roof together. Once you've done that, you can set your UCS and use the FILLET command with a specified radius to create these 3D curves.

Finally, you will turn layer 000-000 off, erase the base plan lines, turn layer 000-000 back on, and use our TOLAYER AutoLISP routine to change the new curved roof lines to layer ROF-070.

Setting Up 3D Roof Curves Drawing

 Enter selection: 2   Edit an existing drawing named 8910FP-P.

Select the [VIEW–RESTORE–SCRATCH] command from AE-MENU.

Enter selection: 1   Begin a NEW drawing named AE-SCR5.

Verify the settings in this chapter's AE-SCRATCH Setup Table.

When you are ready to begin, set layer 000-000 as current and select the Pull Down menu DISPLAY and VPOINT 3D.... Select the icon for the RIGHT FRONT view at +30.

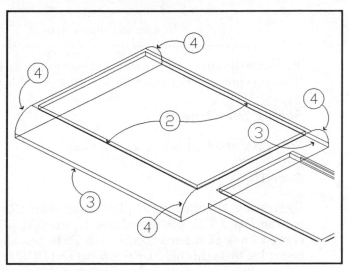

Applying 3D Roof Curve Technique

3D Roof Curves

Command: **OSNAP**   Set a running OSNAP of INTersection.

Command: **3DLINE**   Use to trace roof lines ① that will receive curve.

Command: **3DLINE**   Use to create lines ② that tie top of mansard together.

Command: **SETVAR**
Variable name or ?: **ELEVATION**
New value for ELEVATION <0'-0">: **12'**   Elevation for line ③.

```
Command: SETVAR
Variable name or ? <ELEVATION>: THICKNESS
New value for THICKNESS <0'-0">: 1'          Thickness for line ③.


Command: 3DLINE          Use to create lines ③ that tie the bottom of mansard together.
```

Define a UCS by selecting SETTING from the Pull Down menu, then UCS OPTIONS and the icon for RIGHT. Once you have defined the UCS, use FILLET to add the curves.

```
Command: FILLET                        Use to set radius for ④.
Polyline/Radius/<Select two objects>: R
Enter fillet radius <0'-0">: 5'


Command: FILLET                        Use to define curves ④.
```

To create the curves of the roof that covers the drive through area, repeat the same process, but use a fillet radius of 2'4" instead of five feet.

Saving Your Drawing

When you are done save your drawing.

Benefits

By tracing the roof base plan lines with 3D lines and setting an appropriate UCS, you were able to use FILLET to add the vertical curved lines of the mansards with little trouble. You would not have been able to do this before Release 10 and UCS's came out. UCS's let you change the construction planes of your drawings, and allow you to use all of the commands of AutoCAD in 3D space, just as if you were drawing in 2D space.

Now that you have defined the 3D outline of the roof area, the next big task is to add 3D surfaces to the drawing.

Introduction to 3D Surfaces

This chapter's remaining AutoCAD techniques deal with adding 3D surfaces to the drawing. This is not always a straightforward task. If you were to render your drawing under AutoShade, as it exists now, you would see some hollow areas.

Black Holes and Hollow Areas

When you create an AutoShade rendering of a 3D AutoCAD drawing, you will discover that any areas that you defined without a thickness appear *hollow*. For instance, the ends of your mansard roof (diligently created with 3DLINEs and given curved corners) would show up as gaping holes.

Back in the design drawing phase of the project, you created the first set of lines (including a Z value) for the roof with XYZBOX. Earlier in this chapter, you corrected the thickness for these lines. As long as you have thickness, there should be no problem, right? Well, if you recall, these were the lines that couldn't be adapted for the curve. Now the problem is that AutoShade will treat the new lines as hollow. How do we solve this problem? You add 3D surfaces between these lines.

Horizontal Voids

Other areas that pose problems for AutoShade are horizontal surfaces. They also appear hollow to AutoShade. It is not very comforting to see your building displayed without a roof (especially if you live in a tornado zone). To avoid the *void* you must add a 3D surface so that AutoShade will recognize the area as a solid.

If you want to review the different types of 3D surfaces, please refer to INSIDE AUTOCAD (New Riders Publishing).

RULESURF and EDGESURF Surfaces

For the purposes of the project's presentation drawing, you will be using AutoCAD's RULESURF and EDGESURF commands to add the surfaces that AutoShade requires. The surfaces generated by these two commands appear different in wireframe view, but appear similar when rendered under AutoShade. So for outside of demonstration purposes, you could really make do with just ruled surfaces.

➡ *NOTE! You can use 3DFACE on flat surfaces. However, we find it is sometimes hard to visualize which surfaces have already been defined; and hard to visualize how the final rendering will appear with the surface.*

Basic Rules of Ruled and Edged Surfaces

You define ruled surfaces (RULESURF) by selecting two lines on either side of an area. These lines can be either straight or curved, but be careful — always select your lines from the same end. If you don't, your surface will be twisted.

You define edged surfaces (EDGESURF) by selecting all the entities that form the border of the area you want to surface. After you've selected all the areas, the command generates a mesh type pattern to define the surface.

The variables SURFTAB1 and SURFTAB2 control how dense the lines or mesh appear. SURFTAB1 defines the density of the lines that RULESURF creates. SURFTAB1 and SURFTAB2 govern the density of the mesh that EDGESURF creates.

For this project, use RULESURF to define all the surfaces that are either sloped or warped (such as between the curb and building) or curved (such as the curved portion of the mansard roof). Use EDGESURF to define all the surfaces that are flat, or that are on a single plane, such as the recessed roof areas.

Adding 3D surfaces may sound a little complicated, but if you practice the techniques that we give you in the next sections, we think that you will be able to master the art of adding these surfaces under almost any conditions.

Adding 3D Vertical Surfaces

Let's start by adding vertical surfaces to the ends of the mansard roofs. You'll use RULESURF for this.

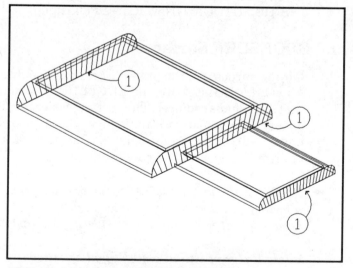

3D Vertical Curve-edged Surfaces Illustration

① ADDING 3D SURFACES. Use the following technique to add the surfaces to the ends of your mansard roofs.

Technique for Defining 3D Vertical Ruled Surfaces

 Did we say this was going to be easy? Well, it is almost easy. The surface for this first area presents a problem. The number of lines that can border an area defined with RULESURF is limited by AutoCAD to two. The vertical surfaces of your mansard roof are defined by four lines and two curves. This problem is easy to solve by simply converting the top 3D line and the two curves into a polyline. After you do this, select that polyline and the bottom 3D line to define the area.

First, set the variables that govern the density of the lines that RULESURF will create. We found that trial and error is the fastest way to set density the way you like it. All our values for SURFTAB1 and SURFTAB2 were obtained by trial and error. Feel free to adjust our values.

Setting Up 3D Vertical Ruled Surfaces Drawing

 Enter selection: 2   Edit an existing drawing named 8910FP-P.

Select the [VIEW–RESTORE–SCRATCH] command from AE-MENU.

Enter selection: 1   Begin a NEW drawing named AE-SCR5.

Verify the settings in this chapter's AE-SCRATCH Setup Table.

When you are ready to begin, set layer ROF-070 as current and select the Pull Down menu DISPLAY and VPOINT 3D.... Select the icon for the RIGHT FRONT view at +30.

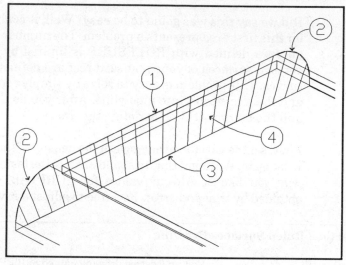

Applying Ruled Surface Technique

Ruled Surfaces with Curved Edges

```
Command: SETVAR
Variable name or ?: SURFTAB1
New value for SURFTAB1 <6>: 24


Command: SETVAR
Variable name or ? <SURFTAB1>: SURFTAB2
New value for SURFTAB2 <6>: 24
```

Command: **PEDIT** Select line ① and fillets ② to convert and join as one Pline.

Command: **RULESURF** Use to create surface ④.
Select first defining curve: Select PLINE of above.
Select second defining curve: Select 3DLINE ③.

Repeat the PEDIT and RULESURF commands in this technique for the other ends of the mansard roofs.

Benefits

Many of the buildings that architects seem to be designing today include a variety of shapes and curves. To create these buildings under AutoCAD, you definitely need to understand how to apply 3D surfaces to your drawing. You can add a surface quickly and with very little effort by using PEDIT and converting a number of different lines into one boundary line.

But, the trick is to know which lines to convert and which surface to use. In this case (and in the majority of cases that you will be faced with), ruled surfaces used with Plines will handle most curved surfaces.

You have defined the 3D vertical surfaces with curved edges, but you have the 3D surfaces of the canopy faces yet to do.

Adding Remaining 3D Curved Surfaces to Canopies

The first ruled surfaces added were flat surfaces bordered with a curve. You can use these surfaces in the next step to help create the *truly* curved surfaces of your mansard roofs. Select two of the curves that you created with FILLET and add a ruled surface to define the area between them.

➥ *NOTE! You need to turn off the last surface you worked on so that you can see the current ruled surface.*

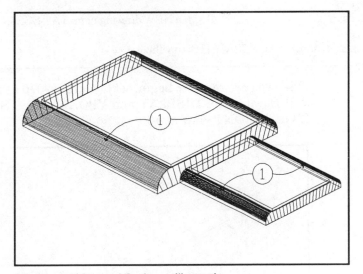

3D Curved Mansard Surfaces Illustration

① ADDING CURVED RULED SURFACES. You can use the following technique to add the ruled surface to the curved areas of our mansard roofs.

Technique for Adding Remaining 3D Curved Surfaces

When you defined the 3D ruled surfaces for the ends of the mansard roof, you had to first convert the curves and the top 3D line into a polyline so that you could select the allowable two lines to represent the borders of your areas.

Now, you are going to add a ruled surface to define the curved area of the mansard roof. It just happens that the same two fillets you created to originally define the curve of the mansard can also be used to define the curve for the ruled surface.

➥ *JUST A REMINDER! When you select lines for RULESURF, always select the same end of each of the two lines that define the border. If you select the top of one and the bottom of the other, you get something that looks like bad modern sculpture.*

Setting Up 3D Curved Ruled Surfaces Drawing

 Enter selection: 2   Edit an existing drawing named 8910FP-P.

Select the [VIEW–RESTORE–SCRATCH] command from AE-MENU.

 Enter selection: 1   Begin a NEW drawing named AE-SCR5.

Verify the settings in this chapter's AE-SCRATCH Setup Table.

When you are ready to begin, set layer ROF-070 as current and select the Pull Down menu DISPLAY and VPOINT 3D.... Select the icon for the RIGHT FRONT view at +30.

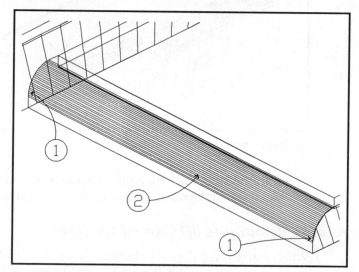

Applying 3D Curved Surface Technique

3D Curved Mansard Surfaces

Command: **RULESURF**   Use to create surface ②.
Select first defining curve: Select fillet ①.
Select second defining curve: Select other fillet ①.
Repeat for three remaining curved surfaces.

As you can see, if you select entities that represent the curved shape of a surface, your ruled surface will conform to that shape.

Benefits

We hope you will agree that being able to create drawings in 3D does have its advantages. With the aid of Release 10 and a UCS, you were able to perform the very difficult task of adding curved surfaces to your drawing by merely selecting two entities.

Now that you have defined the vertical and curved surfaces of the roof, the next areas that you need to define with 3D surfaces are the horizontal surfaces that join the curved surfaces.

Adding 3D Horizontal Surfaces

Since you have already used RULESURF to define the vertical and curved surfaces, let's try using EDGESURF to define the flat horizontal surfaces of the recessed roof and the flat horizontal area that is bordered by the curved mansard surface. Use RULESURF again for the recessed roof curb.

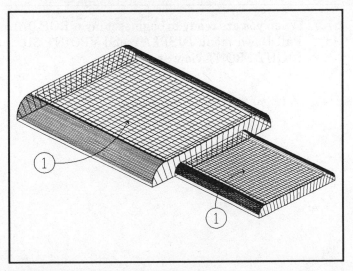

3D Horizontal Surfaces Illustration

① ADDING THE HORIZONTAL SURFACES TO THE RECESSED ROOF AREA. Use the technique below to add the horizontal flat surface to the recess roof area and to the flat surface at the top of the mansard.

Technique for Defining 3D Horizontal Surfaces

Our procedure for defining the 3D horizontal surfaces is similar to that used to define the vertical surfaces. This time, however, you won't have to edit any curves. Use RULESURF to define the surface that joins the curved surfaces of the mansard and the recessed roof area. Then, use EDGESURF to define the flat surfaces of the recessed roof.

You will have to explode the ruled surfaces of the curved areas before you can use them to define one of the border edges. Neither RULESURF nor EDGESURF will allow you to select a surface as a border edge. AutoCAD redefines an exploded surface with lines, so that you can select part of it as an edge.

Setting Up 3D Horizontal Surfaces Drawing

 Enter selection: 2   Edit an existing drawing named 8910FP-P.

Select the [VIEW–RESTORE–SCRATCH] command from AE-MENU.

Enter selection: 1   Begin a NEW drawing named AE-SCR5.

Verify the settings in this chapter's AE-SCRATCH Setup Table.

When you are ready to begin, set layer ROF-070 as current and select the Pull Down menu DISPLAY and VPOINT 3D.... Select the icon for the RIGHT FRONT view at +30.

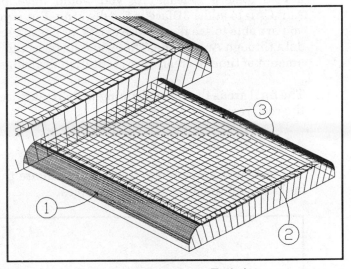

Applying 3D Horizontal Roof Surfaces Technique

3D Horizontal Surfaces

```
Command: EXPLODE
Select block reference, polyline, dimension, or mesh:
```

Repeat for each curved surface ①.

```
Command: EDGESURF
Select edge 1:
Select edge 2:
Select edge 3:
Select edge 4:
```

Use to create ②. Repeat for both areas.
Select all four edges of recessed roof area.

```
Command: RULESURF
Select first defining curve:
Select second defining curve:
```

Use to create ③. Repeat for all surfaces.
Select exploded edge of ruled surface ①.
Select interior edge of curb for recessed roof area.

By using EXPLODE you were able to again use an existing entity (in this case a ruled surface) to help add new entities. And, by using EDGESURF and selecting the four lines that border the recessed roof area, you were able to add a surface that will appear as a solid to AutoShade.

Benefits

One of the most frustrating things that you can run in to when you are creating a presentation drawing, is that after you have spent time to create a perfect 3D drawing, and added yet more time in AutoShade to process the drawing, you may find a portion of your building displayed as a void. Before Release 10, you would have probably gone back to AutoCAD to add a 3Dface to the drawing. But if you use the 3D surfaces, you are able to see if a surface exists (or doesn't exist) before you run the data through AutoShade. These new surfaces can save you a tremendous amount of time in creating presentation drawings.

The final areas that need defining with 3D surfaces are the undersides of the roof, representing the ceiling.

Adding 3D Ceiling Surfaces

The last surfaces to add to the roof system are the ceilings for the main building and the drive through canopies.

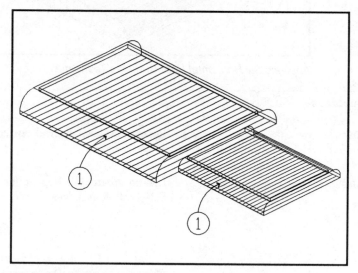

3D Ceiling Surfaces Illustration

① ADDING THE CEILING SURFACES. Use the following technique to add the ceiling surfaces for the building and drive through area canopies.

Technique for 3D Ceiling Surfaces

You could use either EDGESURF or RULESURF to define these last areas of the roof, but we think that RULESURF is a better choice in this case. You only have to select two edges, and since the drawing is becoming dense, a ruled surface will make these areas easier to read.

Setting Up 3D Ceiling Surfaces Drawing

    Enter selection: 2   Edit an existing drawing named 8910FP-P.

Select the [VIEW–RESTORE–SCRATCH] command from AE-MENU.

Enter selection: 1   Begin a NEW drawing named AE-SCR5.

Verify the settings in this chapter's AE-SCRATCH Setup Table.

When you are ready to get started, select the Pull Down Menu DISPLAY and VPOINT 3D.... Select the icon for the FRONT RIGHT view at +30.

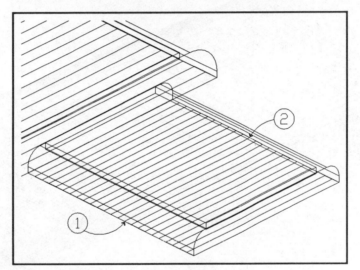

Applying the Ceiling Surface Technique

3D Ceiling Surfaces

```
Command: RULESURF
Select first defining curve:
Select second defining curve:
```

Repeat for both areas ①.
Select as the first border line ①.
Select as the second border line ②.

Repeat for the building ceiling.

By using RULESURF, you added the ceiling as a surface to the building. This will be important when you do the AutoFlix *walk around*, since the ceilings will be displayed as solid surfaces, not voids.

Choosing Surfaces

In the introduction to this section, we told you that you were going to add the horizontal flat surfaces of the drawing with the EDGESURF command. It made sense to change back to the RULESURF command this time for the sake of readability. A ruled surface is composed of half as many lines as an edged surface. We find that when drawings start getting dense, RULESURF is the better choice.

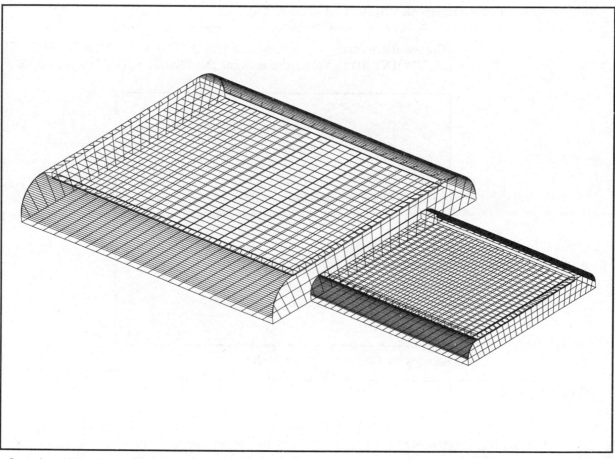

Completed Wire Frame Roof System Illustration

The next command sequences deal with adding 3D surfaces to the areas defined from the exterior walls to the curbs and drive through area. Again we suggest using RULESURF to add these surfaces. Use VPOINT to rotate the building into the best 3D view to add the surfaces.

Defining 3D Site Surfaces

The entrance surfaces act as more horizontal surfaces that AutoShade needs to see defined with 3D surfaces. Use a ruled surface on these areas.

❑ LAYER. Turn off all layers with roof data, and turn on all others.

❑ LAYER. Set layer CON-030 current — AE-MENU Tablet [LAYER–BY ENTITY–SET].

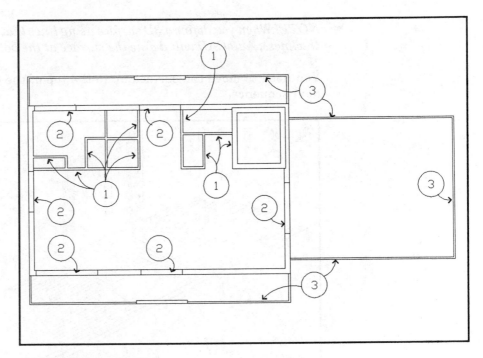

3D Site Surfaces Illustration

① DEFINING WALL LINE. Begin by using the PLINE command with OSNAP INTersection to create a line defining the edge of the entrance slab at the building.

② APPLYING RULED SURFACE TO ENTRANCE SLABS. Define the 3D surface for the entrance slab by using RULESURF and selecting the line you just created and the wind break wall line. Repeat for the other entrance slab.

③ APPLYING RULED SURFACE TO ENTRANCE WALK. Here's where you get to use the 3D entrance walk lines that you created a few minutes ago. To define a 3D surface for these walks joining the entrance slab, simply use RULESURF and select the 3DPOLY line and the 3DLINE opposite to it.

Defining 3D Curb Surfaces

Now that you have your entrance slabs and walks defined, you need to add surfaces to the curbs around the building. You can do this with RULESURF, but you need to consider that these curbs have elevations and thickness.

➡ *NOTE! When you define a 3D surface using lines that have elevations and thickness, AutoCAD will define the surface at the bottom of the line.*

You need to move the surface up to the top of the curb as part of this next sequence.

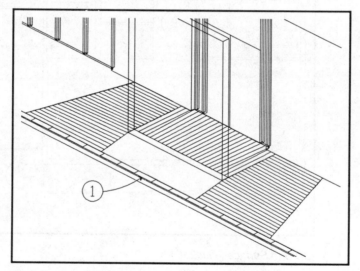

Curb Surfaces Illustration

④ APPLYING RULED SURFACES TO CURBS. You can define the curb surface by using RULESURF and selecting the lines that define the curb section. You will need to repeat this definition for the other curb lines, including those at the corners.

⑤ MOVING 3D CURB SURFACES INTO PLACE. Remember our note from above? Use the MOVE command and an OSNAP of INT to specify the "To point" to move the 3D newly created surfaces to make sure that they are all at the tops of the curbs.

Defining 3D Island Surfaces

The next step is to define 3D surfaces for the drive through islands. There's nothing special about the procedure this time, just use RULESURF.

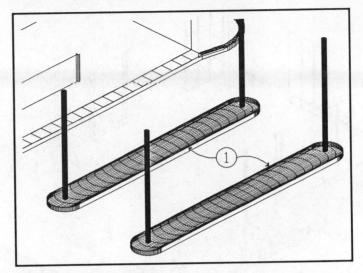

Island Surfaces Illustration

⑥ APPLYING RULED SURFACES TO DRIVE THROUGH ISLANDS. Continue with RULESURF, selecting the arcs at the ends of the islands, to define the surfaces for each.

Defining 3D Planting Area Surfaces

The last 3D surfaces you need to take care of before moving on to AutoShade are for the planting areas. You can use a combination of 3DLINES and 3DPOLY lines to outline these areas, and the now familiar RULESURF to add the 3D surfaces. You will find VPOINT helpful for rotating the building into the best 3D views to add the surfaces.

❑ LAYER. Set layer GRD-020 current — AE-MENU Tablet [LAYER–BY SCREEN–SET].

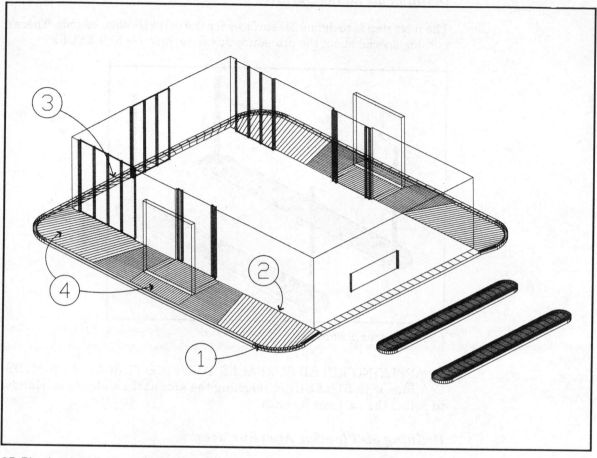

3D Planting Area Surfaces Illustration

① DEFINING PLANTING AREA OUTSIDE CURVE. Let's start with the 3DPOLY command with OSNAP END to define a line parallel to the inside curb line, extending from the walk to the end of the corner curve. Repeat for each of the four curved planting areas.

② DEFINING PLANTING AREA INSIDE LINE. Now, use the 3DLINE command with OSNAP END to define a line parallel to the building and extending from the walk to the curb at the drive through area. For the areas on the other side of the entrances, you can stop the lines at the corner of the building.

③ DEFINING PLANTING AREA ON SIDE OF BUILDING OPPOSITE DRIVE THROUGH. Finally, use the 3DLINE command with OSNAP INT to define a line that is parallel to the building.

④ APPLYING RULED SURFACES TO PLANTING AREAS. Use RULESURF, selecting a building line and the curb line opposite to it, to define each of the 3D planting area surfaces.

❏ SAVE. Now that you are finished adding 3D surfaces to the drawing, it is a good time to save the drawing to disk because you are getting ready to start playing with AutoShade.

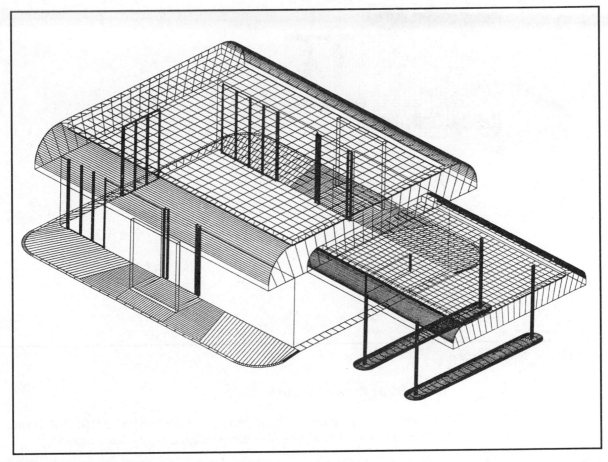

Completed Wire Frame Model

Procedure for Setting Up AutoShade

In the following procedure, we will show you how to set up options in AutoCAD necessary for AutoShade. The commands you'll use will create the data that AutoShade needs to generate a 3D rendering of the exterior of the building. If you use AutoFlix, AutoFlix also uses AutoShade to create a movie file.

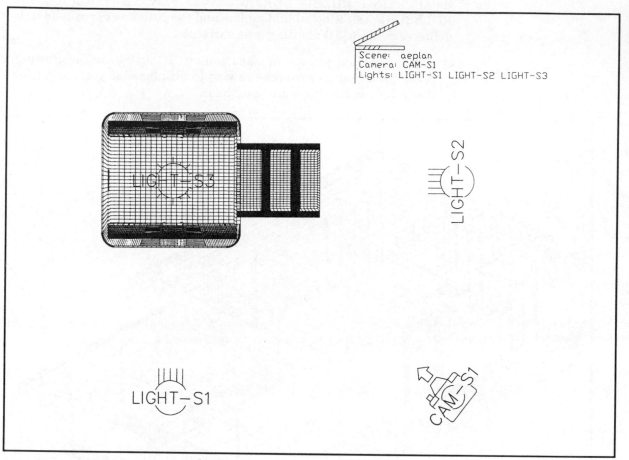

Scene: aeplan
Camera: CAM-S1
Lights: LIGHT-S1 LIGHT-S2 LIGHT-S3

AutoShade Setup Illustration

Setup Steps for AutoShade

You will be using your completed presentation drawing, 8910FP-P. If you are using the AE DISK, the disk comes with a copy of this drawing.

You set the camera and lights that you need to render a drawing, by using commands *within* AutoCAD before you turn the drawing over to AutoShade. You start by setting a plan view, turning all layers turned ON. You then select the AutoShade commands for inserting, naming, and placing the cameras and lights you want to use in the scene from the AutoCAD pull down, tablet, or screen menu.

You can follow the illustration (above) and the command sequence shown below to see how the icons and options are selected for lights and camera settings.

As the illustration shows, you will use DIRECTED and POINT lights, CAMERA, and a SCENE. We will show you how to name the different lights, and where we located them in the drawing. You will also see how to set up the "target" point for the camera.

After everything is set up, select the CREATE FILMROLL icon to create a file of the scene, which we call AEPLAN, to transfer into AutoShade. Then you END the drawing, and exit AutoCAD. We will tell you how to load and operate AutoShade to display the filmroll.

Setting Up AutoShade Drawing

 Enter selection: 2   Edit an existing drawing named 8910FP-P.

Work with the Main Drawing Area.

 Enter selection: 1   Begin a NEW drawing named AE-SCR5.

Verify the settings shown in the AE-SCRATCH Setup Table.

When you are ready to get started, select [OPTIONS] and [ASHADE] from the Pull Down menu area.

AutoShade Setup

```
Command: LIGHT
Enter light name: LIGHT-S1
Point source or Directed <P>: <RETURN>
Enter light location:                    Pick location S1.

Command: LIGHT
Enter light name: LIGHT-S2
Point source or Directed <P>: <RETURN>
Enter light location:                    Pick location S2.

Command: LIGHT
Enter light name: LIGHT-S3
Point source or Directed <P>: D
Enter light aim point:                   Pick location S3.
Enter light location:                    Pick same location.
```

```
Command: CAMERA
Enter camera name: CAM-S1
Enter target point:                       See setup illustration.
Enter camera location:                    See setup illustration.

Command: SCENE
Enter scene name: AEPLAN
Select the camera: CAM-S1
Select a light: LIGHT-S1
Select a light: LIGHT-S2
Select a light: LIGHT-S3
Select a light: <RETURN>
Enter scene location:                     Pick any open area.
Scene AEPLAN included.

Command: FILMROLL
Enter filmroll file name <8910FP-P>: AEPLAN
Creating the filmroll file

COMMAND: END                              Save the drawing and exit AutoCAD.
```

Procedure for Loading and Operating AutoShade

Having created the 3D filmroll file, you are ready to load AutoShade. We assume that you have loaded AutoShade in your AE-ACAD directory.

❑ Start AutoShade by typing **SHADE** at the C:\AE-ACAD> prompt.

❑ Use the Pull Down menu for [FILE] to specify AEPLAN as the filmroll you want to view.

❑ Use the Pull Down menu for [DISPLAY] to create either a FAST SHADE or FULL SHADE rendering of the drawing.

AutoShade Options

There are several options that you can experiment with to enhance the renderings you create. AutoShade provides some default settings for the shading model, light characteristics and many other facets that can make the rendering more realistic. These are found in the [SETTINGS] Pull Down menu. You can also experiment with camera angles and locations while you are in AutoShade, but to permanently change them, you must go back into AutoCAD and change them there. For further information on AutoShade refer to either INSIDE AUTOCAD or the AutoShade Reference Manual.

Benefits

Once you get the hang of defining 3D surfaces in your drawings, it's really not difficult to set up AutoCAD for AutoShade. If you take the time to experiment with options, especially those that control shading, you can get realistic results. The renderings created so far should be important and impressive presentation drawings. Experimenting with the options and creating multiple scenes gives you a lot of flexibility in choosing how to present the design to the client. The time you save by using AutoShade to render the drawings contributes to your flexibility.

Let's go on to produce an animated sequence through the drawing with AutoFlix, the 3D animation enhancement to AutoCAD and AutoShade.

Procedure for Setting Up and Operating AutoFlix

The following steps provide AutoFlix with the data it needs to create a movie of your drawing. Although you can use the same lights as those you specified for the AutoShade rendering, you will need to add some more lighting in AutoCAD to get the drawing ready for AutoFlix.

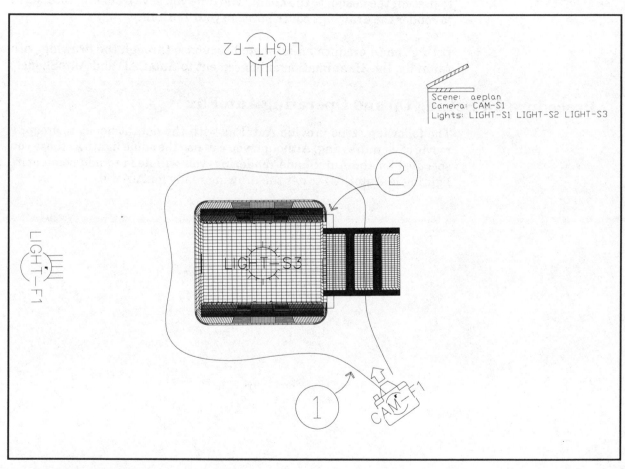

AutoFlix Setup Illustration

AutoFlix Options

AutoFlix creates a movie from multiple AutoShade renderings. It creates these renderings as *frames* along a camera path polyline that you draw in an AutoCAD drawing. You also draw in a path for what the camera *looks at* as it travels along the path. The AutoFlix movie will flow more smoothly if you use the SPLINE CURVE option on the camera path polyline.

We suggest you set an elevation for these lines at six feet, about eye level. With this setup of AutoFlix, the camera will travel 360 degrees around the project. When we did your AutoFlix, we found the original overhead lighting specified for the AutoShade rendering insufficient to illuminate the rear and side of the building so we added more lights to the drawing for the AutoFlix movie. See the illustration above for their locations.

AutoCAD Setup For AutoFlix

```
Command: ELEV                   Set elevation at eye level.
New current elevation <0'-0">: 6'
New current thickness <8'-0">: 0


Command: PLINE                  Use to create camera path ①.
From point:
Current line-width is 0'-0"
Arc/Close/Halfwidth/Length/Undo/Width/<Endpoint of line>:


Command: PLINE                  Use to create look-at path ②.
From point:
Current line-width is 0'-0"
Arc/Close/Halfwidth/Length/Undo/Width/<Endpoint of line>:


Command: PEDIT                  Use to smooth camera path line ①.
Select polyline:
Close/Join/Width/Edit vertex/Fit curve/Spline curve/Decurve /Undo/eXit <X>: S
Close/Join/Width/Edit vertex/Fit curve/Spline curve/Decurve/Undo/eXit <X>: <RETURN>
```

Reload the AutoShade setup in AutoCAD and select the direct light icon for the following.

```
Command: LIGHT                          Add a light a the rear of the building.
Enter light name: LIGHT-F1
Point source or Directed <P>: D
Enter light aim point:
Enter light location:


Command: LIGHT                          Add a light at the side of the building.
Enter light name: LIGHT-F2
Point source or Directed <P>: D
Enter light aim point:
Enter light location:
```

Operating AutoFlix

Now that your setup is complete, you are ready to invoke AutoFlix to animate the drawing. Again, we assume that you are working in the AE-ACAD directory. We assume that you have AutoFlix files copied to that directory.

❑ First load the AutoFlix AutoLISP routine, AFKINET from with AutoCAD.

❑ Next, choose the polyline paths set up in the previous sequence as the camera and look-at paths. Notice in the sequence below how we use a wider angled lens to bring more of the building into the picture.

❑ When the program asks you to specify a base name, answer AEP. The program uses this base name to name the files it creates in the sequence.

❑ When specifying the number of frames you want, the larger the number you choose, the smoother and longer running the animation will be.

Be careful, though, the larger the movie file is, the longer it will take to load. You may need to experiment with the number of frames in your own drawings to produce the best results for the time available.

➡ *NOTE! You also need to make sure that you have plenty of disk space.*

AutoFlix will draw a camera at each frame location along the camera path and will generate a filmroll of each frame. A script file will be created for AutoShade, as well as a movie command file.

OPERATING AutoFlix

```
Command: (load "afkinet")
C:ANIMATE

Command: ANIMLENS
Animation lens focal length in mm <50>: 30

Command: ANIMATE
Choose camera path polyline:                          Select ①.
Choose look-at point (or Path or Same): P
Choose look-at path polyline:                         Select ②.
Initial path elevation <6'>: <RETURN>
Final path elevation <6'>: <RETURN>
Base name for path (1-3 characters): AEP
Number of frames: 20
Initial camera elevation <6'>: <RETURN>
Final camera elevation <6'>: <RETURN>
Twist revolutions <0>: <RETURN>
Layer to move: <RETURN>
Select a light: LIGHT-S1
Select a light: LIGHT-F2
Select a light: LIGHT-F1
Select a light: LIGHT-S2
Select a light: LIGHT-S3
Select a light: <RETURN>

Command: END
```

Generating A Movie

In the next steps, you will invoke AutoShade to create the pictures for each frame and invoke AutoFlix to view the movie.

❑ LOAD AUTOSHADE. At the DOS prompt enter C:\AE-ACAD>**SHADE**.

❑ SELECT AND EXECUTE OUR SCRIPT FILE. Select [SCRIPT] from the FILE Menu and type in AEP to execute your script file. The program creates a rendering file for each frame. This is a lengthy step. As a rough estimate of just how long it takes, multiply the number of minutes it took to create a Full Shade rendering of the drawing by the number of frames you specified for the movie. Exit AutoShade when the program stops processing.

❑ RUN THE MOVIE. At the DOS prompt enter C:AE-ACAD>**AFEGA "AEP"** to invoke AutoFlix and give it the name of your movie file. AutoFlix uses the .MVI file, which contains a list of the rendered frames, to generate and *run* the movie.

Now that you've finished creating your animated view of the exterior of the project design, you can see how the circuit around the building gives you (and the client) a realistic feel for how the whole branch bank building will look and function. The possibilities for enhancing a presentation with other movies are unlimited. You could open the entrance doors, step into the lobby, walk up to the teller counter, and cash your paychecks — well almost.

Summary

AutoCAD offers you many alternatives to presenting a project in the traditional form of hand-drafted drawing. The methods we have shown you in this chapter, which really only scratch the surface of what AutoShade and AutoFlix can do, can significantly enhance the impact of your presentations and increase your design skills.

Setting up your drawings for AutoShade can be tricky, but once you get the hang of applying 3D surfaces, we think you'll find that using AutoShade for 3D renderings gives you a lot more flexibility and speed than manual techniques do. Animating your drawings with AutoFlix is effective for lending realism to your presentations, for showing aspects of your solutions that you couldn't before, and for making your presentations a lot more exciting.

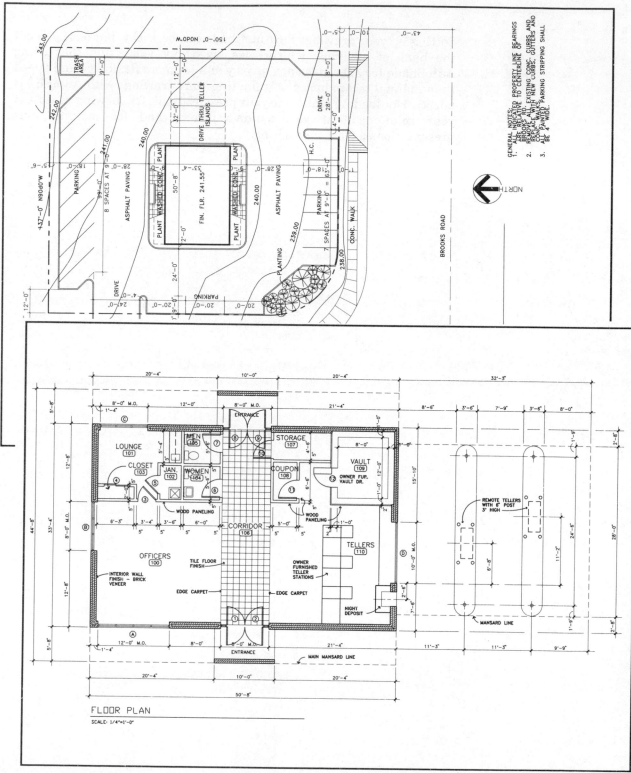

FLOOR PLAN

SCALE: 1/4"=1'-0"

Project Plan Drawings

CHAPTER 6

Plans

Introduction

You have now completed your project planning, design, and presentation efforts for the branch bank project. In this chapter, you will use data from the site and floor plan design drawings that you developed in Chapter Four to create the major components of the working drawings for the project — the site, floor, and reflected ceiling plans. These plans will in turn provide the data you need for developing the remainder of the working drawings — the elevations, sections, and engineering Drawings — in Chapters Seven, Eight, and Nine.

The expanded detail and information that you add to the design drawings will yield a more definitive set of working drawings for the project. In this chapter, we will continue to use the AE-MENU system and give you more techniques for enhancing the features of the site and floor Plans, and for creating a ceiling grid. These techniques will not only apply to the tasks necessary to complete the plans, but will be useful for almost any other drawing.

Site Plan

A finished site plan drawing graphically conveys the objectives that the designer establishes for placing the building on the site, taking into account the client's particular needs as well as the site's physical characteristics.

The design drawing from Chapter Four details some of the items that are specific to the site, such as property lines, streets, building setbacks, and contours, and also indicates the positioning of the building, walks, parking, and other physical improvements. However, to create the actual site plan, a more definitive drawing is needed. We will use the following project sequence to create the site plan.

```
Project Sequence for the Site Plan

Create a working copy of the Design Drawing.
Add Site Plan enhancements.
◆ Technique for Adding Concrete Walk Construction Joints.
◆ Technique for Drawing New Contours.
◆ Technique for Labeling Contour Lines.
◆ Technique for Adding Landscaping.
◆ Technique for Dimensioning.
Add text.
```

Floor Plan

The floor plan serves as the infrastructure for the entire set of working drawings. It graphically describes the building's physical characteristics in a two-dimensional plane. This 2D plan identifies critically related components that control the 3D elements of elevations, sections, and detail drawings.

The floor plan design drawing produced for the project in Chapter Four shows the exterior walls, interior partitioning, plumbing fixtures, structural grid and columns, doors, windows, and a portion of the text required to convey adequate information for construction purposes. To develop the floor plan as a working drawing, follow the project sequence below.

```
Project Sequence for the Floor Plan

Create a working copy of the Design Drawing.
Enhance exterior and interior walls.
◆ Technique for Inserting Room, Door and Window
Annotations.
Indicate corridor floor finish.
Add remote tellers and columns.
Define the plan's dimensions.
◆ Technique for Wall Poucheing.
Add text.
```

As you work with this plan, pay special attention to the layer assignments. Fundamental plan elements, such as walls, partitions, and fixtures will be transferred and reused in the engineering drawings that we will illustrate later in the book.

Ceiling Plan

A ceiling plan depicts information about critical elements of the ceiling, such as the spacing of suspended acoustical systems, where lighting fixtures are located, mechanical penetrations, and the various ceiling materials used. We will use the base plan created in Chapter Four as a drawing base for the reflected ceiling plan for the project. Use the following project sequence to create this plan.

```
Project Sequence for the Ceiling Plan

 Create a new drawing inserting base plan from Chapter
Four.
 Enhance walls and roof lines.
 ◆ Technique for Creating Suspended Ceiling Grid System.
 Indicate porcelain soffit panels.
 ◆ Technique for Inserting Light Fixtures.
 Add text.
```

Drawing Setup for Chapter Six

Again, we have set up two ways for you to approach practicing the concepts and techniques presented in this chapter. First, you can follow each project sequence and complete the drawings as if they were part of a real project. Second, you can read along with the project sequences and use AE-SCRATCH drawings to experiment with the techniques. Either method will provide you with the same benefits — increased productivity and better understanding of how to apply AutoCAD to your work environment.

Method One

If you are using the AE DISK, you have copies of the three working drawings, 8910SP, 8910FP, and 8910CP. The site plan is 8910SP. Select the [VIEW–RESTORE–SCRATCH] command from the AE-MENU and move to the SCRATCH area to experiment with any of the techniques that you want to explore.

If you plan to complete these three plan working drawings from start to finish, load AutoCAD and select Option 1, Begin a New drawing. Since you will be enhancing design drawings, 8910SP-D for example, start the site plan drawing by entering the drawing name 8910SP=8910SP-D. This will create a working copy of 8910SP-D and name it 8910SP for the sit plan drawing. Use the same procedure for the other drawings.

Method Two

If you want to practice the techniques in this chapter, but not complete the drawings, experiment using the AE-SCRATCH drawing, AE-SCR6, shown in the table below. With the exercise sequences that follow, you can set up the technique drawing by using the AutoCAD Setup routine, selecting Architectural units, 1/4"=1" scale, D–24"x36" sheet size, and setting all other options shown in the table.

COORDS	GRID	SNAP	ORTHO	UCSICON
ON	10'	6"	ON	WORLD

UNITS	Set UNITS to 4 Architectural.
	Default the other UNITS settings.
LIMITS	0,0 to 144',96'
ZOOM	Zoom All.

Layer name	State	Color	Linetype
0	On	7 (white)	CONTINUOUS
DIM-010	On	4 (cyan)	CONTINUOUS
NCT-020	On	3 (green	CONTINUOUS
LAN-020	On	3 (green)	CONTINUOUS
CON-030	On	3 (green)	CONTINUOUS
MAS-040	On	1 (red)	CONTINUOUS
GRD-090	On	2 (yellow)	CONTINUOUS
LIG-160	On	3 (green)	CONTINUOUS

AE-SCR6 Setup Table

Site Plan Project Sequence

The site plan for the project is the first working drawing you will produce, enhancing data from the design drawings. We will show you how to create a working copy of the site plan design drawing. With a few modifications, this drawing could serve as a *background* drawing for the engineers. For architectural purposes, this working copy is what you need to augment with specific information necessary to describe the work that would be done on the site.

Adding Site Plan Enhancements

You created the design drawing for the site to convey the concepts of the design to your client. To create the site plan working drawing, you need to add enhancements to spell out the construction methods for the project. First define the sloped areas of the curb cuts and building walks, and highlight these areas with hatch patterns.

❏ LAYER. Set layer CON-030 current — AE-MENU Tablet [LAYER–BY SCREEN–SET].

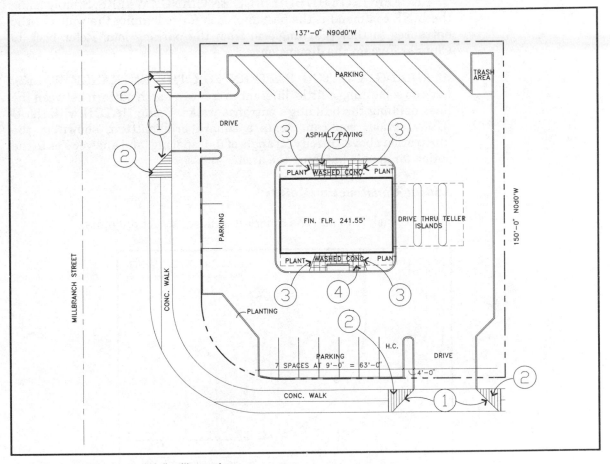

Curb Cuts and Concrete Walks Illustration

① SEPARATING THE SLOPED AREAS OF THE CURB CUTS. To define the sloped areas of the curb cuts, use the [CUT-LINE] command and separate the lines that border these areas, setting a running OSNAP of INTersection to help in the operation.

② ADDING A HATCH PATTERN FOR THE CURB CUTS. Select the lines that border the defined sloped areas. Use a LINE hatch pattern with a scale factor of 60.0, accenting each curb cut.

➥ *NOTE! When you originally set up the design drawing, you selected a scale of 1"=20'-0". These 20 feet (240 inches) represent 240 AutoCAD Drawing Units. So the drawing has a scale factor of 240. To space the LINE hatch pattern at three inches, which is one quarter of a foot, specify a scale factor of 60 (240/4=60).*

③ REDEFINING THE BUILDING ENTRANCE WALKS. Simply using the LINE command is the best approach for redefining the walks at the entrances so that they angle in from the parking lots. Refer back to Chapter Five for the dimensions.

④ ADDING A HATCH PATTERN FOR THE ENTRANCE WALKS. Let's do something a little different to create a hatch pattern between the lines defining the building's entrance walks. Using HATCH with the U (User) option, you can create a square grid pattern, shown in the illustration above. Specify an angle of 0, 1' spacing, and answer Yes to the option for double hatching the area.

Adding Construction Joints

The next task is to add the concrete walk construction joints.

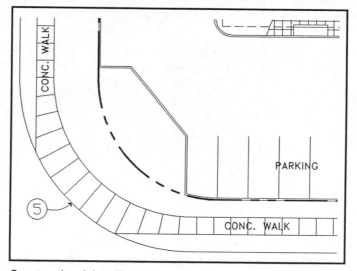

Construction Joints Illustration

⑤ ADDING CONCRETE WALK CONSTRUCTION JOINTS. You can use the following technique to add concrete walk construction joints to your drawing. To apply the technique to your drawing, follow our sequence of commands, but add a step using EXTEND to make the joints on the curve extend to the far walk line.

Technique for Adding Concrete Walk Construction Joints

 Adding construction joints can be a time consuming task, especially if you go about it the long way. You *could* use the LINE command, draw a single joint and then use OFFSET to place some of the joints. But what about the joints for the curved section at the street corner? Should you use ARRAY with the Polar option, then erase the extra lines? What about the vertical walk section — should you use OFFSET again? The answer is no. There's a much faster and simpler way to do this, the key is to convert lines into polylines with PEDIT.

Setting Up New Contours Drawing

 Enter selection: 2   Edit an existing drawing named 8910SP.

Select the [VIEW–RESTORE–SCRATCH] command from AE-MENU.

Enter selection: 1   Begin a NEW drawing named AE-SCR6.

Verify the settings shown in this chapter's AE-SCRATCH Setup Table.

When you are ready to get started, set layer CON-030 as current.

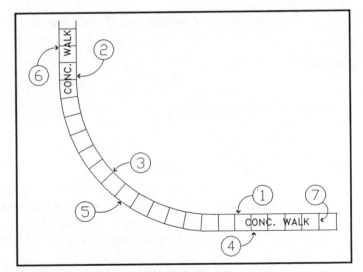

Applying Construction Joint Technique

Concrete Walk Joints

```
Command: LINE                    Draw horizontal line ①.
Command: LINE                    Continue and draw vertical line ②.
Command: FILLET                  Specify a radius of 40' and select ① and ② to create ③.
Command: OFFSET                  Specify a distance of 5' and select ①②③ to create ④⑤⑥.

Command: PEDIT
Select polyline:                 Select walk line ①.
Entity selected is not a polyline
Do you want to turn it into one? <Y> <RETURN>
Close/Join/Width/Edit Vertex/Fit curve/Spline curve/Decurve/Undo/eXit <X>: J
Select objects:                  Select walk curve ③.
Select objects:                  Select walk line ②.
Select object: <RETURN>
2 selected, 2 found.
2 segments added to polyline
Open/Join/Width/Edit Vertex/Fit curve/Spline curve/Decurve/Undo/eXit <X>: <RETURN>


Command: LINE                    Draw a 5' vertical line to represent the construction joint ⑦.

Command: BLOCK
Block name (or ?): JOINT
Insertion base point: END
of                               Select top of vertical line at walk line ①.
Select objects: L
1 selected, 1 found.
Select objects: <RETURN>


Command: MEASURE
Select object to measure:        Select polyline.
<Segment length>/Block: B
Block name to insert: *JOINT
Align block with object? <Y> <RETURN>
Segment length: 5'               Specify construction joint spacing.
```

This construction joint technique uses PEDIT to convert the inside lines of the concrete walk to a polyline. It uses BLOCK on the first joint, and MEASURE to correctly space and insert the remaining joints.

➡ *TIP! If you use an asterisk before the block name, the entity will retain its original characteristics. The line represented by our JOINT block will be inserted as a line, not as a block. This *insertion lets you use EXTEND when you apply this technique to the drawing.*

Benefit

MEASURE, when used with a block, makes the once time-consuming task of inserting and spacing multiple entities quite simple. This command takes on new power when used with AutoCAD's ability to convert multiple pre-drawn lines into polylines. Now, you can not only place blocks at specific distances on straight lines, but also on entities that are made up of both straight and curved lines.

Placing New Contours

The next step in the site plan project sequence is to add lines representing the new contours that the site would require for proper drainage and easy access to the building.

❑ LAYER. Set layer NCT-020 current — AE-MENU Tablet [LAYER–BY SCREEN–SET].

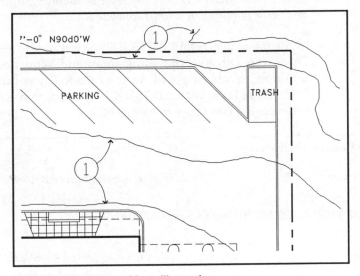

SKETCHED Contour Lines Illustration

① PLACING NEW CONTOUR LINES. We'll use SKETCH to place the new contour lines, as we demonstrate in the following technique; then use PEDIT to smooth them.

Technique for Drawing New Contours

When you create a site plan working drawing, there are generally two types of contour lines that you differentiate. The first type, for those contours that already exist, shows the pre-construction conditions of the site. The second type, for new contours, shows the contours you must add to the site to create the slopes in grade required for drainage and to locate the building's finished floor elevation.

To add these contour lines, you could use the ARC command with the 3Point option, but you would probably spend an enormous amount of time trying to match the surveyor's field notes on the grade elevations of the site. A more effective way is to use the SKETCH command with a large record increment (which controls the distance your pointing device can move across the screen before generating another line segment). Then, by using the PEDIT command with the spline curve option, you can smooth these lines with the required curves.

➡ *NOTE! You'll need to turn ORTHO OFF so that SKETCH can create the contour lines correctly. You'll also have to set the SKPOLY variable to 1 so that the command will create PLINEs instead of LINEs. You should be aware that SKETCH can use a lot of memory, and you may have to stop and save portions of a long contour line.*

In the following technique, we'll show you how to define the new contours. Then, you can use the same command sequence to add the existing contours, except you will want to place those lines on layer ECT-020. Define that layer with a DASHED linetype, so you are able to distinguish between the two types of contours.

Setting Up New Contours Drawing

    Enter selection: 2   Edit an existing drawing named 8910SP.

Select the [VIEW–RESTORE–SCRATCH] command from AE-MENU.

Enter selection: 1   Begin a NEW drawing named AE-SCR6.

Verify the settings shown in this chapter's AE-SCRATCH Setup Table.

When you are ready to get started, set layer NCT-020 as current.

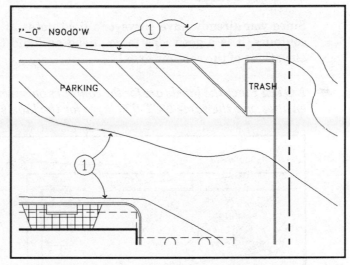

Applying New Contours Smoothing Technique

New Contours

```
Command: SKETCH                  Create contour lines ①.
Command: Record increment <0'-0 1/8">: 5'
Sketch. Pen eXit Quit Record Erase Connect.
<Pen down>  <Pen up>  <Pen down>  <Pen up>  <Pen down> <Pen up>  <Pen up>  <Pen down>
<RETURN>
4 polylines with 282 edges recorded.

Command: PEDIT
Select polyline:
Close/Join/Width/Edit vertex/Fit curve/Spline curve/Decurve/Undo/eXit <X>: S
Close/Join/Width/Edit vertex/Fit curve/Spline curve/Decurve/Undo/eXit <X>: <RETURN>
```
Repeat for each contour line ①.

In the technique for drawing contour lines, you used SKETCH with a large record increment so that you were able to duplicate the grade elevation data. Then, by using PEDIT with the spline curve option, you enhanced the contour lines by smoothing the curves.

Benefits

Using SKETCH, you can establish both new and existing contour lines for a site accurately from information supplied by the surveyor's field notes. You also gain a benefit by being able to then use PEDIT to easily make revisions or additions to the new contours.

Labeling Contours

Since you already have surveyor's field notes at hand for creating the contours, it is a good idea to go ahead and add the text specifying the elevations of the contour lines.

➡ *NOTE! Since the labels are for the contours only, and do not apply to other entities, use the same NCT-020 layer for the labels.*

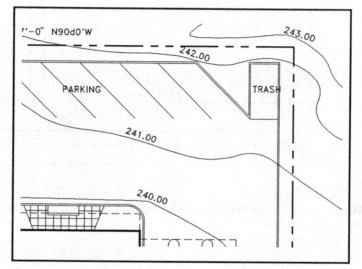

Contour Labels Illustration

① PLACING THE CONTOUR LABELS. Use DTEXT and the following technique to add your text labels specifying the elevations of the contour lines.

Technique for Labeling Contour Lines

We usually place the elevation data for contour lines at angles which follow the curves they are placed on, rather than simply placing them horizontally or vertically.

There are two basic methods you could use for placing the contour labels. You could use TEXT with a running OSNAP of NEAR to get the alignment of the labels with the contours, and then repeat the command at every contour. Or you can use DTEXT and specify one angle of alignment with an OSNAP of NEAR, and place all the labels. We like the second method better because you can locate the starting point and angle of all the labels and not have to restart the whole command sequence over each time you move to a new label location.

Setting Up Labeling Contour Drawing

    Enter selection: 2   Edit an existing drawing named 8910SP.

Select the [VIEW–RESTORE–SCRATCH] command from AE-MENU.

Enter selection: 1   Begin a NEW drawing named AE-SCR6.

Verify the settings shown in this chapter's AE-SCRATCH Setup Table.

When you are ready to get started, set layer NCT-020 as current.

Contour Labels

```
Command: DTEXT
Start point or Align/Center/Fit/Middle/Right/Style: C
Center point: NEAR
to
Rotation angle <E>: NEAR
to
Text: 240.00
Text: 241.00
Text: 242.00
Text: 243.00
Text: <RETURN>
```

```
Command: MOVE
```
Use where required to make sure the text is aligned correctly with the contour lines.

This technique uses DTEXT with the Center option and an OSNAP of NEAR for both the "Center point" and "Rotation angle" of the label. All you have to do then is specify each label location and enter the elevation. You can always use MOVE to adjust a label that doesn't align quite right with a contour line

➥ *NOTE! DTEXT can be tricky. You have to press the <RETURN> key twice (after your last text entry) to commit the text to the drawing. If you enter a Control C (^C) during the operation of DTEXT, all the labels you thought you were adding will be lost.*

Benefits

DTEXT is especially useful for aligning labels with irregular lines. By using this technique you can maintain the standard seen in contour lines in manually produced drawings. As long as you're careful, DTEXT is also nice to use because you can add multiple labels in one operation, something that TEXT won't allow.

Adding Landscaping

We left room for landscaping in this design drawing, but it is not drawn in yet. Let's add some trees to the site plan, filling in the planting area at the street intersection.

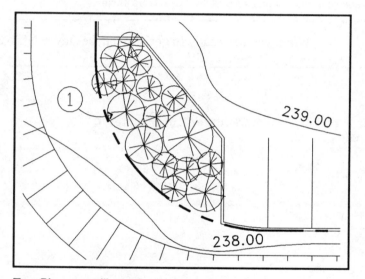

Tree Placement Illustration

① PLACING TREES ON THE SITE PLAN. The illustration titled Tree Placement shows you where to add some trees to the plan. Follow along with the next technique to see how to create the symbol and AutoLISP routine used to make these landscaping additions.

Technique for Adding Landscaping

There are several ways you could go about adding trees. You could first draw a circle with lines radiating from its center and then use the COPY command with the Multiple option to place several of these identically sized trees in the drawing. If you didn't want all the trees to be the same size, you could then draw *another* circle (of a different size), with the same radiating lines, and use COPY again to place these trees. This would require spending a lot of time on repetitive drawing tasks and commands. Another approach would to be create a block of the tree and insert it at different scale factors. But there again, you would have to repeat INSERT over and over, changing the scale factors as you go.

We think the best way to add trees to the drawing is to create a tree block, but use our old friend AutoLISP to vary its size upon insertion. This reduces the number of operations that you have to perform to add landscaping to a site plan. First, we will show you how to create the symbol by using CIRCLE to define the extents of the tree, and LINE to add branches.

If you are using the AE DISK, you already have the tree symbol, called TREE.DWG, and the AutoLISP routine, called TREE.LSP, in your AE-ACAD directory. These support the AE-MENU item called [TREE].

Setting Up Landscaping Drawing

     Enter selection: 2   Edit an existing drawing named 8910SP.

Select the [VIEW–RESTORE–SCRATCH] command from AE-MENU.

     Enter selection: 1   Begin a NEW drawing named AE-SCR6.

Verify the settings shown in this chapter's AE-SCRATCH Setup Table.

➡ *NOTE! We think efficiency is more important than esthetics when it comes to drawing simple landscaping, as we want to do for this site plan. If you are a landscape architect, you obviously will be more creative in your approach to landscaping.*

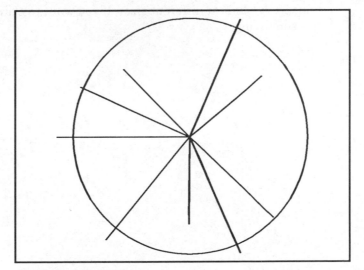

Tree Symbol Illustration

When you are ready to get started, set layer LAN-020 as current.

Tree Symbol

 Check that you have TREE.DWG in your AE-ACAD directory.

Create the TREE block in your current drawing.

Command: **CIRCLE**   Draw a .5 radius circle to represent the tree.

Command: **LINE**   Draw all branches with lines.

Command: **BLOCK**   Select circle and lines for block TREE, specify insertion base at center.

Use the following AutoLISP routine to insert your tree symbol into the drawing. The routine first prompts you for the size of the tree and then for the location where you want to put it.

TREE.LSP AutoLISP Routine

Copy or install the TREE.LSP file in your AE-ACAD directory.

Just read along or create the AutoLISP file.

This routine is a good example how you can use AutoLISP routines to insert symbols at different sizes. If you want to examine the routine, here is the listing:

```
(prompt "Loading TREE utility...")
(defun C:TREE ( / diam pnt)
    (setvar "cmdecho" 0)
    (setq diam (getdist "\nEnter Tree Diameter in Feet:  "))          ;Prompt for tree
    (if diam
        (progn
            (setq pnt (getpoint "\nInsertion Point:  "))              ;Prompt for insertion point
            (while pnt
                (command "INSERT" "/ae-acad/TREE"
                        pnt dia "" ""))                               ;Insert the tree at the specified size
                (setq pnt (getpoint "\nInsertion Point:  "))
            )
        )
)
  (setvar "cmdecho" 1)
)
```

AutoLISP Routine for [TREE]

Now that TREE.LSP is installed, load AutoCAD and test the operation. If you are using the AE-MENU, you can use the [TREE] menu item to insert the tree symbol.

TREE.LSP Operation

```
Command: (load "TREE")
C:TREE
Command: TREE
Enter Tree Diameter in Feet <10'>:        Enter the dimension of the tree.
Insertion Point:                          Pick a point for insertion.
```

This routine will repeat the insertion of the tree symbol until you end with an <RETURN>.

The short AutoLISP routine used in this technique prompted for the symbol's size and then inserted the symbol for you at that size enabling you to display it in different sizes.

We created one symbol, but display it in different sizes.

Benefits

This technique used the by-now-familiar combination of a custom symbol or block (in this case the tree), and a short AutoLISP routine to insert it and change its size. Using AutoLISP's power saves a lot of time and effort. You can enhance the productivity gains, by adding a step to WBLOCK the symbol, as we did in creating symbols for the AE-MENU.

When you are done adding tree symbols, turn to the next step in the site plan. This is to define its dimensions.

Adding Dimensions to Site Plan

To add dimensions to the site plan, you need to indicate the locations of the building, curb cuts, paved areas, and parking striping as well as the location of the site relative to the center lines of the streets.

❑ SETTING DIMENSION VARIABLES. The following are the dimension variables that we suggest you use to follow our techniques for adding dimensions to the site plan.

VARIABLE NAME	SETTING	DESCRIPTION
DIMBLK	SLASH	Arrow block name
DIMDLE	0.1800	Dimension line extension
DIMSCALE	240.0000	Overall scale factor
DIMTAD	On	Place text above the dim line
DIMTIH	Off	Text inside extensions is horiz.
DIMTOH	Off	Text outside extensions is horiz.
DIMTXT	9"	Text height
DIMZIN	2	Zero suppression

Dimension Variable Table

❑ LAYER. Set layer DIM-010 current — AE-MENU Tablet [LAYER–BY SCREEN–SET].

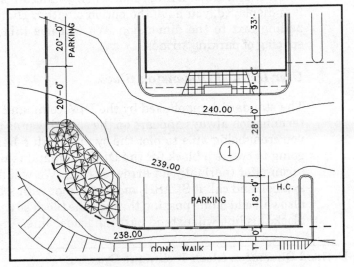

Dimensions for the Site Plan Illustration

① DIMENSIONING THE SITE PLAN. If you use the DIM command with a running OSNAP of NEAR, you can establish the majority of the dimensions displayed in the Dimensions for the Site Plan illustration above without any special consideration for the actual dimension or its placement. We're going to show you how to use a nonstandard dimension block to make a dimension's starting and ending points more obvious.

The parking areas will require the addition of notes with the dimension text to specify the spacing and number of the parking slots indicated on the drawing. Other dimensions that are shown above will have to be moved and a curved leader drawn in to indicate location.

Techniques for Dimensioning

We'll present two dimensioning techniques in this section. One will show you how to define your own dimension block. The second will show you how to add special notes to the dimensions while using the DIM command.

One feature of the DIM command is that the standard symbols it uses can be replaced with user-defined symbols. In our case, one that indicates the termination of the dimension at a leader line. To use our dimension block, you need to specify the block name, called SLASH, as the DIMBLK (Dimension Block) variable. Then, when you place a dimension, this block will appear instead of the standard symbol.

Another feature of DIM is that when you locate a dimension, but before it is displayed, you have the option of either changing the dimension or adding text to the dimension. We will use this feature to specify the spacing of parking stripes.

User-Defined Dimension Block

The standard symbol used by the DIM command for the dimension line termination always appears on the same layer as the leaders. This keeps you from being able to plot the symbol with a heavier pen. But, we are going to create a block for the DIM command to use as a dimension line terminator (normally an arrow or tick). We will add a thickness to this symbol, and call it SLASH, making it plot with a bolder appearance. We also will add a line passing through the block so that when you insert the block this line will extend past the leader line.

Setting Up User-Defined Dimension Block Technique Drawing

Enter selection: 2   Edit an existing drawing named 8910SP.

Select the [VIEW–RESTORE–SCRATCH] command from AE-MENU.

Enter selection: 1   Begin a NEW drawing named AE-SCR6.

Verify the settings shown in this chapter's AE-SCRATCH Setup Table.

When you are ready to get started, set layer DIM-010 as current.

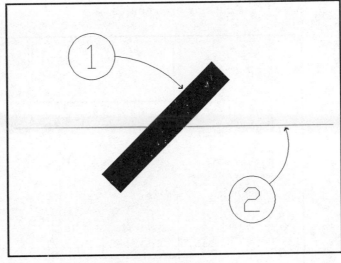

User-Defined Dimension Block Illustration

Dimension Block SLASH and Setting DIMBLK

Command: **PLINE**   Draw slash at 45 degrees, 3/32" long and 1/32" wide ①.

Command: **LINE**   Use to create extension line through center of Pline above ②.

Command: **BLOCK**   Select the Pline and Line, name them SLASH, giving an insertion base
                                  at intersection.

Command: **Dim:**
Dim: **DIMBLK**
Current value   New value: **SLASH**
Dim:

Adding Dimensions with Slash Block

Now that you have set the SLASH DIMBLK, go ahead and use it to place
the dimensions that the site plan requires. Use the Completed Site Plan
illustration (below) as your guide.

When you are done, let's go on to our second technique for adding
dimensions that have special notes attached.

Adding Special Notes to Dimensions

Before AutoCAD commits a dimension to a drawing, the DIM command
gives you the option to either accept the dimension displayed, change the

dimension, or add text to the dimension. We will use this feature to specify the parking area spaces.

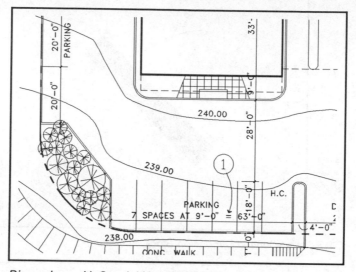

Dimensions with Special Notes Illustration

Dimensions with Notes

```
Command: DIM                           Use to define ①.
Dim: HORIZ
First extension line origin or RETURN to select: PERP
to
Second extension line origin: NEAR
of
Dimension line location:
Dimension text <>: 7 SPACES AT 9'-0" = 63'-0"
Dim:
```

Use this technique to add dimensions for the parking areas. (See the illustration below).

Benefits

The technique of creating a user-defined dimension block shows that the DIM command is another of the many commands that you can customize to fit your own needs. Now when your drawings are plotted, the SLASH marks of your dimensions will appear bold and make your drawings easier to read. The second technique, adding more than just dimensions with the DIM command, shows how you can add notes or special text directly with the dimensions, avoiding unnecessary manual input.

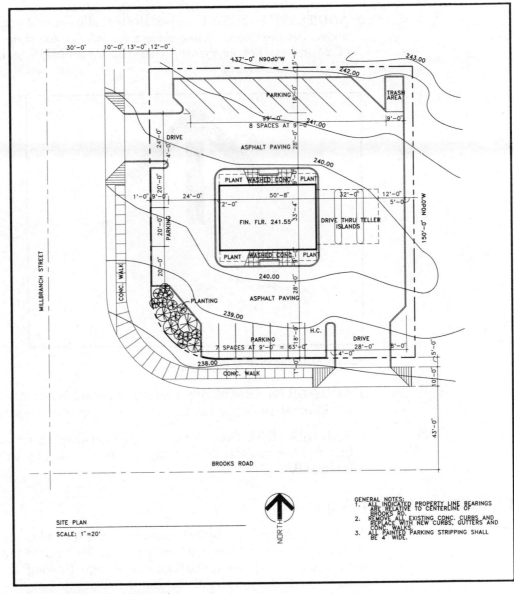

Completed Site Plan Illustration

❑ **ERASE TEXT PLACED DURING DESIGN PHASE.** Because the text
you use in a working drawing is usually more detailed in nature than
that used in a design drawing, it's a good idea to erase the old design
text before you add the new.

❑ ADD TEXT. Use DTEXT to place the text and general notes illustrated above. For the titles, we like using a ROMAND text style and a height of 24 inches. For everything else, use a ROMANS text style and a height of nine inches.

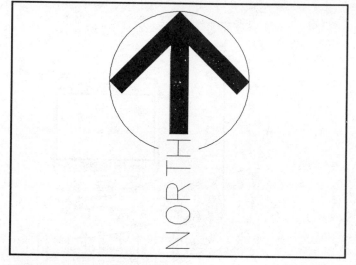

North Arrow Symbol Illustration

❑ ADD NORTH ARROW. We used our standard North Arrow symbol in the illustration — you can use your own or create one similar to ours.

❑ END DRAWING. That's it for the site plan drawing for our simulated branch bank project. Be sure to save the drawing by using the END command.

Floor Plan Project Sequence

The next working drawing for the project is the floor plan. As we did with the site plan, we will show you how to use the design drawing as the base, creating a working copy of the floor plan design drawing. It's a good idea to go through the process of making these working copies rather than adding data directly to an existing drawing. You will be able to verify each of the phases much more easily that way.

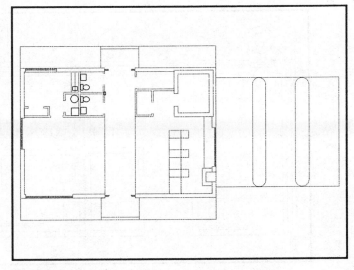

Base Floor Plan Drawing

Use either Methods One to complete the entire working drawing of the floor plan or use Method Two and an AE-SCR6 drawing to experiment with the techniques. Of course, if you are using the AE DISK, you have the completed drawing, called 8910FP.

If you want to do the drawing from start to finish, you will be enhancing the design drawing 8910FP-D. Load AutoCAD and select Option 1, Begin a New drawing. Enter the drawing name 8910FP=8910FP-D. This will create a working copy of 8910FP-D and name it 8910FP.

Enhancing Exterior Walls

The first step in preparing the floor plan working drawing is to add lines to the exterior walls defining the different courses of material you want to use. The wall construction for the project calls for four inches of brick on the exterior and interior walls, with a core of four inches concrete block. The exterior walls are insulated with granular fibrous vermiculite poured into the block cells.

❑ LAYER. Set layer MAS-040 current — AE-MENU Tablet [LAYER–BY ENTITY–SET].

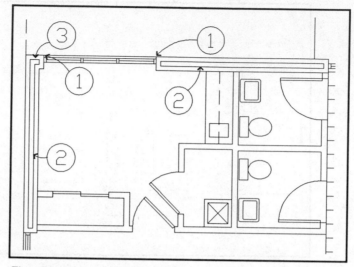

Floor Plan Enhancements Illustration

① SEPARATING DIFFERENT TYPES OF WALL LINES. Try using our [CUT-LINE] command to separate the lines that extend past the intersections defining the wall sections.

② CREATING LINES TO REPRESENT THE DIFFERENT BRICK AND BLOCK COURSES. Use OFFSET, selecting the interior wall, and offset a distance of four inches toward the exterior to define the courses of materials that are illustrated above. Repeat this routine until you have defined all the exterior walls of the building.

③ CLEANING UP EXTERIOR WALL LINES. Use FILLET and TRIM to correct any of your offset lines that need it.

Enhancing Interior Walls

The only interior wall enhancement that you need to make is to define furring strips and wood paneling around the vault concrete walls vault that are exposed to the lobby.

❑ LAYER. Set layer STD-060 current — AE-MENU Tablet [LAYER–BY ENTITY–SET].

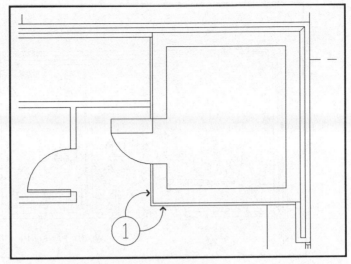

Wood Paneling on Vault Walls Illustration

① ADDING PANELING TO VAULT WALLS. It is a simple operation to represent the wood paneling covering the exposed concrete walls of the vault — just define exterior lines using OFFSET with a distance of two inches. See the Wood Paneling illustration (above) for the locations.

Techniques for Adding Room, Door, and Window Annotations

You may want to review the sections of Chapter Two where we created the annotation symbols and listed the AutoLISP routines used with the symbols. If you are using the AE DISK, or you created the symbols and routines, you can use them in these next sequences. You save a lot of drawing time and effort by letting AutoLISP insert these symbols and keep track of the numbering. If you are not using the AE DISK, just read along to get an idea of how the insertions work.

Adding Room Annotations

The first step to completing the floor plan annotations is to annotate the rooms with numbered symbols.

☐ LAYER. Set layer SYM-010 current — AE-MENU Tablet [LAYER–BY SCREEN–SET].

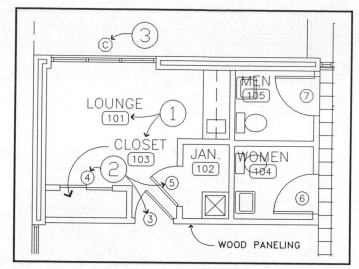

Room, Door and Window Annotations Illustration

① ADDING ROOM ANNOTATIONS. Make use of one of the paired sets of AutoLISP routines and custom symbols that were created in Chapter Two. You can either load the ROOMSYM.LSP routine from the command line or select [ROOM] from the AE-MENU tablet menu. Use this routine to automate the task of placing the incremental room number annotations shown in the illustration above.

Adding Door Annotations

Follow the same sequence as above, using the AutoLISP routine and symbol defined in Chapter Two, to place the door annotations.

② ADDING DOOR ANNOTATIONS. Again, you can either load the DOORSYM.LSP routine or select [DOOR] from the Tablet menu; the illustration above shows you where to place door annotations.

Adding Window Annotations

This sequence is similar to the previous two, using the routine and symbol from Chapter Two. Recall, that the attribute used to indicate each window is a letter instead of a number.

③ ADDING WINDOW ANNOTATIONS. You guessed it! You can either load the WINSYM.LSP routine or select [WINDOW] from the tablet menu. The illustration shows you where the different window annotations are required for the plan.

Drawing Corridor Floor Finish

The specifications for the design called for a surface of quarry tiles, 12 inches square, in the corridor or floor area, spanning between the entrance doors. You can draw in this surface very nicely using the HATCH command. But first, you need to define the hatch area with a polyline. Define this area on the nonplot layer, 000-000.

❑ LAYER. Set layer 000-000 current — AE-MENU Tablet [LAYER–BY SCREEN–SET].

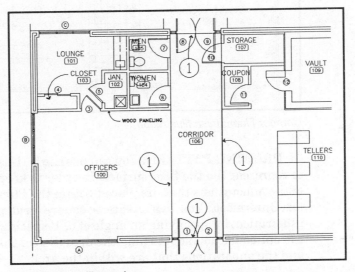

Corridor Area Illustration

① DEFINING AREA TO RECEIVE CORRIDOR FINISH. To define your hatch pattern correctly, you'll have to make sure that you completely outline the area you want to apply the pattern to with PLINE, creating a closed rectangle.

❑ LAYER. Set layer SUR-090 current — AE-MENU Tablet [LAYER–BY SCREEN–SET].

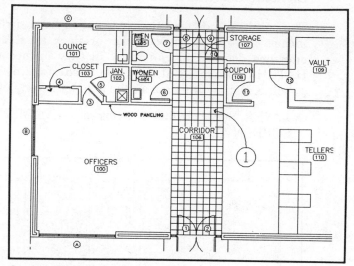

Corridor Floor Finish Illustration

① DRAWING PATTERN. You can use the U (User) option with HATCH to represent the tile floor surface. In order to keep the hatch pattern out of the annotations that are placed within the Plined area, use option "O" for Outermost. You can create a square grid pattern, shown in the illustration, by specifying an angle of 0, 1' spacing, and answering Yes to the option for double hatching the area. Select the PLINE you just drew and the annotations that are within the area.

❑ MAKE ADJUSTMENTS. Use LINE to add any lines left out by the HATCH command.

Adding Remote Tellers and Columns

In Chapter Five, we suggested that you select a remote teller from a manufacture's catalog to use in the presentation drawings for this project. Six inch diameter pipe columns were also added to support the mansard over the drive through. These features are lacking in the design drawing used as a base, so you will have to add these features to the floor plan. You can refer back to Chapter Five to get the data and command sequences that you need, or you can do some estimating by using the following illustration.

❑ LAYER. Set layer SPL-100 current — AE-MENU Tablet [LAYER–BY ENTITY–SET].

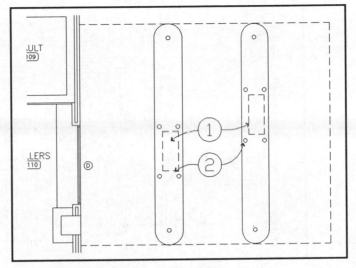

Remote Tellers and Columns Illustration

① ADDING REMOTE TELLERS AND GUARD POSTS. You can draw the remote tellers using LINE; then use CIRCLE to place the six inch diameter guard posts at each corner of the teller islands.

❑ LAYER. Set layer COL-050 current — AE-MENU Tablet [LAYER–BY SCREEN–SET].

② ADDING COLUMNS. You still need to add the support columns defined at the end of each island. Use CIRCLE, and an OSNAP of CEN to select the center of the arc defining each island's curved end. Use a three inch radius.

Defining Floor Plan Dimensions

Once again, we come to dimensioning. We provided some special techniques for dimensioning the site plan earlier in this chapter. You can apply these same techniques to the floor plan, defining the dimensions with the DIM command. You can use the same variables as you did earlier, but you need to set the dimension scale variable (DIMSCALE) to match the scale of this drawing (1/4"=1'-0", or 48 drawing units).

❑ LAYER. Set layer DIM-010 current — AE-MENU Tablet [LAYER–BY SCREEN–SET].

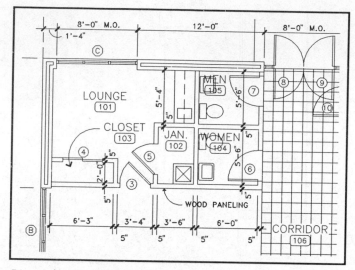

Dimensioned Floor Plan Illustration

❑ DIMENSIONING THE FLOOR PLAN. Use the DIM command with the variables set in the site plan section of this chapter. Except set a DIMSCALE of 48. You will find OSNAPs of INT and NEAR helpful in placing the dimensions shown in the illustration above.

Adding Poucheing to Walls

Our next task is to add a diagonal hatch pattern to the exterior walls to indicate the types of materials we want to use for construction.

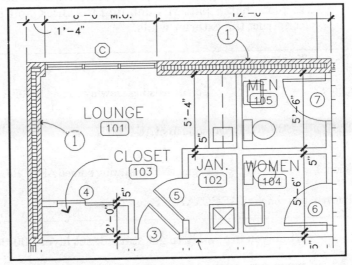

Poucheing Exterior Walls Illustration

① **POUCHEING WALLS.** Use the following technique to add a diagonal hatch pattern to the walls.

Technique for Wall Poucheing

Figuring out how to define different hatch patterns for the materials that make up walls can be difficult. The block and brick courses are separate and distinct entities, but they are located close to each other. You could just use HATCH and select the lines that border one of these areas. You also would have to just hope that all the lines you select are closed at the intersections; and, if not, undo the hatch and use FILLET to close them.

Or, you can use the following technique, which we think shows you the best way to pouche walls (or to apply a pattern to almost any other existing entity that you care to hatch). Use the nonplot layer, 000-000, and the PLINE command with a running OSNAP of END to trace over the lines defining the areas and close the Plines. This will guarantee that you only have to use HATCH once per area, completing the poucheing of all the exterior walls.

➡ *NOTE! To keep from creating large drawing files when you add hatch patterns, don't try to match a standard architectural pattern used in conventional drafting methods. Whenever you can, use an existing AutoCAD hatch pattern that will identify the material without having to add other entities.*

In the following technique, you will see how to use a series of commands to place hatch patterns defining the brick and concrete block used to construct the exterior walls.

Setting Up Wall Poucheing Drawing

    Enter selection: 2   Edit an existing drawing named 8910FP.

Select the [VIEW–RESTORE–SCRATCH] command from AE-MENU.

    Enter selection: 1   Begin a NEW drawing named AE-SCR6.

Verify the settings shown in this chapter's AE-SCRATCH Setup Table.

When you are ready to get started, set layer 000-000 as current.

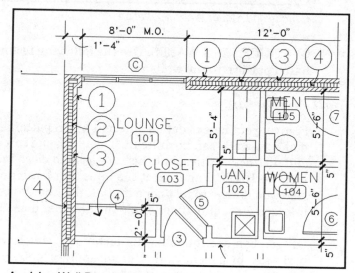

Applying Wall Poucheing Technique

Wall Poucheing

```
Command: OSNAP              Set a running osnap of END.
Command: PLINE              Trace over existing wall lines ① and ②.

Command: LAYER              Set layer MAS-040 as current.

Command: HATCH              Use to define brick in area ③.
Pattern (? or name/U, style): LINE,O
Scale for pattern <1.0000>: 24
```

```
Angle of pattern <0.00>: 45
Select objects:                 Select outside Pline.
Select objects:                 Select inside Pline.
2 selected, 2 found
Select objects: <RETURN>


Command: HATCH                  Use to define concrete block in area ④.
Pattern (? or name/U, style): LINE
Scale for pattern <1.0000>: 24
Angle of pattern <0.00>: 90
Select objects:                 Select inside Pline.
1 selected, 1 found
Select objects: <RETURN>
```

Because you defined the hatch areas with polylines, and because you used the option of Outermost, AutoCAD left the central space between these areas clear. This lets you use a different pattern to define the concrete block core.

Benefits

It may seem like extra work to trace polylines over lines that already exist, or you may question why we didn't just draw with polylines to begin with. There is some logic to tracing. First, polylines are more difficult to edit than lines. As you know from your own experience, designs for architectural projects *do change*. Because of this, we normally do not use polylines to define wall lines. Polylines work well, however, for defining the outline of an area for hatching because they are single entities that are *closed*. Remember, you can always erase or just freeze layer 000-000 to get it out of your way.

Adding Text

There is some additional text, labeling the details you have added and the plan's general conditions, that need to be added to make the floor plan complete. We always use DTEXT to add a large amount of text to a drawing since you can label more than one area at a time.

❑ LAYER. Set layer TXT-010 current — AE-MENU Tablet [LAYER–BY SCREEN–SET].

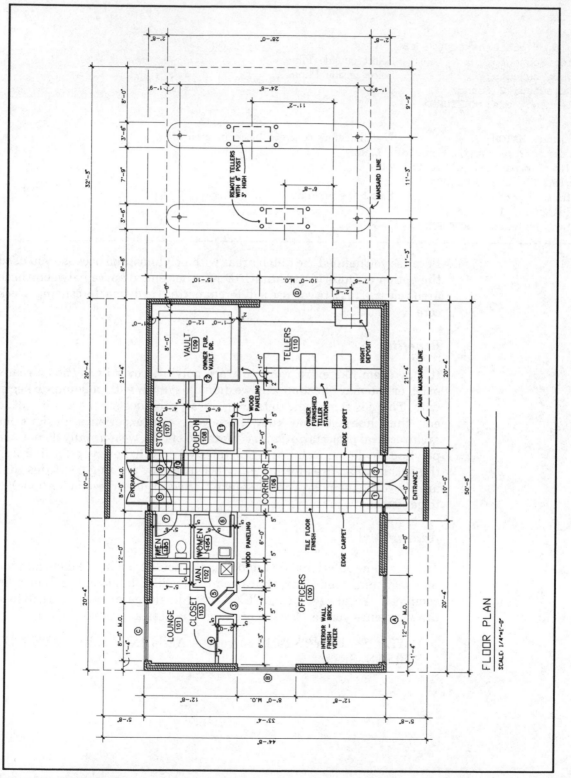

Completed Floor Plan

① ADDING TEXT. If you haven't practiced enough with DTEXT yet, use it to place the text illustrated in the Completed Floor Plan drawing above. For the Titles, use a ROMAND text style, and a height of 18 inches. For everything else, use a ROMANS text style, and a height of six inches.

❑ END DRAWING. You're done with the floor plan! Be sure to save the drawing by using the END command.

Ceiling Plan Project Sequence

If you are using the AE DISK, you have the completed ceiling plan drawing, called 8910CP.

Use Method One to complete the entire working drawing of the ceiling plan, or use Method Two with an AE-SCR6 drawing to experiment with the techniques.

If you plan to complete the entire drawing, load AutoCAD and select Option 1, Begin a New drawing. Enter the drawing name 8910CP.

The ceiling plan has two configurations. For the interior spaces and drive through area, the ceiling is a 2x2 suspended type. The soffits of the main roof are four feet wide porcelain panels. If you recall, you created a ceiling base plan block from the floor plan design drawing in Chapter Four. You can insert this base plan into the new drawing, and enhance it to create the reflected ceiling plan.

Insert Base Plan

Start by inserting the base plan block that you created in Chapter Four.

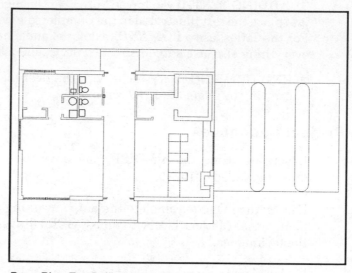

Base Plan For Ceiling Illustration

❑ **INSERT BASE PLAN.** You want to INSERT the base plan block in exploded form so that you can make some additions and modifications. Do this by specifying the block name as *BASEPLAN and insert the drawing at point 0,0.

❑ **CHANGING BASE PLAN LAYER.** The layers used in the base plan block are not what you want to use for the ceiling plan. The easiest way to change layers is to use the CHPROP command and select the entire drawing with the C option. Then, change all entity layers of the base plan to BLD-010.

Enhancing Walls and Roof Lines

Before you start working on the ceiling grid, it a good idea to get rid of some unnecessary lines and entities carried over from the base plan. You'll also want to add the wall sections that span the door openings, because although they are not part of the base plan, they are required for the ceiling plan.

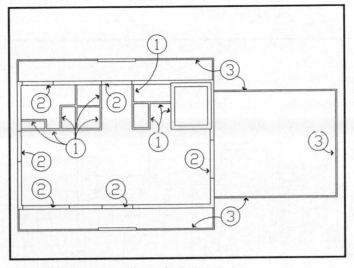

Enhanced Base Plan Illustration

❑ REMOVING LINES. You really don't need the window mullions, frame or glass lines that are part of the base plan. Also, you don't need the doors and swing lines. All of these entities are below the line of the ceiling and don't intersect it in any way. Just ERASE them.

① ADDING DOOR HEADERS. You can use the [LINEMEND] command from the AE-MENU to create the lines representing the door headers. You will need them for defining the different areas of the ceiling plan.

② ADDING STOREFRONT AND WINDOW HEADERS. Just use LINE to add lines to represent the headers of the storefront and windows.

③ ADDING ROOF LINE TURN DOWN. You need to enhance the lines of the drive through roof and soffit areas with a four inch turn down. You can do this easily with OFFSET.

❑ CLEAN UP THE ROOF LINES. You can clean up the corners with FILLET.

Adding Suspended Ceiling Grid System

You can add the ceiling grid system to the base plan by setting an UCS in the center of each area, and using the HATCH command with the U option to place the 2'x 2' grid system.

❑ LAYER. Set layer GRD-090 current — AE-MENU Tablet [LAYER–BY SCREEN–SET].

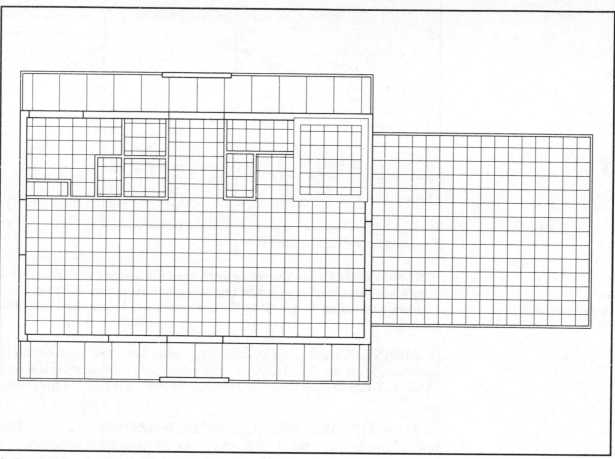

Ceiling Grid System Illustration

① DEFINING THE GRID SYSTEM. You can use the UCS command and XY filters to define a new origin for each area of the floor plan. If you use the HATCH command with the U (User) option, you can create the square grid pattern, shown in the illustration, by specifying an angle of 0, 2' spacing, and answering Yes to the option for double hatching the area.

Technique for Creating Ceiling Grid System

Using AutoCAD to lay out ceiling grid systems is a bit tricky. Most of us have run into problems both in creating grids, and in correctly spacing them. In the following technique we will show you how to use UCS's to help solve these problems.

➡ *TIP! By using OSNAP MID and XY Filters, you can easily find the center of any area and locate a new UCS origin.*

First, you need define each of the areas that require the ceiling grid system with polylines. If you then define a separate UCS for each of these areas, you can use HATCH to create a grid pattern, and use MOVE to adjust the pattern, spacing the grid in each area.

Setting Up Ceiling Grid System Technique Drawing

 Enter selection: 2   Edit an existing drawing named 8910CP.

Select the [VIEW–RESTORE–SCRATCH] command from AE-MENU.

Enter selection: 1   Begin a NEW drawing named AE-SCR6.

Verify the settings shown in this chapter's AE-SCRATCH Setup Table.

When you are ready to get started, set layer GRD-090 as current.

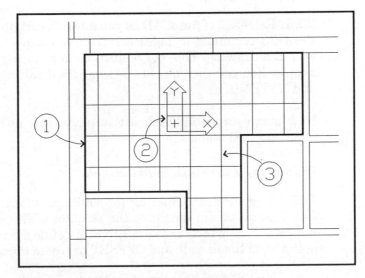

Applying Suspended Ceiling Technique

Ceiling Grid System

```
Command: OSNAP              Use to define a running osnap.
Object snap mode: END
```

```
Command: PLINE                Select any wall line at one end ① and trace walls.

Command: UCS                  Repeat for each area.
Origin/ZAxis/3point/Entity/View/X/Y/Z/Prev/Restore/Save/Del/?/World: O
```
Specify new origin ② by using XY filters and OSNAP MID.
```
Origin point ,0,0: .X
of mid
of (need YZ): .Y
of mid
of (need Z): O
```
Name and save all UCS's with their room names.

```
Command: HATCH
```
Use the U option and specify a 2' pattern each way; repeat for each area ③.

```
Command: MOVE                 Use to adjust the grid.

Command: TRIM                 Use for editing.
```

By tracing the rooms with a closed pline and specifying a UCS with an origin at the center of the area, you can add the grid system with HATCH.

Benefits

While Release 9 of AutoCAD let you move the origin of an entire drawing, the UCS command in Release 10 lets you move the origins of entities within a drawing. This key feature makes placing ceiling grid systems simple. You can then quickly do some final editing with MOVE, TRIM, and EXTEND.

Now that your ceiling grids are in place, look at how to represent the porcelain soffit panels.

Indicating Porcelain Soffit Panels

The ceiling method we used for the soffits calls for the attachment of four foot wide porcelain panels to the structure. The simplest approach to indicate these panels is to use LINE to define a panel line at the center of the wind break wall, and OFFSET to create the remainder.

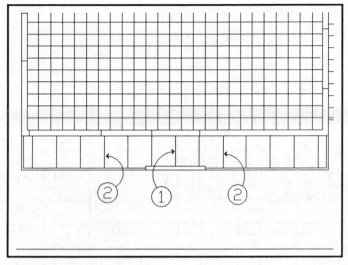

Soffit Panels Illustration

① DRAWING BEGINNING PORCELAIN PANEL LINES. To draw a beginning center panel line on each side of the building, use LINE with an OSNAP of MID, selecting the outside of the wind break wall and drawing the line to the exterior wall line.

② ADDING REMAINING PORCELAIN PANEL LINES. The remaining soffit panels can be added by offsetting the first panel lines four feet' at a time.

Placing Light Fixtures

The last thing you need to do before you add text to finish the drawing, is to place the light fixtures within the grid system. These are shown in the illustration below.

❑ LAYER. Set layer LIG-160 current — AE-MENU Tablet [LAYER–BY SCREEN–SET].

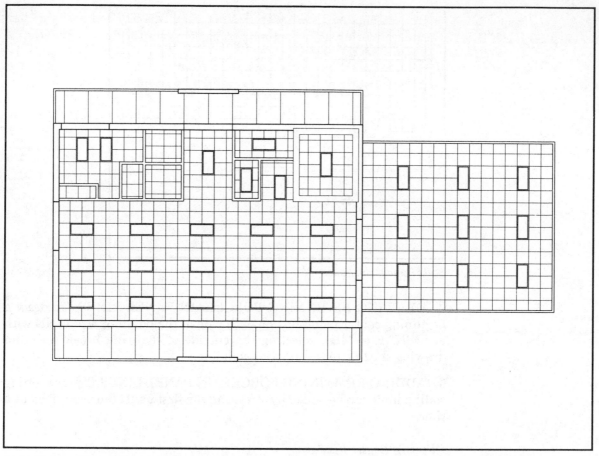

Light Fixtures Layout Illustration

❏ PLACING LIGHT FIXTURES IN GRID SYSTEM. You *could* create a simple 2'x4' fluorescent light fixture block for your drawing; then use INSERT with an OSNAP of INTersection to insert it. You will have to TRIM out the grid lines that pass through the fixtures. But what if you had light fixtures that were of similar shape, but different sizes? Should you create a different fixture block for each of them?

We think the following technique presents a better way to add your light fixtures because it shows how to create a single fixture and vary its size upon insertion with an AutoLISP routine.

Technique for Inserting Light Fixtures

Create a standard 1'x1' light fixture named LIGHT, and use a custom AutoLISP routine named FIXTURE to insert it at the size you want (in this case 2'x 4'). You can use the combined capabilities of a custom block and AutoLISP routine to insert different-sized fixtures over and over again without having to restart the INSERT command.

If you are using the AE DISK, you have the LIGHT.DWG drawing, and the FIXTURE.LSP routine in your AE-ACAD directory. They support the AE-MENU item called [FIXTURE].

Setting Up Light Fixture Technique Drawing

 Enter selection: 2   Edit an existing drawing named 8910CP.

Select the [VIEW–RESTORE–SCRATCH] command from AE-MENU.

 Enter selection: 1   Begin a NEW drawing named AE-SCR6.

Verify the settings shown in this chapter's AE-SCRATCH Setup Table.

When you are ready to get started, set layer LIG-160 as current.

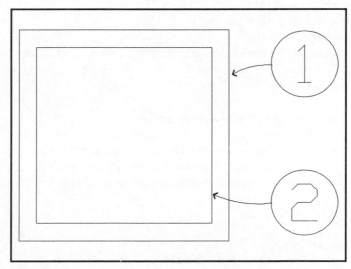

Light Fixture

Fixture

Command: **PLINE**   Draw a 1x1 closed polyline box ①.

Command: **OFFSET**   Specify 1/8" for distance and offset toward inside ②.

Command: **BLOCK**   Specify LIGHT as name and bottom left as insertion base point.

Now that you have the fixture drawn, look at the AutoLISP routine to automate inserting the block.

FIXTURE.LSP AutoLISP Routine

 Check that you have a copy the FIXTURE.LSP file to your AE-ACAD directory.

 Read along, or create the AutoLISP file.

The FIXTURE.LSP routine first prompts you to input the length and width of the fixture you want to insert. The routine assigns this data to the variables that govern the scale of the inserted fixture. Then you are prompted for the insertion point of the fixture, as many times as you require.

```
(prompt "Loading FIXTURE utility...")
(defun C:FIXTURE ( / lngth wdth pnt rot)
    (setvar "cmdecho" 0)
    (setq lngth (getdist "\nEnter Fixture LENGTH in Feet: "))    ;Set fixture length

    (if lngth
      (progn
        (setq pnt (getpoint "\nInsertion point: )
          (while pnt
              (setq rot (getangle "\nRotation angle: "))
              (command "INSERT" "/ae-acad/LGT" pnt lngth wdth rot)   ;Insert
              (setq pnt (getpoint "\nInsertion point: "))
          )
      )
    )
    (setvar "cmdecho" 1)
  )
```

AutoLISP routine for [FIXTURE]

Now that FIXTURE.LSP is installed, load AutoCAD and test the operation. If you are using the AE-MENU, you can use the menu item [FIXTURE].

FIXTURE.LSP Operation

```
Command: (load "FIXTURE")
C:FIXTURE
Command: FIXTURE
Enter Fixture LENGTH in Feet <4'>:        Enter dimension.
Enter Fixture WIDTH in Feet <2'>:         Enter dimension.
Insertion point:                          Pick an insertion point.
Rotation angle:                           Define a rotation angle.
```

This routine will maintain the size, rotation angle and repeat the insertion point prompt until you hit ENTER.

The following sequence shows you how to invoke the routine to place the fixtures and how to edit the ceiling grid. Use the Completed Ceiling Plan illustration (shown below) to guide your insertions.

Applying the Light Fixture Technique

```
Command: OSNAP
Object snap mode: INT

Command: (load "FIXTURE")
C:FIXTURE
Command: FIXTURE
Enter Fixture Length in Feet <4'>: 4'
Enter Fixture Width in Feet <2'>: 2'
Insertion Point:             Select the grid intersection for light and repeat as required.

Command: TRIM                Use to remove excess lines that pass through fixture.
```

This technique centers around the familiar, but very useful, concept of creating a *universal* block and letting an AutoLISP routine automate the insertion at different sizes or scale factors.

Benefits

The benefit of this technique is that you only have to create one block or symbol to satisfy many different situations or tasks. This saves time and effort, and cuts down on the number of symbols you have to maintain.

Adding Text

The last thing to do to complete this drawing is to add the text you see in the illustration below.

❑ LAYER. Set layer TXT-010 current — AE-MENU Tablet [LAYER–BY SCREEN–SET].

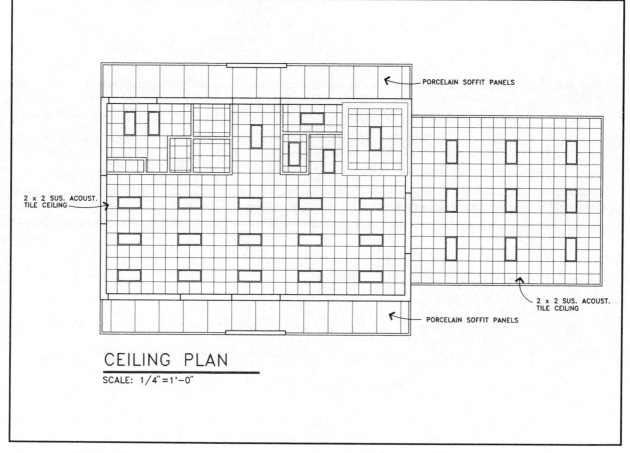

Completed Ceiling Plan

❑ ADD TEXT. Again, you can use DTEXT to place the text. For titles, use a ROMAND text style, and a height of 18 inches. For everything else, use a ROMANS text style, and a height of six inches.

❑ END DRAWING. Now you've got a completed ceiling plan. Be sure to save the drawing with an END.

Summary

You have now finalized the graphic representations of the project's features required for the site plan, floor plan, and reflected ceiling plan. As the primary drawings contained in the working drawing set, they contain extensive information conveying construction purposes. You started out with base information carried over from the design drawings, which saved time and avoided errors.

The techniques in this chapter, especially those using the HATCH command, give us alternatives to drawing each line of the plan and all its related components. The AutoLISP routines and custom symbols that were created (or used from Chapter Two) also helped eliminate line drawing and annotation steps. All contributed to saved time and effort, consolidating drawing tasks for quicker results.

Although these 2D site, floor, and reflected ceiling plans contain the 2D plan information you will need for constructing the project, we still need additional vertical or 3D information (in the form of elevations, sections, details, and engineering drawings) to complete the construction drawings. The drawings that you just created can be distributed to the consulting engineers to help them prepare their respective drawings. Let's go on to Chapter Seven and begin creating the exterior elevations for the simulated project.

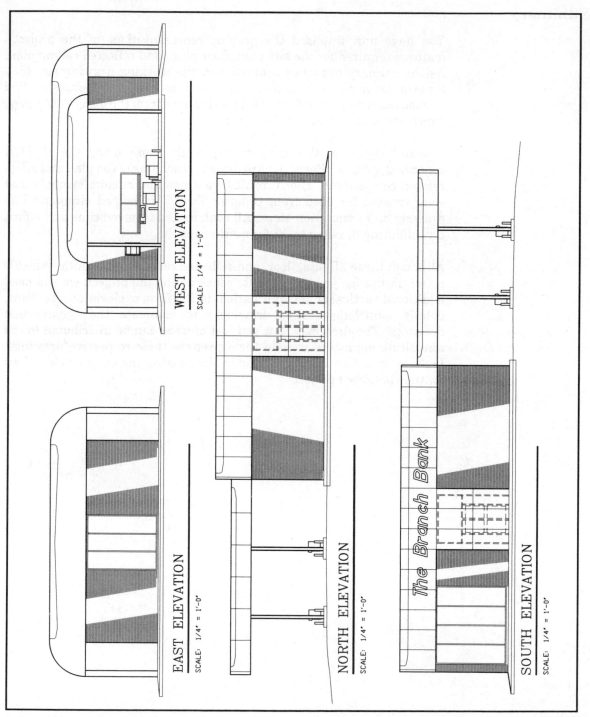

WEST ELEVATION
SCALE: 1/4" = 1'-0"

EAST ELEVATION
SCALE: 1/4" = 1'-0"

NORTH ELEVATION
SCALE: 1/4" = 1'-0"

The Branch Bank

SOUTH ELEVATION
SCALE: 1/4" = 1'-0"

Exterior Elevations Illustration

CHAPTER 7

Elevations

Introduction

The exterior elevation drawings of any project are drawn up and used to exhibit detailed information to the contractor and the owner, or client. With this information, the construction materials that have been selected for the project can be defined. You can determine some of the quantities that it will require, and you have a way to view the *appearance* of the building's exterior.

As you know as designer, it is essential that you have a good working knowledge of construction methods and procedures during this phase of project development. Without such knowledge, even the most ingeniously conceived design could never become a practical reality. In this chapter, we will present a series of techniques and command sequences, creating the elevations of the simulated branch bank project. Hopefully, these techniques will help you enhance your AutoCAD skills for the elevation phase of project development.

Building on Existing Data

In Chapter Five, you created 3D views of a proposed solution for the branch bank project as part of the presentation drawings. In Chapter Six, you created the site, floor, and reflected ceiling plans as part of the working drawings. Now, we will show you how to create accurate exterior elevations by combining the completed elements of the design features, and the detailed plan information from these previous chapters. By extracting the drawing data for the elevations from phases that already are completed, we will not only be able to complete this phase sooner, but will also insure that the drawings are consistent.

Exterior Elevations

The exterior elevation drawings of this project depict the detailed components of the building's exterior facades. This chapter's techniques, are geared to the requirements of four exterior elevation drawings. The project's sequence will produce key portions of each elevation.

Project Sequence

Here is the Project Sequence that you will follow to create the four Elevations of the project — North, South, East, and West:

Project Sequence For The Exterior Elevations
Begin a new drawing.

CREATE EAST ELEVATION
◆ Technique for Defining Major Component Elevations.
Define the exterior limits.
Define mansard curve.
◆ Technique for Defining Sloped Roof Lines.
◆ Technique for Adding Grade Lines and Curbs.
◆ Technique for Creating and Placing Window.
◆ Technique for creating base drawing for West Elevation.

CREATE WEST ELEVATION
Add grade and curbs and drive through.
Define drive through mansard.
Define steel columns.
Create teller window, remote tellers and night deposit.

CREATE NORTH ELEVATION
Define major component elevations.
Define exterior limits.
Define curbs and grade elevations.
Define mansard over drive through area.
◆ Technique for Creating Storefront and Doors.
Place storefront.
Place window.
Define specialty equipment.
Create base drawing for South Elevation.

CREATE SOUTH ELEVATION
Define hatched surfaces.
◆ Technique for Placing Elevation and Annotation Symbols.
Custom lettering of signage.
Add text and titles.

We will derive the data for the components of these drawings from the drawings completed in previous chapters.

Drawing Setup for Chapter Seven

Again, we have set up two ways for you to approach practicing the concepts and techniques that we will present in this chapter. First, you can follow the project sequences and complete the elevation drawings as if they were part of a real project. Second, you can read along with the project sequences and use the AE-SCRATCH drawings to experiment with the techniques. Either method will provide you with the same benefits — increased productivity and better understanding of how to apply AutoCAD to your work environment.

Method One

If you are using the AE DISK, you have the completed elevation drawings, called 8910EL.

If you plan to complete the four component exterior elevation drawings of this chapter, load AutoCAD and select Option 1, Begin a New drawing. You need to create the exterior elevation drawings, called 8910EL, so enter 8910EL as the drawing name. Select the [VIEW–RESTORE–SCRATCH] command from the AE-MENU and move to the SCRATCH area to experiment with any of the techniques before you apply them to the drawing.

Method Two

If you want to practice the techniques in this chapter, but not complete the drawings, experiment using the AE-SCRATCH drawing, AE-SCR7, that we have defined in the table below. With the exercise sequences that follow, you can set up the technique drawing by using the AutoCAD Setup routine, selecting the Architectural unit type, 1/4"=1" scale, D—24"x 36" sheet size, and setting all the other options defined in the table.

COORDS	GRID	SNAP	ORTHO	UCSICON
ON	10'	6"	ON	WORLD

UNITS	Set UNITS to 4 Architectural.
	Default the other UNITS settings.
LIMITS	0,0 to 144',96'
ZOOM	Zoom All.

Layer name	State	Color	Linetype
0	On	7 (white)	CONTINUOUS
000-000	On	7 (white)	CONTINUOUS
MIS-010	On	1 (red)	CONTINUOUS
GRA-020	On	3 (green)	CONTINUOUS
ROF-070	On	5 (blue)	CONTINUOUS
WIN-080	On	5 (blue)	CONTINUOUS

AE-SCR7 Setup Table

AutoLISP Drawing Utilities

Before we get started with the drawing sequences, we want to show you two AutoLISP utilities that help you speed your production by enhancing the functions of AutoCAD's EXTEND and TRIM commands.

DULXEXTD.LSP and DULXTRIM.LSP AutoLISP ROUTINES

Check that you have copies of the DULXEXTD.LSP and DULXTRIM.LSP files in your AE-ACAD directory.

Read along, or create the AutoLISP files.

The AutoLISP utility, called DULXEXTD, lets you select multiple lines to extend at once, instead of having to extend them one at a time. This routine supports the menu item called [DULXEXTD] on the AE-MENU.

Testing the DULXEXTD.LSP Utility

Now that DULXEXTD.LSP is installed, or you have used your text editor to create the routine, load AutoCAD and test the operation.

DULXEXTD.LSP Operation

```
Command: (load "DULXEXTD")
C:DULXEXTD
Command: DULXEXTD
Select boundaries to extend to:        Select line to extend to.
Select lines to extend:                Select lines for extension.
```

The second utility, DULXTRIM.LSP, is similar. It allows you to select multiple lines to trim at once, instead of having to trim them one at a time. It supports the menu item, called [DULXTRIM], on the AE-MENU.

Testing the DULXTRIM.LSP Utility

Now that DULXTRIM.LSP is installed, load AutoCAD and test the operation.

DULXTRIM.LSP OPERATION

```
Command: (load "DULXTRIM")
C:DULXTRIM
Command: DULXTRIM
Select edge for cutting:        Select line for cutting edge.
Select lines to cut:            Select lines to trim.
Select side to cut:             Select side to trim.
```

These two routines illustrate one of the major advantages of using custom AutoLISP routines — you can easily modify and enhance the operation and function of existing AutoCAD commands to improve their usability and speed. These utilities let you cut down the number of selections and keystrokes needed to trim or extend multiple lines by using these two routines.

Here are the listings for the two routines:

```
(defun C:DULXEXTD (/ bou ent pt len cou)
  (setvar "cmdecho" 0)
  (setq cou -1)                                   ;set counter
   (prompt "Select boundary edge(s)...")          ;prompt user
  (setq bou (ssget)) (terpri) (terpri)            ;boundry edge
  (prompt "Select object(s) to extend...")
  (setq ent (ssget)) (terpi) (terpri)             ;entities
  (setq len (sslength ent))                       ;number of entities
   (setq pt (getpoint "Pick side to extend..."))  ;reference point for extend
  (command "EXTEND" bou "")                        ;command
  (repeat len
```

```
      (setq cou (1+ cou))
      (command (list (ssname ent cou) pt))
   )
   (command "")
   (setvar "CMDECHO" 1)
   (prin1)
 )                                                          ;defun
```

AutoLISP Routine for [DULXEXTD]

```
(prompt "Loading DULXTRIM utility...")
(defun C:DULXTRIM ( / edge linset side setsiz looper)
   (setvar "cmdecho" 0)
   (prompt "\nSelect edge for cutting:  ")
   (setq edge (ssget))                                      ;get the edge
    (prompt "\nSelect lines to cut:  ")
   (setq linset (ssget)                                     ;get the lines
         side (getpoint "\Select side to cut:  ")           ;get the side
         setsiz (sslength linset)                           ;size of line set
         looper -1                                          ;initialize loop ptr
    )                                                       ;setq calculations
   (command "TRIM" edge "")                                 ;trim it
   (repeat setsiz                                           ;for all lines
      (setq looper (1+ looper))                             ;increment loop ptr
      (command (list (ssname linset looper) side))
    )                                                       ;repeat
   (command "")

   (setvar "cmdecho" 1)
 )                                                          ;defun
```

AutoLISP Routine for [DULXTRIM]

Exterior Elevation Project Sequence

Having looked at these two routines, let's begin the project sequence for the exterior elevations with the east elevation, defining its major dimensions, or construction component elevations. We will employ different techniques for the special features of each of the elevations, so you want to wait until after the last sequence to add the text to the drawings.

East Elevation

We will start by defining the major component elevations of the building. Next, we will add the building extents by defining the grade, walls and roof as they relate to the major component elevations. Finally, we will create the window.

The data for the areas displayed in the illustration (below) are derived from the drawings created in previous chapters.

Defining Major Components of the East Elevation

Start by reusing data, such as the finished floor elevation and the ceiling and roof heights, to serve as the starting lines for this drawing. Then, you will offset these lines to create the major building components.

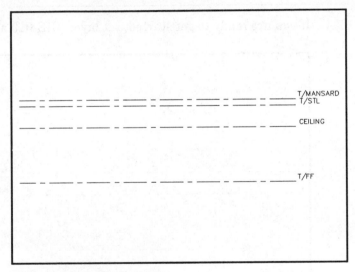

Major Components Illustration

① DEFINING MAJOR BUILDING COMPONENTS. Use the following technique to define the elevations of the major building components.

Technique for Defining Major Components

First, to create this exterior elevation, define a series of reference lines. These lines represent the locations of major building components and are related to the finished floor elevation. In Chapter Six, you defined the finished floor elevation as 241.55'. In Chapter Four, you used the XYZBOX routine to specify a ceiling height of 12 feet. In Chapter Five, you defined the top of the mansard roof at 18 feet.

Using this data, establish the finished floor, ceiling, top of steel and top of mansard elevations, or reference lines. To define these lines, use OFFSET and select the finished floor reference line as the object to offset, and specify the distance from that line.

Setting Up Major Components Drawing

 Enter selection: 2   Edit an existing drawing named 8910EL.

Select the [VIEW–RESTORE–SCRATCH] command from AE-MENU.

 Enter selection: 1   Begin a NEW drawing named AE-SCR7.

Verify the settings shown in this chapter's AE-SCRATCH Setup Table.

If you are ready to get started, set layer MIS-010 as current.

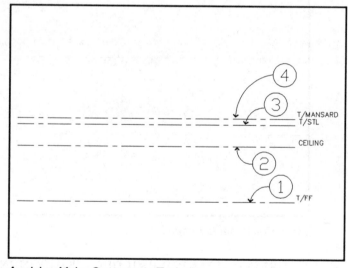

Applying Major Component Technique

Major Components

Command: **LINE**	Use to define finished floor elevation ①.
Command: **OFFSET**	Define ceiling, specify distance 12', for ②.
Command: **OFFSET**	Define top of steel, specify distance 16'8, for ③.
Command: **OFFSET**	Use to define top of mansard, specify distance 18', for ④.

OFFSET gets you started with the reference lines that will be helpful for locating the rest of the building components.

Benefits

Reference lines are nothing new for the draftsperson who has used conventional drafting methods. But with AutoCAD, these lines can take on a whole new meaning. Certainly reference lines enhance your productivity in locating major building components elevations, but they also give you a quick way to make design changes. You can move large areas of a drawing either up or down, according to an elevation or structural change.

Now that you have defined the major components , the next task is to define the building extents or exterior limits.

Defining Exterior Limits

You need to define the extents of the building from wall to wall, and from grade to top of mansard. You can refer to Chapter Six for the dimensions if you need to.

❑ LAYER. Set layer MAS-040 current — AE-MENU Tablet [LAYER—BY SCREEN—SET].

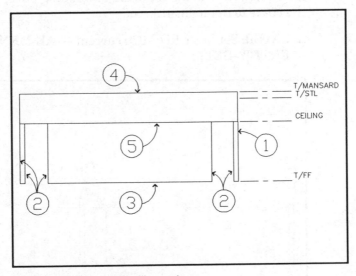

Building Exterior Limits Illustration

① DRAWING VERTICAL BASE LINE. Start by using LINE to draw a vertical line extending from the finished floor elevation line up to the top of mansard roof line.

② DEFINING VERTICAL WALL LINES. Now, use OFFSET to create lines representing the exterior walls.

③ DRAWING GRADE OR FLOOR LINE. Use LINE again to create a line from the vertical base line to the last wall line.

④ DRAWING ROOF LINES. Next, use LINE with OSNAP END to draw a horizontal line extending from the vertical base line to the last vertical wall line.

⑤ DEFINING ROOF AND CEILING LINE. Use OFFSET with a distance of 12 feet to create the bottom roof or ceiling line.

❑ CLEAN UP LINES. It's a good idea to use TRIM at this point to remove any excess wall lines.

❑ CHANGE ROOF LAYERS. The lines representing the mansard should be on layer ROF-070. Use CHPROP to change the layer.

Defining Mansard Curve

Now that you've defined the major building lines, the next step is to add the curve to the mansard roof.

❑ LAYER. Set layer ROF-070 current — AE-MENU Tablet [LAYER–BY ENTITY–SET].

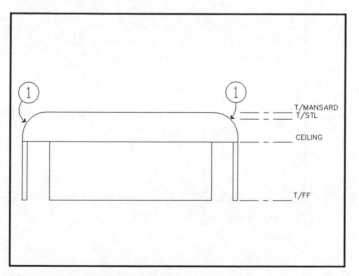

Mansard Curve Edge Illustration

① CREATING CURVED EDGE. To create this curve, use FILLET, specifying a radius of five feet, and select the vertical and top of mansard lines. You'll need to do this for both sides.

Defining the Slope of the Roof

Now, you are ready to add to the elevation drawing lines that represent the recessed sloped roof. For detailed information on the roof slope, you may want to skip ahead to Chapter Eight for a quick look at this area in the building section.

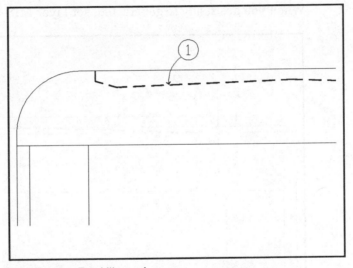

Slope of the Roof Illustration

① DEFINING THE RECESSED SLOPED ROOF LINES. Use the following technique to define the slope roof lines for this elevation.

Technique for Defining Sloped Roof Lines

This technique illustrates how to define the sloped roof lines that are bordered by the mansard. Your recessed roof slopes to the outside edge of the mansard to allow rain water to escape through the roof drains. Using the following command sequence, you can define reference lines and then tie them together to create the roof lines. These roof lines will have ridges in the centers, and valleys at the outer edges which will run perpendicular to the vertical curved edges of the mansard.

Setting Up Sloped Roof Line Drawing

  Enter selection: 2   Edit an existing drawing named 8910EL.

Select the [VIEW–RESTORE–SCRATCH] command from AE-MENU.

  Enter selection: 1   Begin a NEW drawing named AE—SCR7.

Verify the settings shown in this chapter's AE-SCRATCH Setup Table.

When you are ready to get started, set layer MIS-010 as current.

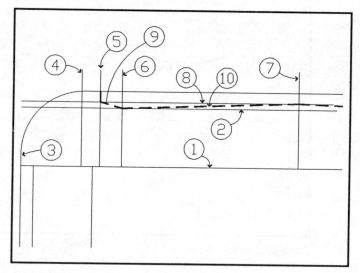

Applying Sloped Roof Line Technique

Slope of Roof

Command: **OFFSET**   Select ①, specify 4'-6, to create ②.

Command: **OFFSET**   Select ③, specify 5', to create ④.

Command: **CHANGE**   Use to increase length of ④.

Command: **OFFSET**   Select ④, specify 1'-6, to create ⑤.

Command: **OFFSET**   Select ⑤, specify 2', to create ⑥.

Command: **LINE**   Select MID of ① to create ⑦.

Command: **OFFSET**   Select ②, specify 6", to create ⑧.

Command: **OSNAP**
Object snap mode: **INT**

Command: **LINE**   Use to create lines ⑨ and ⑩.
From point: Select the insertion of top of mansard and ⑤.
To point: Select the intersection of ② and ⑥.
To point: Select the intersection of ⑦ and ⑧.
To point: **<RETURN>**

COMMAND: **ERASE**   Pick ①②④⑤⑥⑧.

Command: **CHPROP**   Select ⑨ ⑩ and vertical at ⑤, specify LT HIDDEN.

COMMAND: **MIRROR**   Use to create other half of roof.

By creating reference lines from existing entities, you were able to *connect the dots* to create the recessed sloped roof line. Then you used MIRROR to create the other half of the area.

Benefits

As we show in this technique, editing predrawn entities to create reference lines or points, and then connecting the lines, or points, to create a new entity, is faster than trying to create the entity from scratch. We frequently find that doing things this way — editing instead of inputting — can take one tenth the amount of time.

➡ *TIP! It always helps to try to think ahead. Ask yourself, "Can I use any of these existing entities to create, or help create, a new entity or even a series of new entities?"*

Adding the Concrete Curbs and Grade Lines

Now, you can add the concrete curbs and show the grade in elevation.

❑ LAYER. Set layer CON-030 current — AE-MENU Tablet [LAYER–BY SCREEN–SET].

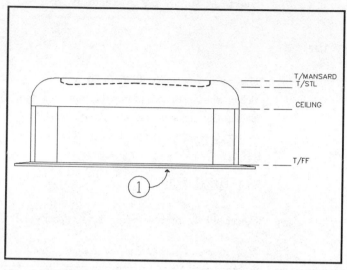

Concrete Curbs Illustration

① ADDING GRADE LINES AND CONCRETE CURBS. Use the following technique to add the concrete curbs that encompass the building, and lines to represent the grade.

Technique for Adding the Grade Lines and Curbs

 Since you have defined the exterior building limits, you can refer back to the building and site plan drawings (shown in Chapter Six) if you feel you are missing any information for the next step.

Again we make some reference lines. You can add these lines (shown below) by using OFFSET. Then, you merely have to use EXTEND, TRIM and LINE to complete this portion of the drawing.

Setting Up for Grade and Curb Drawing

 Enter selection: 2   Edit an existing drawing named 8910EL.

Select the [VIEW–RESTORE–SCRATCH] command from AE-MENU.

Enter selection: 1   Begin a NEW drawing named AE-SCR7.

Verify the settings shown in this chapter's AE-SCRATCH Setup Table.

When you are ready, set layer CON-030 as current.

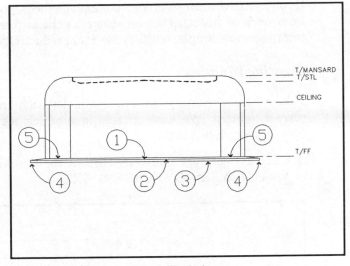

Applying the Grade and Curb Technique

Grade and Curbs

Command: **OFFSET**	Select ①, specify 6, to create ②.
Command: **OFFSET**	Select ②, specify 6, to create ③
Command: **OFFSET**	Select exterior wind break wall, specify 3'4, to create ④.
Command: **MOVE**	Move ④ down so ② and ③ will extend to them.
Command: **EXTEND**	Select ④ as edge, select ② and ③ for extension.
Command: **TRIM**	Select ② and ③ as edges and trim ④.
Command: **LINE**	Use OSNAP INT to create line ⑤.

Again, this technique shows you how to use existing entities to create an entire new group of entities. You created the curbs and defined their extents by offsetting the finished floor and exterior wind break wall lines. To add the grade lines, you simply use LINE with OSNAP INTersection. EXTEND and TRIM were handy for clean up.

Benefits

If you started drafting with conventional methods, you can see how reference lines work in AutoCAD to create new entities from existing ones. These techniques are simple, but they are real productivity boosters.

Creating Windows

Take a look at creating the window in elevation and positioning it in the correct location. We have a technique which we think simplifies this sometimes difficult task.

❑ LAYER. Set layer 080-WIN current for the windows — AE-MENU Tablet [LAYERS–BY SCREEN–SET].

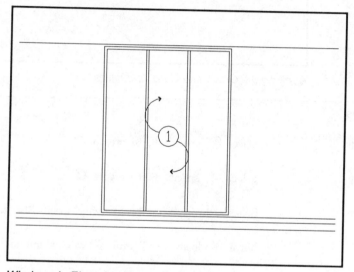

Windows in Elevation Illustration

① ADDING THE WINDOW. You can use the following technique to create and position in the drawing a window in elevation.

Technique for Creating and Placing Windows

Creating a window in elevation can sometimes be a problem. Defining the opening is easy, but creating a balanced frame is not. In the following technique, you will create and locate a window unit for the east elevation. (If you need to, refer to Chapter Six for the location of the exterior opening.)

Begin by defining the center of the window and half of the vertical members. Then, create the sill and head and complete the window by using MIRROR. The last thing you do is convert the window drawing into a block so that it can be inserted in another elevation.

Setting Up Window Drawing

 Enter selection: 2   Edit an existing drawing named 8910EL.

Select the [VIEW–RESTORE–SCRATCH] command from AE-MENU.

 Enter selection: 1   Begin a NEW drawing named AE-SCR7.

Verify the settings shown in this chapter's AE-SCRATCH Setup Table.

When you are ready to get started, set layer WIN-080 as current.

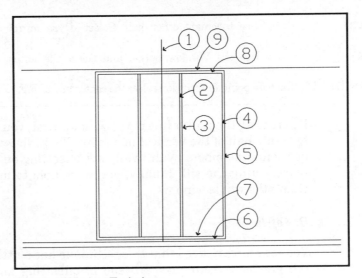

Applying Window Technique

Window

Command: **LINE**	Create an 8' line to represent the center of the window ①.
Command: **OFFSET**	Select ①, specify 1'-3, to create ②.
Command: **OFFSET**	Select ②, specify 2", to create ③.
Command: **OFFSET**	Select ③, specify 2'-5, to create ④.

Command: **OFFSET**	Select ④, specify 2", to create ⑤.
COMMAND: **LINE**	Use to create ⑥.
Command: **OFFSET**	Select ⑥, specify 2", to create ⑦.
Command: **OFFSET**	Select ⑦, specify 11'-8, to create ⑧.
Command: **OFFSET**	Select ⑧, specify 2", to create ⑨.
COMMAND: **TRIM**	Use to clean up frame.
Command: **FILLET**	Use as required to complete clean-up operation.

Repeat TRIM and FILLET commands until all corners are correct.

COMMAND: **MIRROR**	Use to create other half. Select ① as mirror line.
COMMAND: **BLOCK**	Use the Window option and select the entire drawing.

Name the block WIND and for the base point for insertion select the center line at F/F.

By creating the center line of the window first, you were able to OFFSET to create half of the vertical members. Then, you were able to create the horizontal members by drawing and offsetting only one horizontal line, representing the sill. Finally, you were able to mirror the first half to create the whole window.

Benefits

You may have noticed that this technique has a little different twist than the previous techniques. Instead of offsetting an existing entity, you created new lines — a center line and a sill line — which were then offset to complete the window. Creating a complete window in elevation from these single vertical and horizontal members is an accomplishment.

Saving Your Drawing

When you are done, save your drawing. You will use it next to create a base drawing for the west elevation.

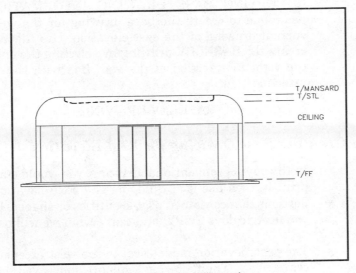

Partially Completed East Elevation Illustration

Creating Base Drawing for West Elevation

To create the west elevation, we can use some base information from the east elevation. We can also use this same data to create a BASELEV, base elevation drawing. In addition, we will reuse this same drawing yet again as a guide for creating the building section drawing in Chapter Eight.

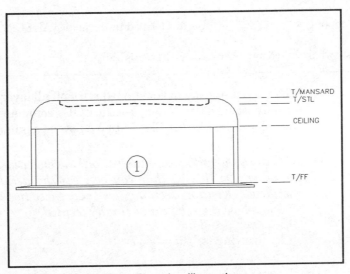

Base Drawing for West Elevation Illustration

① COPYING BASE DRAWING INFORMATION. Follow the next technique to create the base drawing for the west elevation from the current drawing of the east elevation. Use the WBLOCK command to create the BASELEV drawing by selecting the entire elevation. Specify the right intersection of the wall line with the floor slab line as the insertion point.

❑ CREATING BASELEV DRAWING.

Technique for Creating Base Drawing for West Elevation

In the days of conventional drafting, you would start with one elevation, print it, then use it as guide for the next elevation by tracing over the portions that contained usable entities or shapes. But with AutoCAD all you need to do is COPY, and that's what we will be doing next.

The east elevation is identical to the west except for the drive through components. The following sequence shows you how to use the existing east elevation to create new data for the west elevation.

Setting Up Base Drawing

Enter selection: 2   Edit an existing drawing named 8910EL.

Select the [VIEW–RESTORE–SCRATCH] command from AE-MENU.

Enter selection: 1   Begin a NEW drawing named AE-SCR7.

Verify the settings shown in this chapter's AE-SCRATCH Setup Table.

When you are ready to get started, turn off all layers that contain data only for the east elevation. Then, use the COPY command to create a base for the west elevation. Use the illustration below as a guide for your COPY.

➥ *NOTE! In the following illustration, the elevations are shown one above the other. Do this so that you can see the details more easily. In the completed elevation drawing, we show these two elevations side by side, which is how they should be finally located.*

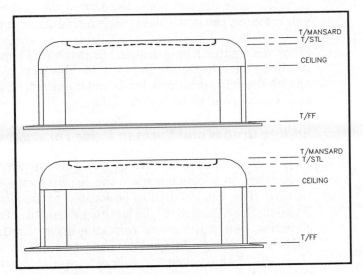

Applying Base Drawing Technique

Base Drawing

Command: **LAYER**   Turn off layers WIN-080 and CON-030.

Command: **COPY**   Use the Window option and select the entire elevation.

That was pretty easy. COPY gives you a base drawing for the west elevation in one fell swoop.

➥ *TIP! When you specify a COPY displacement point between the two elevations, always remember to leave yourself enough room for text.*

Benefits

If the building had the same entities on both of these elevations (one a mirror image of the other) you could have used MIRROR with an OSNAP of MID, selecting either the roof or finished floor line, to create the other elevation. With either of these methods, you can create new elevations from old in fraction of the time it would take to draw the new elevation from scratch.

West Elevation

Reusing and modifying information defined in other drawings is one of the most effort-saving advantages that AutoCAD offers. The west elevation is really just a modification of the east elevation. And, as you will see later, the north elevation is a modification of the south elevation.

Since you created a base drawing from the east elevation in the last step, it will be easy to modify the data for the west by changing the grade, adding the drive through islands and mansard, adding the remote tellers and window, and doing a little cleanup.

Adding Grades and Curbs at Drive Through

As you recall, in Chapter Five we made an adjustment to the grade of the west elevation to accommodate the requirements for the drive through window. The window had to be located at elevations of 3'-0" from the finished floor, and 3'-6" from the pavement. To complete the west elevation, you will have to raise the drive through area immediately adjacent to the building. You need to raise it six inches, since the elevation of the pavement everywhere else is -12 inches (related to finished floor).

If you need to, refer to Chapter Five for the required dimensions for this change in elevation; then follow along with the command sequence below.

❑ LAYER. You set the correct layer, CON-030, in the technique above.

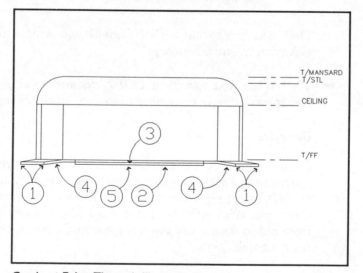

Grade at Drive Through Illustration

① OFFSETTING OUTSIDE CURB LINES. Use OFFSET with a distance of five feet, select the outside vertical curb lines, and offset them toward the inside.

② TRIMMING EXISTING CURB LINES. You'll need to use TRIM to remove the curb lines between the two offset curb lines.

③ DEFINING NEW TOP OF PAVEMENT. To define the new top of pavement, use OFFSET with a distance of six inches, select the finished floor line, and offset toward the bottom of the drawing.

④ TYING THE TWO CURB LINES TOGETHER. To draw the sloped top of the curb line at each side of the building, use LINE with an OSNAP of END. Then, use OFFSET again with the default six inches and offset these lines. You can use FILLET to clean up the intersection.

⑤ DEFINING DRIVE THROUGH ISLANDS. To define the top of the outside island, select the top of the pavement at the building and OFFSET it the default of six inches toward the bottom of the drawing. To represent the ends of the islands, select the exterior wall lines and give them a distance of 2'-8" toward the inside. You'll need to use MOVE to move these lines down a minimum of 12 inches. TRIM the two vertical lines by selecting the finished floor line and the last offset line. Finally, use the AE-MENU [TOLAYER] command and change the last lines to layer CON-030.

Defining Drive Through Mansard

The mansard roof that covers the drive through area will be constructed in the same manner as the roof of the main building, but it differs in the depth and radius of its curves. Follow along with the command sequence below to create the drive through mansard.

❑ LAYER. Set layer ROF-070 current — AE-MENU Tablet [LAYER–BY SCREEN–SET].

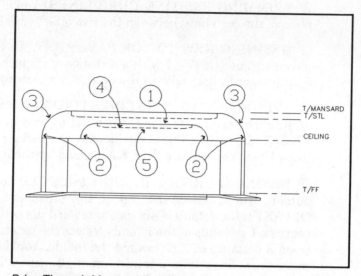

Drive Through Mansard Illustration

① DEFINING TOP OF DRIVE THROUGH MANSARD. To define the top of the drive through mansard, you can OFFSET the line that represents the main mansard down a distance of 2'-8".

② DEFINING VERTICAL LIMITS OF MANSARD. Again, use OFFSET, but this time select the 12-inch vertical line of the outside edge of the main mansard, and offset it a distance of 7'-8" toward the inside.

③ DEFINING CURVE OF MANSARD. To create the curve of the roof, select the vertical and horizontal lines that you offset in the last steps, and use FILLET with a radius of 3'-4".

④ DEFINING INTERIOR ROOF LINE. Here you can make use of COPY and an OSNAP END; select one half of the interior hidden lines of the main roof and specify the "to point" on the drive through mansard. Repeat the same operation for the other side.

⑤ CLEANING UP INTERIOR ROOF LINES. Since the slopes of the main and drive through roofs are the same, use FILLET with a radius of 0 and simply select the two lines to join them at the center.

Defining Steel Columns

Now, you can add the columns that support the mansard roof over the drive through area.

❑ LAYER. Set layer STL-050 current — AE-MENU Tablet [LAYER–BY SCREEN–SET].

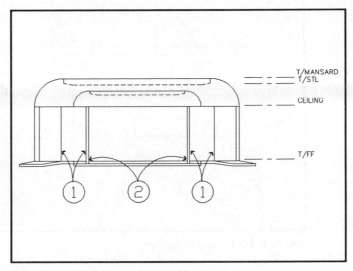

Drive Through Columns Illustration

① DEFINING COLUMNS. First OFFSET the vertical exterior wall lines of the building a distance of 4'-2" toward the inside. Repeat for both columns. Next, OFFSET these lines a distance of six inches toward the inside to define the columns.

② CHANGING LAYER AND CLEANING UP. Use CHPROP and change the layer from MAS-040 to STL-070. Try using the AE-MENU [DULXEXTD] command, select the top of the outside island and extend the column lines down.

Creating Teller Window, Remote Tellers and Night Deposit Box

As we discussed before when we talked about specialty items, most of the time you get the information you need about specialty items from solicited cut sheets, or from catalogues such as *SWEETS*. Here again, we will leave it up to you to represent the banking equipment items in the next sequence. However, feel free to approximate what you need from our illustrations (below).

❑ LAYER. Set layer SPL-100 current — AE-MENU Tablet [LAYER–BY SCREEN–SET].

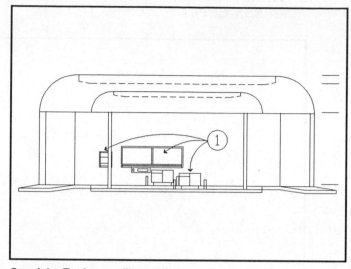

Specialty Equipment Illustration

① ADDING SPECIALTY EQUIPMENT. If you create the equipment items, we recommend that you create them in the scratch area. (Use the [VIEW–RESTORE–SCRATCH] command to the display the scratch area.) After you have duplicated the manufacturer's specifications for the teller window, remote tellers, and night deposit box, you can use MOVE, or BLOCK, to relocate these items in the MAIN area at their correct locations.

❑ CLEAN UP. To remove any extra lines that pass through the columns, try using the AE-MENU [DULXTRIM] routine, and select the column lines.

❑ SAVE your drawing.

North Elevation

Just as we did with the east elevation, we are going to first define the elevations of the major components for the north drawing. Then, we will create lines to represent the building extents. Next, we will add the curbs (displaying the changes in grade), the drive through canopy and islands, the entry and concealed storefront, and the specialty equipment.

You can finish the north elevation by adding the window, and creating a copy that you can mirror for the south elevation.

Defining Major Component Elevations

Continue working in the same drawing. First, you need to add lines that represent the elevation of major building components.

❑ LAYER. Set layer MIS-010 current — AE-MENU Tablet [LAYER–BY SCREEN–SET].

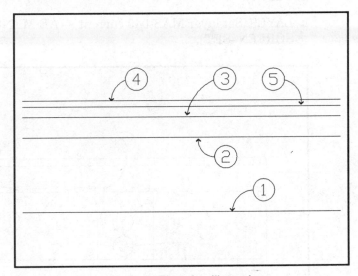

Base Information for North Elevation Illustration

① DRAWING FINISHED FLOOR LINE. Use LINE to create a line to represent the finished floor.

② CREATING CEILING LINE. To define the ceiling line, select the finished floor line and offset it a distance of 12'-0" toward the top of the drawing.

③ DEFINING DRIVE THROUGH MANSARD. To define the top of the mansard over the drive through area, select the ceiling line and OFFSET it a distance of 3'-4" toward the top of the drawing.

④ DEFINING MAIN MANSARD. To define the top of the main mansard, OFFSET the top of the mansard line a distance of 2'-8" toward the top of the drawing.

⑤ DEFINING MAIN MANSARD ROOF LINE. To define the roof line at the main mansard, OFFSET the top of the main mansard a distance of 12 inches toward the bottom of the drawing.

Defining Exterior Limits

Next, you need to define the extents of the north elevation. By defining a vertical line to represent the right exterior wall, you will be able use OFFSET to create the remaining vertical lines. Then, you can use OFFSET again to create the horizontal lines by using the finished floor line as a base.

❑ LAYER. Set layer MAS-040 current — AE-MENU Tablet [LAYER–BY SCREEN–SET].

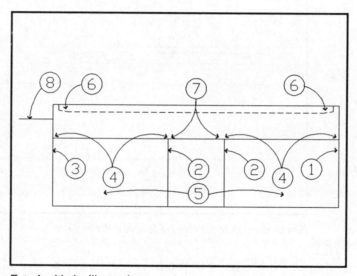

Exterior Limits Illustration

① DRAWING BUILDING LINE. Use LINE and draw a line from the finished floor line up 18 feet or to the top of the main mansard.

② CREATING WIND BREAK WALL AT ENTRANCE. To define the wall line, use OFFSET with a distance of 20'-4", selecting the building line, and offsetting toward the left. Now, define the other side of the wall by reusing OFFSET with a distance of 10'-0", selecting the last line, and offsetting it to the left again.

③ DEFINING THE LEFT BUILDING LINE. Again, use OFFSET with a distance of 20'-4". Select the left wind break wall line and offset it to the left to define the left building line.

④ CUTTING LINES. You can use our AutoLISP routine [CUT–LINE] to cut each of the horizontal elevation lines approximately 12 inches to the left of the building line. If you use the [CUT–LINE] routine again with an OSNAP of INTersection, you can cut the vertical building lines at the ceiling line. This will let you assign different layers to the two pieces of the vertical line.

⑤ DEFINING FLOOR, CEILING, AND MAIN ROOF LINES. Use TRIM to remove excess lines that extend past the building lines. Use CHPROP and change these elevation lines to the correct layers and linetypes as illustrated above.

⑥ DEFINING RECESS MAIN ROOF PARAPET WALL. Use the same technique as you used with the east elevation to define the roof lines. Please note that the roof line displayed in this elevation is flat.

⑦ TRIMMING WIND BREAK WALL LINES. Just use TRIM to remove the excess lines that are displayed above the ceiling line.

⑧ DEFINING THE DRIVE THROUGH TOP OF MANSARD. Use MOVE to relocate the offset line that you created in the major component sequence. Move the line so that it's right end is flush with the far left building line.

Defining Curb and Grade Elevations

The grade of the pavement at the north elevation, as illustrated above, is basically flat except at the drive through area, where it gradually slopes down. If you look at the east elevation, you see that the pavement is located at an elevation of -12 inches from the finished floor, with the top of the curb located at -6 inches.

You can also see that the top of the first drive through island is at the finished floor elevation; and that the second is at -6 inches. Both are six inches above the adjacent pavement. Follow this next command sequence to see how to use LINE, OFFSET, COPY and EXTEND to define these paved areas, curbs, islands, and changes in elevation.

❏ LAYER. Set layer CON-030 current — AE-MENU Tablet [LAYER–BY SCREEN–SET].

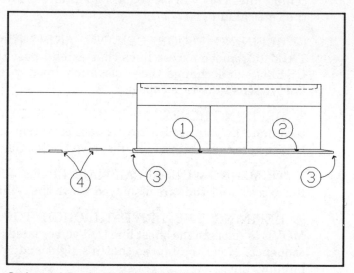

Curbs and Grade Change Illustration

① DEFINING TOP OF PAVEMENT. To define the top of the pavement, OFFSET the finished floor line a distance of 12 inches toward the bottom of the drawing. This line also needs to be changed to layer CON-030. Use CHPROP.

② DEFINING TOP OF CURB. To define the top of the curb, simply OFFSET the top of pavement line that you just created a distance of 6 inches toward the top of the drawing.

③ DEFINING ENDS OF CURBS. You can use LINE and draw lines to represent the ends of the curbs on both sides of the building; then EXTEND the curb lines to the ends.

④ DEFINING PAVEMENT AND ISLANDS AT DRIVE THROUGH. Represent the first island by using LINE to draw a 3'-6" wide by 6" deep rectangle, with the top of the island at the same elevation as the finished floor. Then use COPY to locate the second island to the left 11'-3" and down 6".

To complete the pavement lines, just use LINE with an OSNAP INTersection to draw a line from the bottom of the curb at the building to the bottom of each island. Extend the line past the last island as you see in the illustration above.

Defining the Mansard Over the Drive Through Area

Defining the mansard should be familiar ground. You're going to define the lines that make up the mansard and columns of the drive through area by using existing entities.

❑ LAYER. Set layer ROF-070 current — AE-MENU Tablet [LAYER–BY SCREEN–SET].

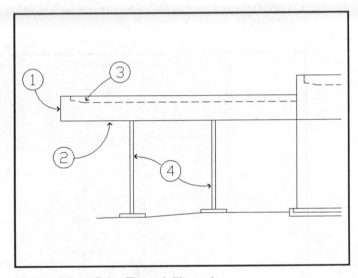

Mansard Over Drive Through Illustration

① DEFINING OUTER EDGE OF MANSARD. If you select the mansard line that is adjacent to the drive through area and OFFSET it 32'-3" toward the left, you will have placed the outer edge correctly. You'll have to clean up a bit by using TRIM to remove the excess line above the elevation line that indicates the top of the drive through mansard.

② DEFINING BOTTOM LINES OF MANSARD. You can OFFSET the line you created and moved into place a distance of 3'-4". You'll need to use TRIM to remove the excess line that sticks out past the outer edge of mansard line just created.

③ DEFINING ROOF PARAPET WALLS. You can use the same technique for these roof lines as that used for the east elevation. Note, however, that the roof line displayed in this elevation is flat.

④ ADDING COLUMNS. To define the pipe columns of the islands, start by using LINE with an OSNAP of MID, selecting the top of each island, and drawing a line up to the mansard by using an OSNAP of PERP. You can then OFFSET each of these center lines three inches to the left and right. Finally, erase the center lines to complete the columns.

Creating Store Front and Doors

Now, you need to add the store front and doors at the entrance to the branch bank.

❑ LAYER. Set layer WIN-080 current — AE-MENU Tablet [LAYER–BY SCREEN–SET].

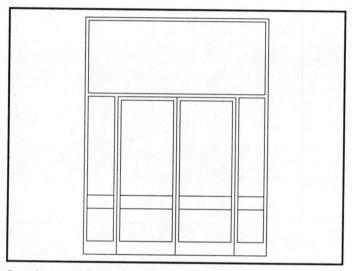

Storefront and Doors Illustration

① CREATING STORE FRONT AND DOORS. By using the following technique, you can create and add the store front and doors for the entrances to the elevation.

Technique For Creating Store Front and Doors

Our next technique is similar to the window technique that we described earlier in this chapter. It is a little more complex. Try to stay with us. Practice always *makes perfect*.

Again, we will use the flexibility of OFFSET. With this one command, you can essentially create the entire elevation section, illustrating the store front and doors of the entrance. When you define the horizontal members, you can use TRIM and FILLET to clean-up the intersections with the frame. Then, you can use MIRROR to create the other half of the store front and doors.

Finally, we will show you how to use the CHANGE command to move the lines representing the doors to the correct layer.

Setting Up Store Front and Doors Drawing

     Enter selection: 2   Edit an existing drawing named 8910EL.

Select the [VIEW–RESTORE–SCRATCH] command from AE-MENU.

     Enter selection: 1   Begin a NEW drawing named AE-SCR7.

Verify the settings shown in this chapter's AE-SCRATCH Setup Table.

When you are ready to get started, set layer WIN-080 as current and either move to an empty portion of the drawing editor, or use the SCRATCH area to create the store front drawing.

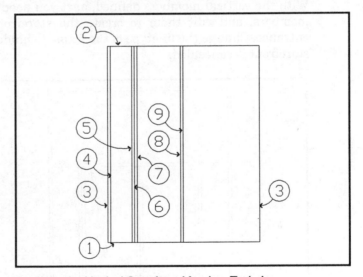

Applying the Vertical Storefront Member Technique

Vertical Frame Members

COMMAND: **LINE**   Create a line 8' long to represent the frame sill ①.

Command: **OFFSET**   Define frame head: select ①, specify 12', for ②.

COMMAND: **LINE**   Create vertical line ③ that extends from ① to ②.

Command: **OFFSET**   Define opposite side of frame: select ③, specify 8', for ③.

Command: **OFFSET**   Define frame member: select ③, specify 2", for ④.

Command: **OFFSET**   Define glass opening: select ④, specify 1'-4, for ⑤.

Command: **OFFSET**   Define frame and door: select ⑤, specify 2", for ⑥ and ⑦.

Command: **OFFSET**   Define door frame: select ⑦, specify 2'-8, for ⑧.

Command: **OFFSET**   Define center of frame: select ⑧, specify 2", for ⑨.

COMMAND: **MIRROR**

Use to create other half of frame, select ⑨ as mirror line.

With the vertical members defined, next you need to add the horizontal members; and edit them to create the storefront and doors for the entrance. Change the linetype of the frame to hidden to indicate that the storefront is concealed.

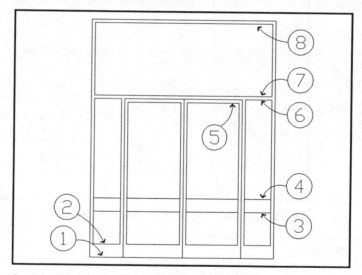

Applying Horizontal Frame Members Technique

Horizontal Frame Members

Command: **OFFSET**   Define kick plate: select ①, specify 8", for ②.

Command: **OFFSET**   Define top of guard rail: select ②, specify 1'-8, for ③.

Command: **OFFSET**   Define bottom of guard rail: select ③, specify 8", for ④.

Command: **OFFSET**   Define door head: select ①, specify 8', for ⑤.

Command: **OFFSET**   Define door frame: select ⑤, specify 2", for ⑥.

Command: **OFFSET**   Define frame: select ⑥, specify 2", for ⑦.

Command: **OFFSET**   Define frame head: select top horizontal line, specify 2", for ⑧.

Command: **TRIM**   Use to clean up all corners.

Command: **FILLET**   Use as required to complete clean-up operation.

COMMAND: **BLOCK**

Select entire drawing, name the block STFT, and use center of frame at sill for insertion point.

By drawing two lines and using OFFSET, we were able to create the store front entrance frames and doors.

Benefit

You have used the OFFSET command frequently throughout these first chapters (and in your own work). This command is one of the simplest to use, but its capabilities are often overlooked. After drawing a single vertical and a single horizontal line, you can use OFFSET to practically create an entire drawing to absolute dimensions.

Placing Storefront

Now that you've finished creating the storefront in the SCRATCH area, you can move this detail into the MAIN drawing area and place it in the appropriate location.

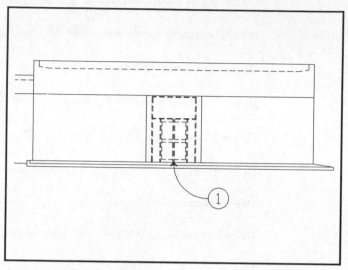

Storefront Placement Illustration

① PLACING STOREFRONT. First, select one of the wind break wall lines and OFFSET it a distance of 4'-0" toward the inside. Now, use INSERT, specify the block name as STFT. Using an OSNAP of INTersection, insert the block at the intersection of the offset line and the finished floor. You can erase the offset line once you've inserted the storefront.

Placing Window

The size of the window that you created for the east eLevation matches that of the window required for the north elevation. Insert this window block (called WIND) again at the correct location for this elevation.

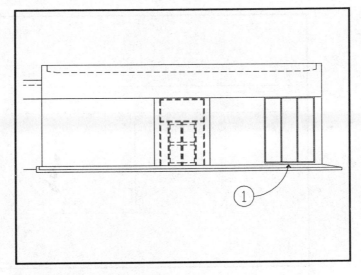

Window Placement Illustration

① PLACING WINDOW. First, select the right building line and OFFSET it 5'-4" (as derived from the Floor Plan of Chapter Six) toward the left. Then, use INSERT, specify the block name as WIND. Using an OSNAP of INTersection, insert the block at the intersection of the offset line and the finished floor. You won't need the offset line after you finish inserting the window.

Defining Specialty Equipment

Put in the specialty equipment, using whatever data you used in the east elevation (whether you obtained the manufacturer's data or approximated our drawing). Start with the remote tellers on this elevation. Our illustration is shown below.

❑ LAYER. Set layer SPL-100 current — AE-MENU Tablet [LAYER–BY SCREEN–SET].

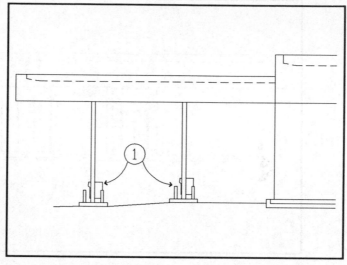

Special Equipment Illustration

① SPECIALTY EQUIPMENT. Use the appropriate commands to duplicate manufacturer's specifications for the remote tellers in the SCRATCH area. When you are ready to relocate the tellers to the appropriate locations in the MAIN area, use the BLOCK command.

❑ CLEAN UP. Try using the [DULXTRIM] routine to select the column lines and remove any excess lines that pass through the columns.

❑ SAVE. Save your drawing.

Creating Base Drawing for South Elevation

The final exercise for this north elevation is to create a copy of this elevation and mirror it to serve as a base drawing for the south elevation. This operation is similar to what you did for the west elevation.

After you create this base drawing, you need to increase the width of the window size for the south elevation. You need to make this adjustment before you mirror the base drawing. Once it's mirrored, you won't be able to explode the window block to make the changes.

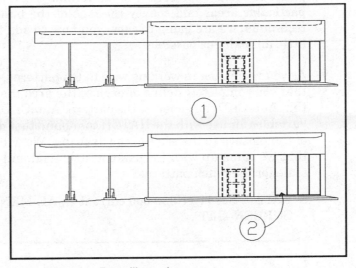

South Elevation Base Illustration

① COPYING NORTH ELEVATION. First, use COPY, select the north elevation, and place the copy below it.

② MAKING WINDOW ADJUSTMENT. Now, EXPLODE the window. You need to add one pane to the left side of the window. Use COPY, select the inside head, sill, and mullion lines, and use OSNAP END to select the base point. Use OSNAP PERP and select the outside mullion. Finally, FILLET to clean up the corners.

❑ MIRROR BASE DRAWING. To finalize the south elevation base drawing, use MIRROR, select the copied elevation, and create a mirror image of it.

❑ SAVE your drawing.

South Elevation

The south elevation is the last drawing required to complete the exterior elevation working drawings for this simulated project. In this last elevation, use a hatch pattern to represent the brick veneer, insert the elevation and section annotations, and add the text and titles.

Defining Hatched Surfaces

Usually when we create exterior elevations we define the different surface characteristics of the building with hatch patterns. These patterns help represent the materials that we have specified for particular areas and convey the *look* of the building. In the following technique, we are going to show you how to add these patterns with as little difficulty as possible.

As you remember in working with hatch patterns, you have to make sure that you completely define or enclose the area that you want to hatch so that AutoCAD can generate the pattern. AutoCAD does not yet to provide variables for use with the HATCH command that directly relate the scale of the pattern to the scale of the drawing. We will provide a scale factor that we came up with (by using a little trial and error) to get the right scale for the hatch patterns.

❑ LAYER. Set layer 000-000 current — AE-MENU Tablet [LAYER–BY SCREEN–SET].

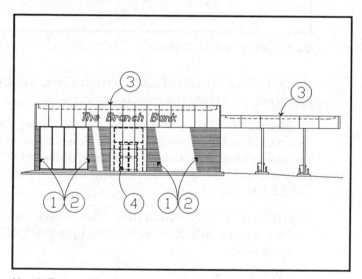

Hatch Pattern Illustration

① ENCLOSING AREA WITH PLINE. Pick one of the areas in the illustration hatched with the LINE pattern to trace over with the PLINE command.

❑ LAYER. Set layer MAS-040 current — AE-MENU Tablet [LAYER–BY SCREEN–SET].

② HATCHING THE AREA. Now you can select this pline using HATCH, and specify the LINE pattern with a scale factor of 16. Repeat these steps for all the other like-patterned areas.

③ ADDING SEAMS TO ROOF. Use the LINE command with OSNAP MID and select the top of mansard. Draw a line down to the bottom of mansard to represent a seam. Now, use OFFSET, first with a distance of two feet, and then with a distance of four feet, to define the seams. Erase the center line to complete the main mansard. Repeat the same process for the mansard roof over the drive through area.

❑ ERASE PLINE. When all areas are completed, use ERASE to remove the plines.

Technique for Placing Elevation and Annotation Symbols

You are almost done. You need to add the elevation symbols and an annotation to designate the section cut for Chapter Eight (Building Sections).

❑ LAYER. Set layer SYM-010 current — AE-MENU Tablet [LAYER–BY SCREEN–SET].

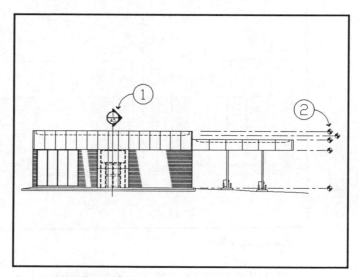

Annotations Illustration

① PLACING ANNOTATIONS. Back in Chapter Two, we created two separate symbols that we could superimpose on one another (using an AutoLISP routine) to create a section annotation. We called the symbols DETAIL and EELEV, and their insertion routine is called EELEV.LSP. You can use these symbols and the associated AutoLISP routine to place your section cut symbol. Answer the prompts for annotation number and sheet number with **1** and **A5**.

② PLACING ELEVATION SYMBOL. Use the CIRCLE, LINE, and HATCH commands to create an elevation symbol and insert it at the end of the major component elevation lines.

Custom Lettering the Signage

If you have third party font files available to you, use a Helvetica-type text font for the "The Branch Bank" lettering shown in the illustration below. Add a 15 degrees angle, and specify a height of 24 inches.

If you don't have extra fonts, choose another font from AutoCAD, use the same angle and height. Then add the signage to the south elevation.

❑ LAYER. Set layer TXT-010 current — AE-MENU Tablet [LAYER–BY SCREEN–SET].

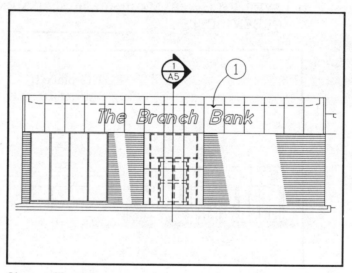

Signage Illustration

① PLACING THE SIGN. Use TEXT with a custom Helvetica font (or a font of your choice) to place the sign. Use the illustration above as a guide.

Adding Text and Titles

The last thing you need to do to complete the drawings in this chapter is to add the text and titles. Some text and titles are illustrated below in the Completed South Elevation illustration shown below. You can refer back to the complete drawing at the start of the chapter for any additional text and titles that you need.

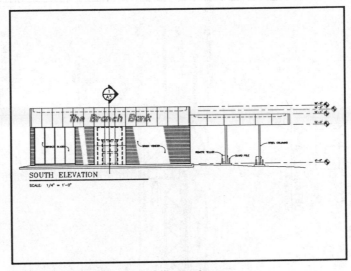

South Elevation Completed Illustration

❑ ADD TEXT. Again, use DTEXT to place the text. For the titles, use a ROMAND text style, and a height of 18 inches. For everything else, use a ROMANS text style, and a height of six inches.

❑ END DRAWING. The exterior elevations are now complete. Be sure to save your drawing by using the END command.

Summary

If you followed along with us in this chapter, you have learned how to exploit AutoCAD's capabilities to reuse and expand data from existing drawings. Reusing data saved a lot of time and effort in creating the exterior elevations for the simulated project. These capabilities give you a major performance improvement over conventional drafting methods. This holds true whether you're an architect responsible for initiating much of the detailed work for a project, or an engineer who benefits from *framework* drawings created from the same data.

Now that you've completed these components of the working drawings, let's move on to developing the building sections.

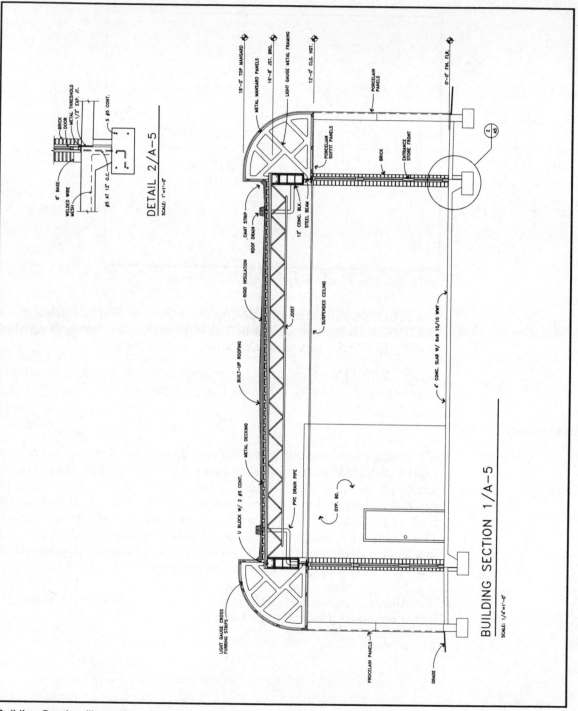

Building Section Illustration

CHAPTER 8

Building Sections

Introduction

Building sections and details are included in construction documents to indicate the internal, sometimes hidden, details and construction connections of a building. Generally speaking, details are the third dimension descriptions that further expand the plan and elevation drawings. They show the contractor, through drawings of various scales, just how the different materials and components should fit together. Details provide information about an enormous range of subjects, from the foundation footings to the roof system, from the exterior to the interior walls, and include critical dimensions and notations. In addition, special requirements pertaining to the building, such as fire proofing, are sometimes detailed in the graphic rendering of each wall section and detail drawing.

In this chapter, we will show you techniques using AutoCAD's editing commands that will save you drawing time. As you've seen in other chapters, if you can manipulate existing data rather than input it from scratch, you'll be using your time more effectively. The AutoLISP routines that we will show in this chapter also will save you from time-consuming drawing input because they prompt you for necessary information.

Building Sections and Details

Building sections are *cut* along specified lines, indicated on the floor Plan, through different parts of the building and from different directions. We usually indicate, using detail annotations, those special areas that we need to define further with large scale detail drawings. We will show you how to reuse and further enhance existing data to create the drawings for this chapter, just as we did for the drawings in previous chapters. You can refer back to Chapters Six and Seven for the information and cut locations that you need to create the single building section and the corresponding detail for this chapter.

Project Building Section

You will create a representative wall section, 1/A5, for the simulated project in this chapter's project sequence. The same information used to create this section, that is cut at the center of the building, could easily be reused to create other sections.

Project Detail

You also will create a representative detail, called 2/A5, to further define the footing and foundation area of the building section at the entrance. Create this detail at the scale defined in the setup routine, but later we will show you how to scale it to a larger size for plotting. Any special conditions you have indicated will be displayed more clearly.

```
Project Sequence for the Building Section Drawing

 Begin a new drawing.
 Define the major component elevation lines and insert
BASELEV drawing.
 Define the exterior wall lines.
◆ Technique for Defining the Foundation, Floor Slab, and
Wall Components.
 Correct the layers.
 Enhance the exterior walls.
 Define the porcelain wall panels and store front.
 Define the suspended ceiling.
 Define the brick surface.
 Define the mansard framing.
 Define the concrete block and structural members.
 Define metal mansard panels.
 Define the structural roof framing.
◆ Technique for Creating Joist Web Members.
 Define the roofing system.
◆ Technique for Creating a Mirror Image for the Other
Half of the Section.
 Define the interior wall features.
 Add the text, titles and Detail Annotation.
◆ Technique for Creating the Detail Base Drawing.
 Enhance the brick surface.
 Enhance the door section.
 Add the reinforcing.
 Changing the scale of the Detail.
 Add the text and title.
```

Drawing Setup for Chapter Eight

Once again, we have set up two ways for you to approach practicing the techniques that we present in this chapter. First, you can follow each project sequence and complete the drawings as if they were part of a real project. Second, you can follow along with the project sequences and use the AE-SCRATCH drawings to experiment with the techniques. Either way, you will gain some benefit through increased productivity.

Method One

If you plan to complete the building section drawing with the detail, load AutoCAD and select Option 1, Begin a New drawing. Enter the drawing name 8910BS, including the prototype drawing SECT-PT (8910BS=SECT-PT). Select the [VIEW–RESTORE–SCRATCH] command from the AE-MENU and move to the SCRATCH area to experiment with any of the techniques listed before you apply them to the drawing.

If you are using the AE DISK, you already have a completed copy of 8910BS.

Method Two

If you want to practice the techniques in this chapter, but not complete the drawings, experiment by using the AE-SCRATCH drawing, AE-SCR8, that we have defined in the table below. With the exercise sequences that follow, you can set up the technique drawing by using the AutoCAD Setup routine, selecting Architectural units, 1/4"=1" scale, D–24"x36" sheet size, and setting all other options defined in the table.

COORDS	GRID	SNAP	ORTHO	UCSICON
ON	10'	6"	ON	WORLD

UNITS	Set UNITS to 4 Architectural
	Default the other UNITS settings.
LIMITS	0,0 to 144',96'
ZOOM	Zoom All.

Layer name	State	Color	Linetype
0	On	7 (white)	CONTINUOUS
MIS-010	On	1 (red)	CONTINUOUS
JST-050	On	2 (yellow)	CONTINUOUS

AE-SCR8 Setup Table

AutoLISP Drawing Utilities

Again before we start the actual drawing, we will show you some AutoLISP utilities that are designed to help you create the structural shapes of the drawing. They prompt you for the dimensions and starting point of an angle, and a wide flange shape, then they draw the shape to your dimensions. These routines support the menu items, called [ANGLE] and [BEAM] on the AE-MENU.

ANGLE.LSP and BEAM.LSP AutoLISP Utilities

Check that you have a copy of the ANGLE.LSP and BEAM.LSP files in your AE-ACAD directory.

Read along, or create the AutoLISP files.

The first routine, ANGLE.LSP, creates an angle shape. When you use it, you will be prompted for the vertical and horizontal leg lengths, the thickness of the legs, and for the lower left corner point of the shape. The routine uses this point as a reference point to create the shape.

Testing the ANGLE.LSP Routine

Now that you have installed the program, load it and type in ANGLE for a test. If you are using the AE-MENU, test the [ANGLE] menu item.

ANGLE.LSP Operation

```
Command: (load "ANGLE")
C:ANGLE
Command: ANGLE
Vertical Leg Length:                    Specify dimension.
Horizontal Leg Length <same>:           Specify dimension.
Leg Thickness:                          Specify dimension.
LL Corner Pt:                           Pick the reference point.
```

The next utility, called BEAM, is designed to create a wide flange structural shape, beam or column, to your specified dimensions. You specify the height, width, and line width or thickness. The routine draws the beam or column shape at the reference point you select. If you are using the AE-MENU, test the [BEAM] item.

Testing the BEAM.LSP Routine

Now that the program is in place, load it and type in BEAM to test its operation.

BEAM.LSP Operation

```
Command: (load "BEAM")
C:BEAM
Command: BEAM
Overall Width <web>:                    Specify dimension.
Overall Height <legs>:                  Specify dimension.
Leg Thickness:                          Specify dimension.
Web Thickness:                          Specify dimension.
Start Point <Left Side>:                Pick the reference point.
```

Again, these routines are good examples of how you can use AutoLISP to draw the symbols with dimensions that you input.

Here are the listings for the two routines.

```
(prompt "Loading ANGLE utility...")
(defun C:ANGLE ( / vdist hdist lth px v1 vx v2 h1 hx h2 mx m1 e1 e2 e3)
    (setvar "cmdecho" 0)
    (setq vdist (getdist "\nVertical leg length:  ")
        hdist (getdist "\nHorizontal leg length <same>:  ")
)                                                      ;setq calculations
    (if (not hdist) (setq hdist vdist));               ;default = vert dist

    (setq lth (getdist "\nLeg thickness:  ")
       px (getpoint "\nLL corner point:  ")
       v1 (polar px 1.5708 vdist)
       vx (polar px 1.5708 (- vdist lth))
       v2 (polar vx 0.0000 lth)
       h1 (polar px 0.0000 hdist)
       hx (polar px 0.0000 (- hdist lth))
       h2 (polar hx 1.5708 lth))
       mx (polar px 1.5708 lth)
       m1 (polar mx 0.0000 lth)
);setq calculations
  (command "PLINE" v1 "W" "0" "" px h1 "")
  (command "ARC" v2 "C" vx v1)
  (setq e1 (entlast))
  (command "ARC" h1 "C" hx h2)
  (setq e2 (entlast))
```

```
 (command "PLINE" h2 m1 v2 "")
 (setq e3 (entlast))
 (setvar· "FILLETRAD" lth)
 (command "FILLET" "p" m1)
 (setvar "FILLETRAD" 0.0)
 (command "PEDIT" px "J e1 e2 e3 "")

 (setvar "cmdecho" 1)
 (gc)  (princ)
);defun
```

AutoLISP Routine for [ANGLE]

```
prompt "Loading BEAM utility...")
(defun C:BEAM ( / hdist vdist vth hth px a1 ax a2 mx1 m1 mx2 m2 b1 bx b2
                hx h1 h2 e1 e2 e3 e4 e5)
   (setvar "CMDECHO" 0)
   (setq hdist (getdist "\nOverall width <web>:  ")         ;get user parameters
         vdist (getdist "\nOverall height <legs>:  ")
         vth   (getdist "\nLeg thickness:  ")
         hth   (getdist "\nWeb thickness: ")
         px    (getpoint "\nStart point <Left Side>:  ")
  )                                                         ;setq calculations
  (setq a1  (polar px 1.5708 (* vdist 0.5))                 ;various calculations
        ax  (polar a1 4.7124 vth)
        a2  (polar ax 0.0000 vth)
        mx1 (polar px 1.5708 (* hth 0.5))
        m1  (polar mx1 0.0000 vth)
        mx2 (polar px 4.7124 (* hth 0.5))
        m2  (polar mx2 0.0000 vth)
        b1  (polar px 4.7124 (* vdist 0.5))
        bx  (polar b1 1.5708 vth)
        b2  (polar bx 0.0000 vth)
        hx  (polar px 0.0000 (* hdist 0.5))
        h1  (polar hx 4.7124 (* hth 0.5))
        h2  (polar hx 1.5708 (* hth 0.5))
  )                                                         ;setq calculations

  (command "PLINE" h2 "W" "0" "" m1 a2 "")                  ;set new line width
  (setq e1 (entlast))                                       ;save name
  (setvar "FILLETRAD" vth)                                  ;reset system var
```

```
(command "FILLET" "P" h2)                      ;construct arc
(command "PLINE" b2 m2 hl "")                   ;draw polyline
(setq e2 (entlast))                             ;save name

(command "FILLET" "P" h1)                       ;construct arc
(command "ARC" a2 "C" ax al)                    ;draw arc
(setq e3 (entlast))                             ;save name

(command "ARC" b1 "C" bx b2)                    ;draw arc
(setq e4 (entlast))                             ;save name

(command "PLINE" al bl "")                       ;draw polyline
(setq e5 (entlast))                             ;save name

(command "PEDIT" el "J" e2 e3 e4 e5 "" "")      ;edit 2D polyline
(command "MIRROR" "L" "" h1 h2 "N")             ;mirror image
(command "PEDIT" "L" "J" px "" "")              ;edit 2D polyline

(setvar "CMDECHO" 1)
(gc) (princ)
)                                               ;defunc
```

AutoLISP Routine for [BEAM]

Building Section Project Sequence

Let's start this project sequence as we did in Chapter Seven, by defining the major dimensions, or construction component elevation lines. If you insert the base elevation drawing (BASELEV) created in Chapter Seven, you will have a guide to help you locate different entities. These are defined by the section ut annotation "1/A5" from Chapter Seven.

While we have created special techniques and command sequences which are designed to speed your production of both this building section and working drawings, we are going to concentrate only on defining the *right half* of the building section. Bear with us, it will all come out right in the end. As you might guess — it's done with mirrors!

Defining the Major Component Elevation Lines

The major building component elevations involved in the building section are the top of footing, finished floor, ceiling height, joist bearing, and top of the main mansard. By defining these elevation lines, you will be able to use them as references for creating the rest of the drawing.

You begin by creating a line extending from the outside of one wind break wall to the outside of the other wind break wall. You'll be able to offset this line to create the other elevation lines. But first, you need to insert the BASELEV drawing on layer 000-000, the *non-plot*, or scratch layer.

❑ LAYER. Set layer 000-000 current — AE-MENU Tablet [LAYER–BY SCREEN–SET].

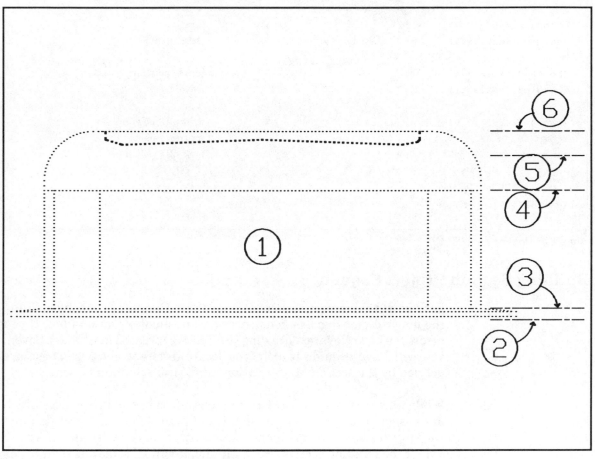

Major Component Elevation Lines Illustration

① INSERTING BASELEV DRAWING. Use INSERT to place the BASELEV drawing, created in Chapter Seven, into the current drawing to locate the elevation lines.

❑ LAYER. Set layer CON-030 current — AE-MENU Tablet [LAYER–BY SCREEN–SET].

② ADDING TOP OF FOOTING LINE. First, use LINE to draw a horizontal line 22'-4" in length to represent one half of the section width and the top of the building footings.

③ ADDING FINISHED FLOOR LINE. Now, use OFFSET with a distance of 2'-0", select the top of footing line, and pick an offset point toward the top of the drawing to create the finished floor elevation.

④ DEFINING CEILING LINE. Again, OFFSET the finished floor line a distance of 12'-0" toward the top of the drawing to define the top of ceiling elevation line.

⑤ CREATING JOIST BEARING LINE. OFFSET the finished floor line again, this time with a distance of 16'-8" toward the top of the drawing, to create the joist bearing elevation line.

⑥ ADDING TOP OF MAIN MANSARD LINE. If you OFFSET the finished floor line a distance of 18'-0" toward the top of the drawing, you will define the our last elevation line, the top of main mansard.

Now that you've defined the major component elevations, you can move on to defining the exterior wall lines.

Defining the Exterior Wall Lines

After you add the exterior wall lines that extend from the top of footing to the joist bearing elevations, you can use the wall lines and the existing elevation lines as reference guides for creating the rest of the building section.

❑ LAYER. Set layer MAS-040 current — AE-MENU Tablet [LAYER–BY SCREEN–SET].

Adding Foundation, Floor Slab and Wall Components

This next step is rather lengthy, so we've broken it down into pieces. You need to add the foundation, floor slab, and wall components to the building section drawing.

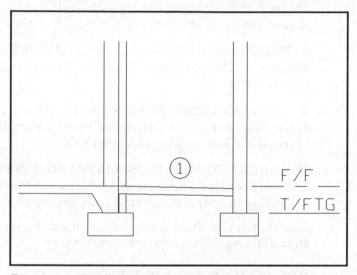

Foundation, Floor Slab and Wall Components Illustration

① ADDING BUILDING SECTION COMPONENTS. Take a look at the following series of techniques to add these components.

Technique for Defining Foundation, Floor Slab and Wall Components

With the following series of commands, we will show you three techniques to create an outline of the footings, foundation, floor slab, and exterior wall lines of the section.

The first technique, called Wind Break Wall, treats the section through the wind break wall, footing, and entrance slab.

The second technique, called Exterior Wall, creates the lines representing the wall exterior, interior, and brick.

The third technique, called Exterior Wall Foundation, creates the outline of the footing and foundation wall for the exterior wall.

Use the reference lines created earlier in the chapter (shown in the major component illustration) to help create these new components. Since you will be using editing commands and existing data, these techniques will save you from having to separately enter each line with specific coordinates.

Setting Up Components Drawing

    Enter selection: 2   Edit an existing drawing named 8910BS.

Select the [VIEW–RESTORE–SCRATCH] command from AE-MENU.

Enter selection: 1   Begin a NEW drawing named AE–SCR8.

Verify the settings shown in this chapter's AE-SCRATCH Setup Table.

When you are ready to get started, use the ZOOM W command and define a window to include the last two vertical lines shown in the Exterior Wall Lines illustration (above), and the reference lines F\F (for finished floor) and T/FTG (for top of footing).

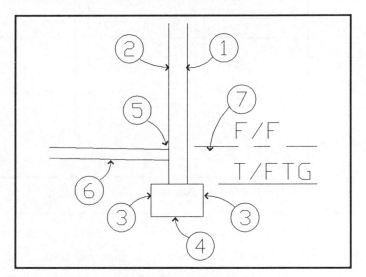

Applying Wind Break Wall Outline Technique

Wind Break Wall

Command: **OFFSET**   Select ①, specify 8" distance, to create ②.

Command: **OFFSET**   Select ① and ②, specify 8", to create ③.

Command: **OFFSET**   Select T/FTG, specify 12", to create ④ bottom of footing.

Command: **EXTEND**	Extend ③ to ④.
Command: **TRIM**	Select T/FTG as cutting edge and remove excess lines of ③ above.
Command: **OFFSET**	Select F/F, specify 2", to create ⑤ top of entrance slab.
Command: **OFFSET**	Select ⑤, specify 4", to create ⑥ entrance slab.
Command: **PLINE**	Create a sloped polyline ⑦ to define grade.

Now that you have defined the wind break wall, move on to the exterior wall.

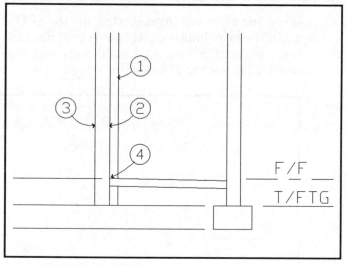

Exterior Wall Outline Illustration

Exterior Wall

Command: **OFFSET**	Select ①, specify 4", to create ② brick.
Command: **OFFSET**	Select ②, specify 8", to create ③ interior wall line.
Command: **TRIM**	Select ② as cutting edge and remove excess slab lines.
Command: **CHANGE**	Select top of entrance and change point to F/F line ④.

The last area that you need to add is the foundation for the exterior wall.

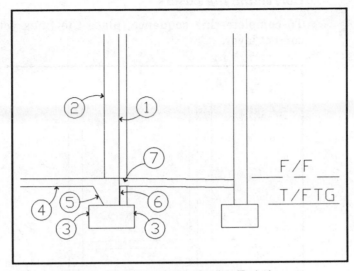

Applying Exterior Wall Foundation Outline Technique

Exterior Wall Foundation

Command: **OFFSET**	Select ① and ②, specify 8", to create ③ footing.
Command: **OFFSET**	Select F/F, specify 4", to create ④ bottom of floor slab.
Command: **OFFSET**	Select ②, specify 4" to create reference for ⑤.
Command: **LINE**	From the top of footing to intersection ④ at reference line.
Command: **TRIM**	Use this command to remove all excess lines at ⑤ and ⑦.
Command: **OFFSET**	Select ①, specify 1", to create ⑥.

This technique shows how you can use reference lines as guides to add other entities; it also shows you how to use those new entities, as a group, to create even more entities.

Benefits

You just created the outline of the wind break wall, footing, exterior wall, and foundation for the building section by drawing only two lines. You added the grade and the turn down section of the exterior foundation with Plines. This is just another good example of why editing is faster than inputting.

Correcting the Layers

To complete this sequence, place the lines you just created on the correct layer.

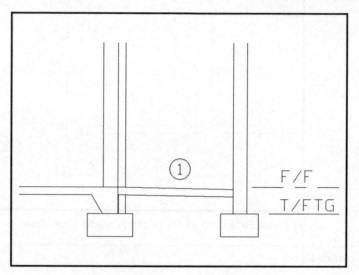

Layer Corrections Illustration

① CHANGE LAYERS. Use CHPROP to move floor slab, top and bottom of footing, and the bottom portion of the foundation wall to layer CON-030.

Enhancing the Exterior Walls

Now, we can continue by making some enhancements to the exterior walls. First, remove the line that defines the edge of the floor slab and then change the outside wind break wall linetype.

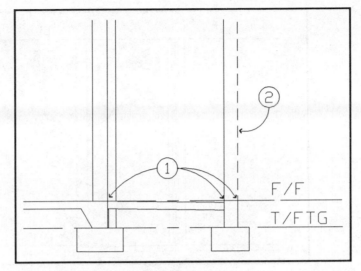

Exterior Wall Enhancements Illustration

① CUTTING WALL LINES. Start by using the [CUT] command (CUT.LSP) from the AE-MENU. Select the line that defines the edge of slab and cut it at its intersection with the finished floor. You can use the same command to cut the exterior wind break wall lines at their intersections with the finished floor line. Since you don't need the top portion of the edge of slab line, simply ERASE it.

② CHANGING WIND BREAK WALL LINES. Change the linetype of the top portion of the wind break wall lines to HIDDEN. Use CHPROP to do this.

Defining the Porcelain Wall Panels and Store Front

Next, add the lines that represent the exterior wind break wall panels in elevation, and the store front frame and door in section. Use LINE and OFFSET to create these entities. Then, clean up the lines and change them to the correct layer.

❑ LAYER. Set layer SPL-100 current — AE-MENU Tablet [LAYER–BY SCREEN–SET].

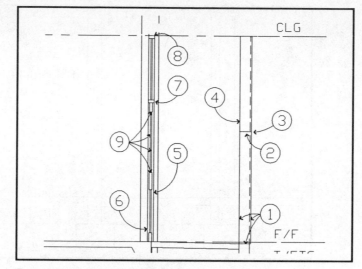

Porcelain Wall Panels and Store Front Illustration

① DRAWING EXTERIOR PANEL LINE. Start with the LINE command with an OSNAP of INT and select the intersection of the slab and the inside wind break wall line. Draw a line extending to the right 10 inches, turn up, and connect it with an OSNAP of PERP to the line that represents the ceiling elevation line.

② CREATING PANEL HORIZONTAL JOINT LINE. You can create this joint line by using OFFSET with a distance of six feet. Select the 10 inch long horizontal line you just created, and offset it toward the top.

③ CLEANING UP. Use TRIM to remove the excess from the joint line.

④ CHANGING LAYERS. Use CHPROP to change the top portion of the inside wall line to layer SPL-100.

Now, you are ready to start working on the store front and doors.

❏ LAYER. Set layer DRS-080 current — AE-MENU Tablet [LAYER–BY SCREEN–SET].

⑤ CREATING STORE FRONT FRAME. To create the exterior frame line, OFFSET the outside line of the wall a distance of three inches toward the left. Use CHPROP to change this line to layer DRS-080.

⑥ DEFINING FRAME WIDTH. To create the interior frame line, OFFSET the exterior frame line a distance of four inches toward the left.

⑦ DRAWING DOOR HEAD MEMBER. Create a rectangle to represent the door head by using LINE. This member should sit eight feet above the finished floor and should be four inches wide and two inches thick.

⑧ ADDING FRAME HEAD MEMBER. Use COPY and select the door head member that you just created to place the frame head member top at 12 feet above the finished floor.

⑨ ADDING DOOR AND GLASS LINES. Use OFFSET to create the door frame lines with a thickness of two inches, and the glass line with a thickness of one inch.

❏ CLEAN UP. You will have some frame and door lines left over that you don't need. Use TRIM and select the ceiling line as the cutting edge to clean up.

Defining the Suspended Ceiling

The suspended ceiling is located at an elevation of 12 feet. Use the ceiling elevation line as the bottom of the suspended ceiling and offset that line to represent the ceiling's thickness.

❏ LAYER. Set layer GRD-090 current — AE-MENU Tablet [LAYER–BY SCREEN–SET].

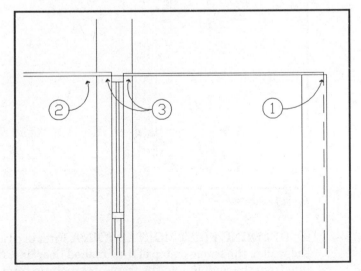

Suspended Ceiling Illustration

① PREPARING CEILING ELEVATION LINE. To start off, you will need to use BREAK to separate the ceiling line at the outside wind break wall line. Then, use CHPROP to change the ceiling elevation line to layer GRD-090.

②DEFINING CEILING THICKNESS. If you OFFSET the line from the last step a distance of one inch toward the top, you easily define the suspended ceiling thickness.

❑ CLEAN UP. Use TRIM to remove the wall lines that pass through the ceiling.

Defining the Brick Surface

The final task to perform, before you move to the mansard and structural framing, is to add lines to define the brick surface at the store front. You only have to draw the first few masonry joints; then you can manipulate those lines with the ARRAY command to add more graphic detail.

❑ LAYER. Set layer MAS-040 current — AE-MENU Tablet [LAYER–BY SCREEN–SET].

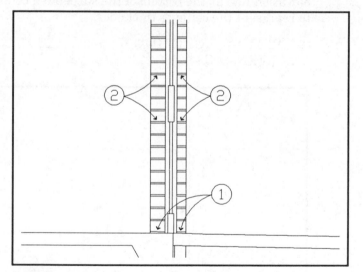

Brick Illustration

① DEFINING FIRST MORTAR JOINT. First use LINE to draw a line from the wall to the frame on top of the finished floor line. Next, draw a horizontal line from the wall line to the frame 1/2 inch up from the finished floor line. Repeat these lines for the other side of the frame.

② CREATING REMAINING MORTAR JOINTS. Now, use the ARRAY command on the lines you just created. Start by selecting the first mortar joint line that lies on top of the finished floor line and specify a rectangular array. Use 1 column and 36 rows (F/F to CLG 12'-4" per joint). This will create the first joint line. Repeat the array sequence for the second joint line, but this time specify the line that is 1/2 inch above the finished floor line.

Repeat again for the other side to complete the brick surface.

Defining the Mansard Framing

The mansard framing consists of light gauge metal studs with metal clips to attach the metal mansard panels. Add the support for the framing later when you add the structural members and roof joists. You create this framing system by using the LINE command with the cross hairs rotated to 45 degrees. Add the curve by using FILLET with the specified radius.

❑ LAYER. Set layer FRA–050 current — AE–MENU Tablet [LAYER–BY SCREEN–SET].

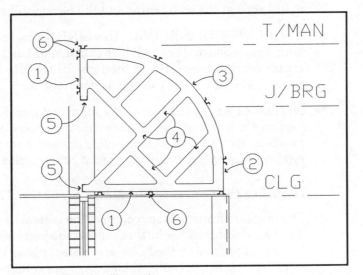

Mansard Framing Illustration

① DEFINING EXTERIOR EXTENTS OF FRAME. Start by using LINE to draw a horizontal line extending from the center of the exterior wall at the store front, 1 1/2 inch above the suspended ceiling panels, to within two inches of the exterior wind break panel line.

Use LINE again and draw a vertical line from the end of the horizontal line at the center of the wall up to within two inches of the top of mansard elevation line.

② DEFINING BASE POINT OF MANSARD CURVE. Now, use LINE with relative coordinate input. Select the end of the horizontal line that you just created and input @12<90 to create the vertical line that the curve attaches to.

③ DEFINING MANSARD CURVE. Using the ARC command with the C,S,E option, you can define the mansard curve by selecting the intersection of the offset line and the vertical line for the arc center point. Use the outside end point of the horizontal line for the start point, and end the arc at the vertical line.

④ DEFINING FRAMING MEMBERS. To create the first framing member, OFFSET the arc you just created a distance of five inches toward the inside. Now use LINE with the cross hairs rotated 45 degrees; then use OFFSET to define the remaining framing members.

⑤ DEFINING BEARING MEMBERS. Use LINE to define the vertical metal stud at the wall center and the horizontal stud at the top.

⑥ CREATING PANEL CLIP. Use LINE to create a clip that has one inch long legs, is one and one half inches tall, and has a one- and one-half inch center span distance. If you then use BLOCK to convert this clip into a symbol, you can insert the clip as you need it.

➡ *NOTE! We have just created the mansard panel clips at drawing scale and on the framing layer, disregarding our own advice on how to create symbols. We did it because the clips are not a standard drawing symbol type, and we only use them in special construction methods.*

Defining the Concrete Block and Structural Members

The mansard framing is supported at the bottom by a structural steel beam. The 12 inch concrete blocks that support the joists (which will be added later) are also supported by the beam and a steel plate. The top concrete block is a "U" block that has two #5 reinforcing bars running the continuous length of the wall. This block is filled with concrete. Mounted in the concrete fill of the "U" block are the base plates, spaced according to the joists, that you will use to attach the joists to the wall.

You can create these entities, using the LINE, OFFSET, and ARRAY commands. Use the following illustration and sequences as a guide.

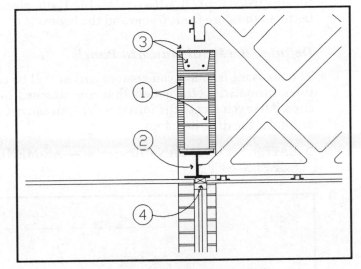

Concrete Block and Structural Framing Illustration

① DEFINING CONCRETE BLOCK. The block that rests on the beam is at 12'-8" above the finished floor elevation line or eight inches above the ceiling elevation line. Use LINE and draw a line from wall line to wall line to define the first mortar joint.

To define the remaining blocks and mortar joints, you can use either OFFSET or ARRAY with a distance of eight inches for the block and 1/2 inch for the joint.

Use OFFSET with a distance of 1 1/2 inches to define the concrete block wall thickness. Use the same distance to define the "U" block. Use HATCH with the LINE pattern to hatch the block walls.

❑ LAYER. Set layer STL-050 current — AE-MENU Tablet [LAYER–BY SCREEN–SET].

② DEFINING STEEL BEAM. Try our [BEAM] command (BEAM.LSP) and specify a wide flange beam dimension of 6"x 6". You can use PLINE to define a steel plate 1/2 inch thick centered on the beam. If you need to, you can also use MOVE to place the beam and plate in the center of the wall at the correct elevation.

③ DEFINING STEEL MEMBERS FOR "U" BLOCK. Use PLINE to define a 1/2 inch thick steel plate for the joist that rests on top of the block. To represent the reinforcing bars, use DONUT to place two donuts with inside diameters of 0 and outside diameters of 5/8 inch.

④ DEFINING BLOCKING. Use LINE to place the wood blocking between the store front frame and the beam.

Defining the Metal Mansard Panels

The mansard framing you created earlier will be covered with two inch thick insulated metal panels that are attached to the frame by metal clips. After you create the initial curve, you can use editing commands to minimize the drawing steps.

❏ LAYER. Set layer CAN-070 current — AE-MENU Tablet [LAYER–BY SCREEN–SET].

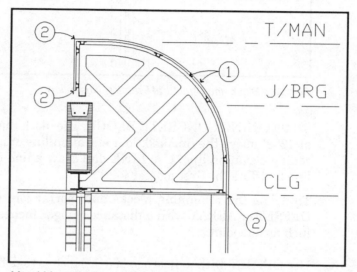

Metal Mansard Panels Illustration

① DEFINING PANELS. Start by using OFFSET, first with a distance of 1 1/2 inches to define the inside panel edge, and again with a distance of two inches to define the panel thickness. Select the framing curve as your base entity.

② CLEANING UP. Use FILLET to clean up all corners.

Defining the Structural Roof Framing

The framing for the roof of the project consists of bar joists and intermediate rib metal decking. A steel angle is used to close the metal decking edges at the mansard. You use the joist bearing elevation line as your reference. For this sequence, use a combination of LINE and OFFSET to add the structural roof framing.

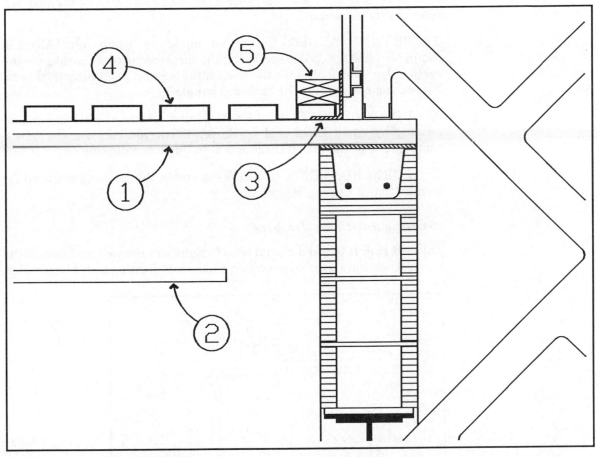

Structural Roof Framing Illustration

① DEFINING TOP JOIST CHORD. Start by using BREAK and select the joist bearing elevation line at the outside wall line. Break the portion of the line that passes through the mansard. Use CHPROP to change the elevation line to layer STL-070. Now, you can OFFSET the elevation line a distance of 1 1/2 inches toward the top; and offset that line again to define the joist top chord. Use LINE to close the joist end at the wall.

② DEFINING THE BOTTOM CHORD. To define the bottom chord, OFFSET the bottom line of the top chord a distance of 24 inches toward the bottom of the drawing. You can then OFFSET that line a distance of 1 1/2 inches toward the bottom to define the thickness of the bottom chord. Now, use LINE and draw a line 12 inches to the left of the inside wall line to define the end of the bottom chord. Use TRIM to clean up the lines.

➡ *NOTE! We will show you how to use a technique for defining the joist web later in this sequence.*

③ CREATING ANGLE. Create an angle by using our [ANGLE] command (ANGLE.LSP), specifying the dimensions of six inches for the vertical leg, four inches for the horizontal leg and a 3/8 inch thickness. Place the angle against the mansard panel.

④ ADDING DECKING. Start by creating a section of 1 1/2-inch metal decking by using LINE and OFFSET. Then, you can use the COPY command with the MULTIPLE option to finish the decking.

⑤ ADDING BLOCKING. Use LINE to create the blocking required for termination of the roofing system.

Creating Joist Web Members

The next task is to add diagonal lines to represent the web members of the bar joist. We have a technique which makes this drawing task a lot simpler.

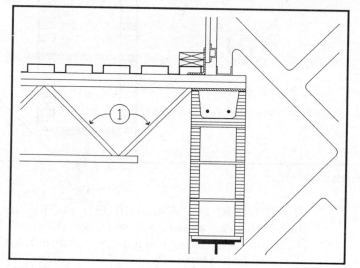

Joist Illustration

① CREATING THE JOIST WEB MEMBERS. Take a look at the following technique for creating the joist web members. You'll first draw a pattern and then use it to create the rest of the members.

Technique for Creating Joist Web Members

Creating the web members of joists can be quite time consuming, especially if you have to create one line at an angle, and then flip the crosshairs to create the next. This presents another good opportunity to create a portion of a drawing, then manipulate the data with editing commands. By first creating a pair of web members, you can use COPY to complete the joist.

Setting Up Joist Web Drawing

   Enter selection: 2   Edit an existing drawing named 8910BS.

Select the [VIEW–RESTORE–SCRATCH] command from AE-MENU.

   Enter selection: 1   Begin a NEW drawing named AE-SCR8.

Verify the settings shown in this chapter's AE-SCRATCH Setup Table.

When you are ready, set layer JST-070 as current.

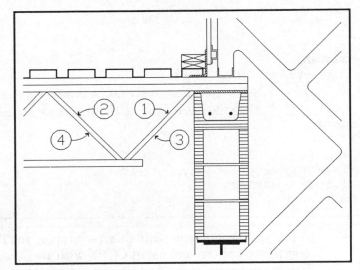

Applying Joist Web Members Technique

Joist Web Members

```
Command: SNAP
Snap spacing or ON/OFF/Aspect/Rotate/Style <0'-1">: R
Base point <0'-0",0'-0">: <RETURN>
Rotation angle <0>: 45

Command: LINE                Use to create lines ① and ②.
From point:
```

```
To point: PERP
to                          Select the top line of the bottom chord.
To point: PERP
to                          Select the bottom line of the top chord.
To point: <RETURN>

Command: OFFSET
Offset distance or Through <Through>: 1
Select object to offset:    Select ①.
Side to offset:             Pick a point to the right to create ③.
Select object to offset:    Select ②.
Side to offset:             Pick a point to the left to create ④.
Select object to offset: <RETURN>

Command: OSNAP
Object snap mode: INT

Command: COPY
Select object: C
First corner:
Other corner:
4 found.
Select object: <RETURN>
<Base point or displacement>/Multiple: M
Base point:                 Select the intersection of ③ with the top chord.
Second point:               Select the intersection of ④ with top chord and repeat until complete.
Second point: <RETURN>

Command: SNAP
Snap spacing or ON/OFF/Aspect/Rotate/Style <0'-1">: R
Base point <0'-0",0'-0">: <RETURN>
Rotation angle <45>: 0
```

After you experiment with this technique, you'll be able to see how creating a pattern and using COPY with the M option can make tasks like creating joist web members much simpler.

Defining the Roofing System

The roofing system consists of 1 1/2-inch rigid insulation with tapered sections, built-up roofing, cant strips and roof drains. The roof drains are PVC types and have 3-1/2 inch PVC drain pipes that turn and go down inside the exterior walls. For this sequence, use OFFSET to create lines. Their intersections, together with other lines will serve as reference points.

❑ LAYER. Set layer ROF-070 current — AE-MENU Tablet [LAYER–BY SCREEN–SET].

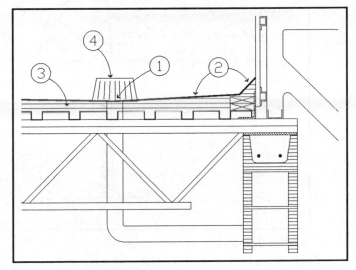

Roofing System Illustration

① DEFINING REFERENCE POINTS. For your first reference point, OFFSET a section of decking, which is the closest to the roof drawing location, a distance of 1 1/2 inch toward the top. Then, move to the center of the building section, select a section of decking and OFFSET it a distance of six inches.

② DRAWING ROOF LINE. Use PLINE, setting a thickness of 1/2 inch for the pline's width, to create a line passing above the wood blocking. This line should meet the reference point at the roof drain and then end at the center reference point. You also need to place a 45-degree line to represent a 4-inch cant strip.

③ INDICATING INSULATION. To represent the rigid insulation , use LINE to draw a line 1/2 inch above the roof decking. You can then OFFSET this line a distance of 1/2 inch to create the lines required to indicate the rigid insulation in section.

④ ADDING ROOF DRAIN. Select the detail of the roof drain from a catalogue, and use the LINE command to draw and place it three feet to the left of the wall.

Making a Mirror Image

That takes care of half of the building section. To complete the other half, you simply need to make good use of the MIRROR command.

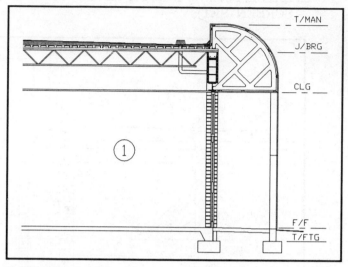

Building Section Illustration

① MIRRORING THE BUILDING SECTION. Use the following technique to create the other half of the building section.

Technique for Creating the Other Half of the Section

 In the first part of this chapter, you defined a line to represent one half of the finished floor elevation line. Since then, you have been using that reference line to create half of the building section. Now that you have completed the right half, you can mirror it to complete the other half.

Setting Up Creating Other Half of Section Drawing

 Enter selection: 2   Edit an existing drawing named 8910BS.

Select the [VIEW–RESTORE–SCRATCH] command from AE-MENU.

Enter selection: 1   Begin a NEW drawing named AE-SCR4.

Verify the settings shown in this chapter's AE-SCRATCH Setup Table.

When you are ready, zoom in on the completed half of the building section.

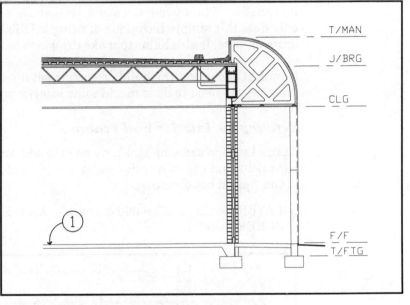

One Half of Completed Building Section Illustration

Building Section

```
Command: MIRROR
Select objects: W
First corner:              Form window around entire section.
Other corner:
196 found.                 This number may vary greatly.
Select objects: <RETURN>
First point of mirror line: END
of                         Select finished floor slab line.
Second point: <Ortho on>   Pick any point toward the top of the drawing.
Delete old objects? <N> <RETURN>
```

If you wish, try using our [LINEMEND] command from the AE-MENU to join the two halves of the section. The drawing does not require it.

Benefits

While it may not have been easy to follow every step we outlined to create the first half of the building section (building sections, by nature, are complicated), we feel sure that you appreciate the fact that you *didn't* have to follow along to create the whole thing. When you apply this technique of mirroring drawings to your own projects, you'll appreciate its benefits. When two halves of a section are essentially identical, not only does this simple technique of using MIRROR save you half of your drawing time, it also helps to make drawings more consistent.

Now that you've got the building section set up as a *whole*, the last bit of drawing you need to do is to add some interior wall lines and a door.

Defining the Interior Wall Features

As final section drawing tasks, we need to add some lines to represent, in elevation, changes in interior walls and to locate any doors that appear in the section cut direction.

❑ LAYER. Set layer GYP-090 current — AE-MENU Tablet [LAYER–BY SCREEN–SET].

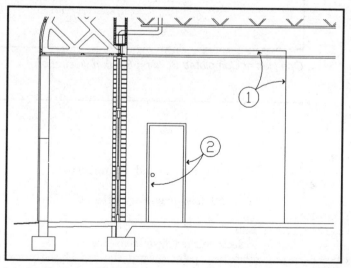

Interior Features Illustration

① CREATING WALL LINES. Simply use LINE to create lines to represent the interior walls in elevation.

❑ LAYER. Set layer DRS-080 current — AE-MENU Tablet [LAYER–BY SCREEN–SET].

② PLACING DOOR. You have an interior door to consider. Again, just use LINE to both place the three foot door, and to indicate its two-inch frame.

Adding the Text, Titles and Detail Annotation

The final additions to this building section will be the text, title, and the detail annotation that you see illustrated below. As you may recall, we created an annotation symbol called DETAIL in Chapter Two. You'll be able to use it in this drawing to label the detail for the next project sequence.

❑ LAYER. Set layer TXT-010 current — AE-MENU Tablet [LAYER–BY SCREEN–SET].

❑ ADD TEXT. Use DTEXT with a height of three inches and a ROMAND style to place all the general text for this drawing.

❑ ADD TITLE. For the title, just use TEXT with a height of six inches.

❑ ADD DETAIL ANNOTATION. To flag the area of the building section that you are going to use for your detail, use CIRCLE and place a circle around the footing and foundation portion. Use the illustration above as a guide.

Then, use our DETAIL command to place the detail annotation symbol with a number of 2/A5.

❑ SAVE DRAWING. You are finished with the building section. This is a good time to save the drawing before you move on to enhancing the detail.

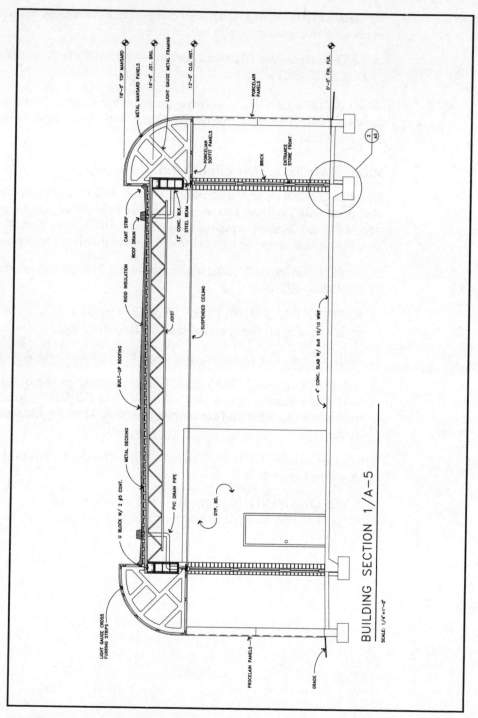

Completed Building Section Illustration

Creating a Detail

The detail we have chosen as an example is the building's foundation and footing. In the following steps, we will show you how to actually cut and copy this section out of the building section; then move it to its own space in the drawing. There you can make enhancements to the brick, the door and frame, and add reinforcing to the foundation.

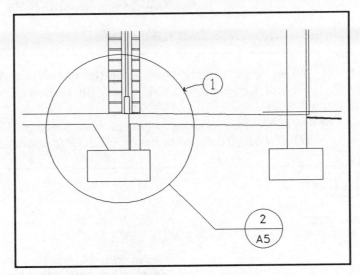

Foundation Detail Illustration

① CREATING FOUNDATION DETAIL. Take a look at the following technique to see how to create the detail.

Technique for Creating the Detail Base Drawing

One of AutoCAD's major benefits is that it lets you copy portions of drawings so that you can enhance them. In the following technique, you will take a portion of the building section, defined by the circle and detail annotation; cut the lines at the circle; and copy that portion to another area of the drawing where you can use it as a detail.

Setting Up Creating Detail Drawing

 Enter selection: 2   Edit an existing drawing named 8910BS.

Select the [VIEW–RESTORE–SCRATCH] command from AE-MENU.

 Enter selection: 1   Begin a NEW drawing named AE-SCR8.

Verify the settings shown in this chapter's AE-SCRATCH Setup Table.

When you are ready, zoom in on the foundation portion of the building section. Refer to foundation detail illustration above to help you.

➥ *NOTE! The following sequence also use our [CUT–LINE] command (CUT.LSP). You need to have CUT.LSP in your AE-ACAD directory.*

Detail

```
Command: OSNAP
Object snap mode: INT
```

```
Command: CUT                    Repeat for each line that crosses circle ①.
Select objects:                 Select a line that passes through the circle.
1 selected, 1 found.
Select objects: <RETURN>
Enter Cut Point:                Select intersection of circle and line.
Select object: <RETURN>
```

```
Command: MOVE
Select object: W
First corner:
Other corner:
26 found.
Select object: <RETURN>
Base point or displacement:     Pick a point within circle.
Second point of displacement:   Pick a point at the top right of the drawing.
```

By using this series of OSNAP, CUT, and MOVE commands, you were able to copy a specific area of one drawing to use as a detail drawing.

Benefits

Throughout this book our motto has been: "If you've already got it, reuse it." This technique for creating your detail base drawing provides another example of why this approach is worthwhile. You save a lot of time and redrawing effort in being able to copy a portion of the section, and enhancing it to perform another project function.

Now that you have moved your detail base drawing, enhance it to create a detail of the footing and foundation.

Enhancing the Brick Surface

The first areas you need to enhance are the brick mortar joints. Offset the joints inward to make them more realistic.

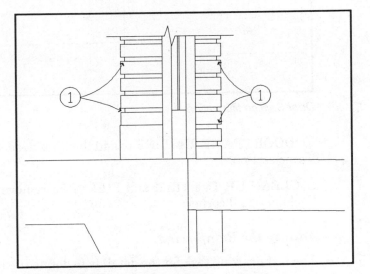

Enhanced Detail Drawing Illustration

① MORTAR JOINTS. OFFSET the outside brick line on each side a distance of 1/2 inch toward the center of the wall.

❑ CLEAN UP. Use TRIM to remove all excess lines.

Enhancing the Door Section

Next, you can enhance your detail by adding lines to define the door and frame in greater detail.

❑ LAYER. Set layer DRS-080 current — AE-MENU Tablet [LAYER–BY SCREEN–SET].

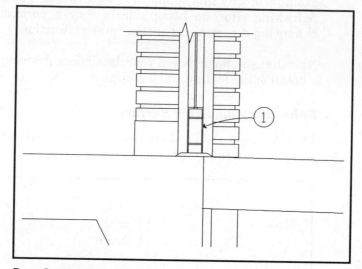

Door Section Illustration

① DOOR FRAME. Use LINE to add the lines needed to enhance the door frame section.

❑ CLEAN UP. Use TRIM and FILLET to remove all excess lines and clean up all corners.

Adding the Reinforcing

The last drawing task for the detail is to define the reinforcing of the foot and foundation. You can do this by drawing in the reinforcing forcing bars, both in section and in elevation.

❑ LAYER. Set layer STL-050 current — AE-MENU Tablet [LAYER–BY SCREEN–SET].

① REINFORCING IN SECTION. To place the five #5 reinforcing bars that you want to show in the footing, use DONUT (as you did when you created the "U" Block earlier in the chapter) and specify a center diameter of 0 and an outside diameter of 5/8 inches.

② REINFORCING IN ELEVATION. To indicate the reinforcing bars that extend out of the slab up into the foundation wall, use LINE.

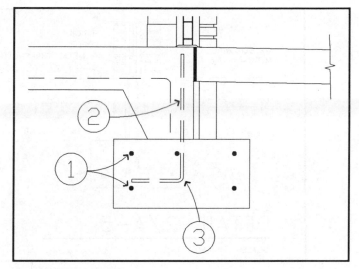

Reinforcing Illustration

③ CLEAN UP. Now, use the FILLET to add the curve of the reinforcing bar. Use CHPROP to change these lines into HIDDEN linetype.

Changing the Scale of the Detail

The general rule of thumb is to draw or display details at a larger scale than the rest of the drawing so that it is easier to see. For this detail, you want to change the scale from its current 1/2"=1'-0" to 1"=1'-0". (This doubles its size).

➡ *NOTE! If you don't add the text and dimensions now, you can change the drawing scale, adding the text and dimensions later at their appropriate size. Alternately, you can set the dimension variables now to 1/2 their default value for the current scale.*

❑ CHANGE SCALE. To change the scale of your Detail to 1"=1'-0", use the SCALE command and specify a scale factor of 2.

Adding the Text and Title

The last thing you need to do for the detail is to add the text and title to describe it. These are shown in the illustration below.

❑ LAYER. Set layer TXT-010 current — AE-MENU Tablet [LAYER–BY SCREEN–SET].

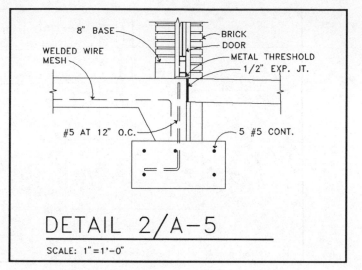

Completed Detail Illustration

❑ ADD TEXT. To place the text you need to describe the drawing, use DTEXT, a three inch text height, and the default text style.

❑ ADD TITLES. Change the text height to six inches with TEXT, and add the title.

❑ END DRAWING. You are done! Be sure to save the completed building section and its detail with an END.

Summary

Building sections and details are generally project specific. Therefore, they require a great deal of time to create. The techniques, such as mirroring the identical side of the drawing, that you used in this chapter to create the building section drawing for the simulated project, cut down on drawing time. We also showed you some ways to retrieve, reuse, and manipulate existing data by using *short-cut* editing commands. In addition, you used our AutoLISP routines, such as the BEAM routine to create the steel beams, saving unnecessary and tedious line input.

With the plans, elevations, building section and detail drawings now finished, we are ready to pass these drawings on to the engineers. In the next chapter, we will show you how to use the data from these drawings to generate the structural, mechanical, plumbing, and electrical plans for the branch bank project.

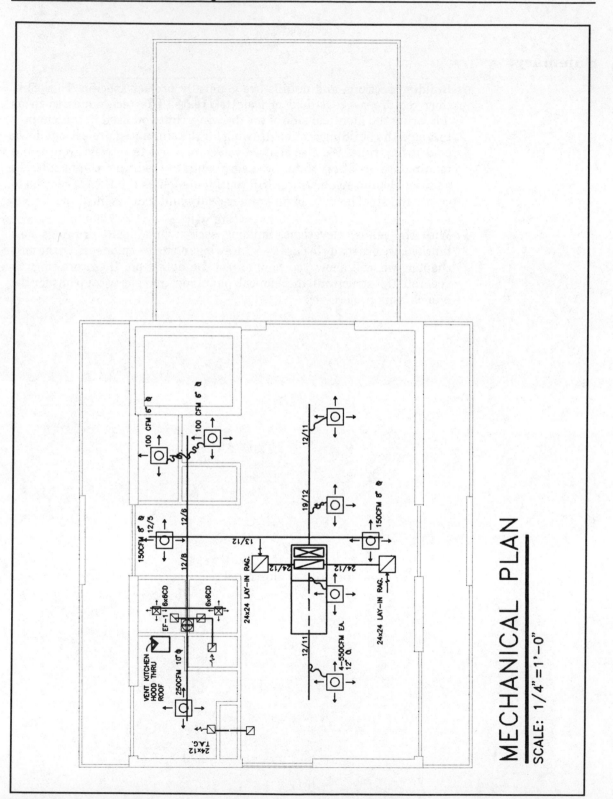

Mechanical Plan Illustration

CHAPTER 9

Engineering Drawings

Introduction

As you know, architects hire consulting engineers to perform critical functions. Engineers make sure that the buildings the architects design will stand up, will be cool in the summer and warm in the winter, that they will be supplied with water and that waste will be removed, and that all the lights and equipment that architects propose using will function the way they are intended to. In short, engineers are responsible for putting down on paper how the building designs will function and operate.

Architects have the responsibility to supply all the design and working drawing base information to the project engineers as quickly as possible so that they can get to work designing the structural, mechanical, plumbing and electrical systems. Since the engineers on a project are usually working within the same time frame as the architects, and since the architects are the only source of project data, it is also important to pass along new or revised data to the engineers as soon as it becomes available. Otherwise, as we all know, the project may not come in on time.

Saving time is one of the best reasons why architects who use AutoCAD should do business with engineers who use AutoCAD (or at least a compatible CAD software package). Being able to share drawing data directly is what makes this time savings possible.

Engineering Drawings

As we did earlier with the architectural working drawings, we will use only the major plan drawings to illustrate the techniques and project sequences for this chapter. The engineering drawings that we will use are:

- Roof Framing Plan   8910ST/S—1

- Mechanical Plan   8910ME/M—1

- Plumbing Plan   8910PL/P—1

- Electrical Lighting Plan   8910EE/E–1

As you may recall, a BASEPLAN drawing of the floor plan design was created back in Chapter Six. In this chapter, you will be able to use this drawing again to help generate the engineering drawings.

Roof Framing Plan

The roof framing plan for our simulated project will contain information that a contractor would need to be able to estimate and erect the framing system that supports the main and drive through roofs. We will define the structural system, by size and by placement of it's members with their associated elevations, using the following project sequence:

```
Project Sequence for the Roof Framing Plan

Begin a new drawing.
◆ Technique for BASEPLAN Framing Modifications.
Define the joists and bridging.
Define the beams.
Define the framing members at the roof openings.
Add text and titles.
```

Mechanical Plan

The mechanical plan for the project will describe a roof top HVAC unit and supply-and-return duct systems. Mechanical engineers usually begin their design process by calculating heat loss and gain for the building. Then, they select the mechanical equipment accordingly. They use the equipment they specify and the required CFM of air movement to size the duct systems and the supply and return air grilles. Let's assume that all the necessary calculations are complete. This project sequence outlines drawing tasks required for completing the mechanical plan:

```
Project Sequence for the Mechanical Plan

Begin a new drawing.
◆ Technique for Supply and Return Grilles.
Define equipment and ducts.
Add text and titles.
```

Plumbing Plan

The branch bank contains data on the routing and sizes for water supply and waste removal piping. An engineer typically sizes the pipes by the requirements of the plumbing fixtures indicated on the architect's floor plan. While the project sequence for this drawing is quite simple compared to the other engineering drawings, it is just as important to the function of the building.

```
Project Sequence for the Plumbing Plan

Begin a new drawing.
◆ Technique for BASEPLAN Plumbing Modifications.
Draw piping layout.
Add text and notes.
Create the sanitary riser.
Add remaining text and titles.
```

Electrical Lighting Plan

The electrical working drawing that we have chosen for you to create in this chapter shows how AutoCAD lets an engineer take one of the architect's plans, in this case, the reflected ceiling plan (from Chapter Six), and use it to define electrical requirements. We use it to define the wiring for the lights. Here is the project sequence for this drawing.

```
Project Sequence for the Electrical Lighting Plan

Begin a new drawing.
Modify Reflected Ceiling Plan 8910CP.
Define and add fixtures.
Create wiring layout.
Add text, Fixture Table and titles.
```

Drawing Setup for Chapter Nine

Once again, we have set up two ways for you to approach the techniques shown in this chapter. First, you can follow the project sequences and complete the drawings as if they were part of a real project. Second, you can read along with the project sequences and use the AE-SCRATCH drawings to experiment with the techniques. Either method will provide you with a better understanding of applying AutoCAD to your work environment.

Method One

If you plan to complete the four engineering drawings of this chapter, load AutoCAD and select Option 1, Begin a New drawing. Refer to the Engineering Drawing Table below to answer the "Enter drawing name:" prompt.

If you are using the AE DISK, you have a copy of each completed drawing.

DRAWING NAME	DESCRIPTION
8910ST	Roof Framing Plan
8910ME	Mechanical Plan
8910PL	Plumbing Plan
8910EE	Electrical Lighting Plan

Engineering Drawing Table

Select the [VIEW–RESTORE–SCRATCH] command from the AE-MENU and move to the SCRATCH area to experiment with any of the techniques listed before you apply them to the drawing.

Method Two

If you want to practice the techniques in this chapter, but not complete the drawings, experiment using the AE-SCRATCH, AE-SCR9, that we have defined below. You can set up the technique drawing by using the AutoCAD Setup routine, selecting Architectural units, 1/4"=1" scale, D–24"x36" sheet size, and setting all other options shown in the table.

COORDS	GRID	SNAP	ORTHO	UCSICON
ON	10'	6"	ON	WORLD

UNITS	Set UNITS to 4 Architectural.
	Default the other UNITS settings.
LIMITS	0,0 to 144',96'
ZOOM	Zoom All.

Layer name	State	Color	Linetype
0	On	7 (white)	CONTINUOUS
000-000	On	7 (white)	CONTINUOUS
MIS-010	On	1 (red)	CONTINUOUS
JST-050	On	2 (yellow)	CONTINUOUS
PLU-150	On	5 (blue)	CONTINUOUS

AE-SCR9 Setup Table

AutoLISP Drawing Utilities

The following AutoLISP routine, called RXHAIR, lets you rotate the cross-hairs to an angle so that you can create lines to that angle. It supports the AE-MENU item called [RXHAIR]. You will find it helpful later on in the drawing sequences.

RXHAIR AutoLISP Utilities

 Check to see that you have a copy of the RXHAIR.LSP file in your AE-ACAD directory.

Read along, or create the AutoLISP file.

This utility first prompts you for an angle, then starts the LINE command. When you have completed the lines you want, the routine will return the angle setting to your last setting.

Testing the RXHAIR.LSP Routine

Now that RXHAIR.LSP is installed, load AutoCAD and test the routine. If you are using the AE-MENU, test the [RXHAIR] menu item.

RXHAIR.LSP Operation

```
Command: (load "RXHAIR")
C:RXHAIR
Command: RXHAIR
All angles are measured 0-Horizontal-To-Right      Message.
Enter rotation angle:                              Enter angle of line.
```

This routine will automatically return to the previous rotation angle after each use.

You will find the RXHAIR routine helpful when you are adding plumbing lines that tie at angles into others. If you want to examine the routine, here is the listing for RXHAIR:

```
(prompt "\nLoading RXHAIR utility...")
(defun C:RXHAIR ( / angin)
   (setvar "cmdecho" 0)
                                                    ;don't echo command
   (prompt "\nAll angles are measured 0-Horizontal-To-Right...")
   (setq angin (getint "\nEnter rotation angle:  "))
   (command "SNAP" "R" "" angin)                    ;set to user value

   (setvar "cmdecho" 1)
)                                                   ;defun
```

AutoLISP Routine for [RXHAIR]

Roof Framing Plan Project Sequence

Begin this chapter's project sequence for the roof framing plan by modifying the BASEPLAN drawing you created earlier to match the requirements for the structural plan. Once you have set this base, add lines to represent the joists, bridging, and beams. Next, add the framing required for the mechanical openings. To complete the drawing, you add the framing sizes and elevations and any special notes required.

Modifying the BASEPLAN

The first step is to insert the BASEPLAN drawing created in Chapter Six and modify it to serve as the base of the roof framing plan.

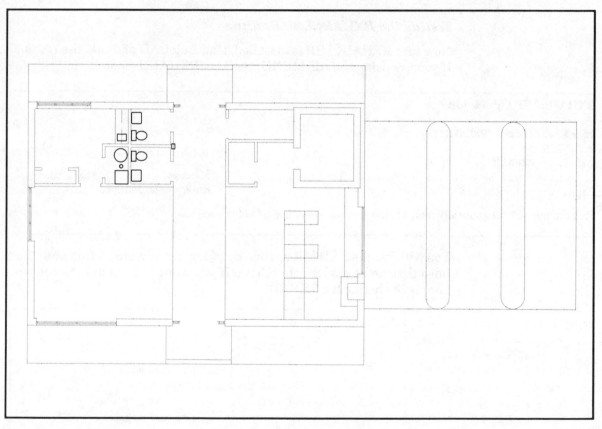

BASEPLAN Illustration

① INSERTING AND MODIFYING BASEPLAN. Use the following technique to insert and modify the BASEPLAN drawing.

Technique for BASEPLAN Framing Modifications

The first modifications to make to the BASEPLAN drawing will enhance the readability of the framing plan. Change all of the wall and roof lines to a DOT linetype. This will create a *screen* effect for the BASEPLAN. The structural framing members that you are going to add will stand out more.

➡ *CAUTION! While this technique is effective in making drawings easier to read, it will cause your drawing file size to increase. If you are using a pen plotter, you can also expect your plotting time to increase.*

We will leave the plumbing fixture lines as CONTINUOUS lines. You will need these fixtures later when you create the plumbing drawing. Also, we want to WBLOCK this drawing so that you can use it for the other engineering drawings.

After you have made these initial modifications, we will ask you to erase all the interior wall lines, doors, and windows. You can then define the true wall thickness at the joist bearing, and add the columns at the drive through area.

Setting Up BASEPLAN Drawing

Enter selection: 2   Edit an existing drawing named 8910ST.

Select the [VIEW–RESTORE–SCRATCH] command from AE-MENU.

Enter selection: 1   Begin a NEW drawing named AE-SCR9.

Verify the settings shown in this chapter's AE-SCRATCH Setup Table.

To get started, set layer MIS-010 as current. Use the following sequence.

➡ *NOTE! This sequence uses our [LINEMEND] command (LINEMEND.LSP) to clean up the lines at the end.*

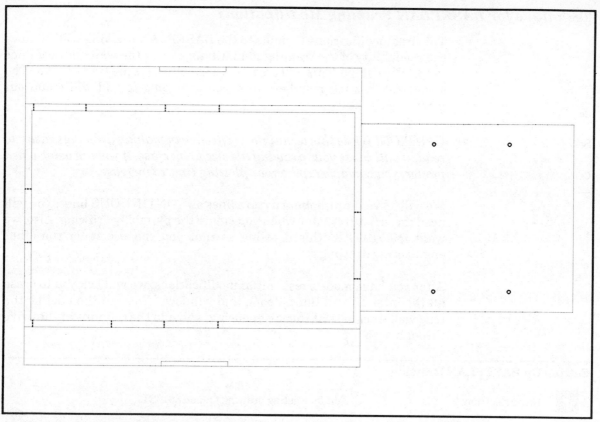

Applying the BASEPLAN Modification Technique

BASEPLAN Framing Modifications

```
Command: INSERT
Block name (or ?): *BASEPLAN
Insertion point: 0,0
```
Use the default for the scale factors and rotation.

```
Command: LAYER
?/Make/Set/New/ON/OFF/Color/Ltype/Freeze Thaw: OFF
Layer name(s) to turn Off: PLU-150
?/Make/Set/New/ON/OFF/Color/Ltype/Freeze Thaw: <RETURN>
```

```
Command: CHPROP                 Change linetype to DOT.
```

```
Command: LTSCALE
```

Change the scale factor to 10 for screened appearance of BASEPLAN drawing.

Command: **WBLOCK**

Use to create BASEPLAN for later use in other engineering drawings.

Command: **OOPS**   Use to return drawing to display.

Command: **CIRCLE**   Repeat for columns at each end of each island, 6" diameter.

Command: **ERASE**

Select interior walls, doors, windows, islands and plumbing fixtures to prepare for framing.

Command: **[LINEMEND]**   Correct all exterior wall lines.

You inserted the BASEPLAN drawing, enhanced its appearance, and created a new BASEPLAN drawing that can be used for the other engineering drawings later in this chapter. You also have simplified the base drawing by removing all the lines that weren't required for the roof framing plan.

➡ *NOTE! If your drawing contains different linetypes, the LTSCALE setting will effect these lines also.*

Benefits

This insertion technique saves everyone time. The architect can prepare BASEPLAN drawings (like you did from a floor plan *working drawing*) with little trouble. The engineer saves time because he or she doesn't have to start from scratch — the background data has already been taken care of. By adding the *screened* appearance to the base, you make sure that the engineer's data won't be diminished by the architect's data.

Defining Joists and Bridging

The next task is to add lines to represent the joists, bridging, and mansard framing members. As you may remember from Chapter Eight, the mansard framing system attaches to the joists at each end. Because of this attachment, you will be unable to distinguish between these two framing systems. This is where structural details help to clarify situations.

Your first drawing step will be to use the DIVIDE command to indicate nodes which will you can use with OSNAP to locate the framing member lines.

❑ LAYER. Set layer MIS-010 current — AE-MENU Tablet [LAYER–BY SCREEN–SET].

■ SET NODE INDICATOR. Use the SETVAR command to set the PDMODE to 3. This will change the point indicator to an "X".

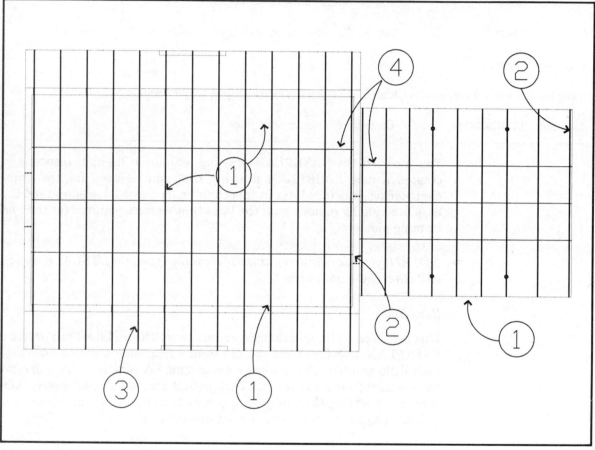

Joist and Mansard Layout Illustration

① SETTING JOIST REFERENCE POINTS. To define the locations of the joists, use the DIVIDE, specifying 12 spaces, and select the top inside exterior wall line and top mansard edge line of the drive through area.

② SETTING BRIDGING REFERENCE POINTS. To define the bridging locations, reuse DIVIDE, specifying four spaces, and select the left inside exterior wall line. Use DIVIDE once more and select the right mansard line of the drive through area.

❑ LAYER. Set layer JST-070 current — AE-MENU Tablet [LAYER–BY SCREEN–SET].

❑ SET OSNAP. Use OSNAP and set a running NODE mode.

③ DEFINING JOISTS. You can use LINE to create the joists by selecting the NODEs and extending the lines to the roof overhang lines. Use the AE-MENU [DULXEXTD] command to complete lines.

④ DEFINING BRIDGING. Use LINE to create the bridging by selecting the nodes and extending the lines to the opposite inside wall line and the exterior wall line of the building for the drive through area.

Adding Beams

Typically, you use steel wide flange beams as lintels over window and store front wall openings, and employ special framing considerations to support the framing members above. Place lines to represent the lintels over the wall openings and to show support beams for the mansard at the corners of the building and over the drive through area from the building to the outside edge.

❑ LAYER. Set layer STL-070 current — AE-MENU Tablet [LAYER–BY SCREEN–SET].

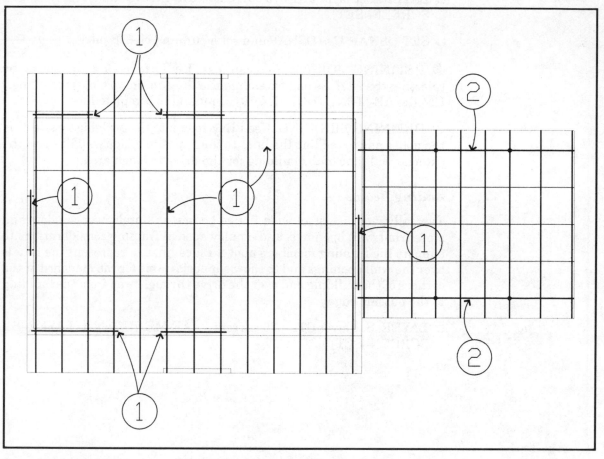

Beam Layout Illustration

① DEFINING LINTELS. Use LINE to place the required lintels spanning eight inches beyond both sides of all our wall openings.

② DEFINE BEAMS. Use the LINE again to add the beams at the corners of the building and over the pipe columns of the drive through.

Defining Framing Members at Roof Openings

When rooftop mechanical units are used, the openings have to be defined to allow the duct system entry into the interior of the building and to provide access to the roof for maintenance. In the following sequence, we will show you how to define these openings.

❑ LAYER. Set layer FRA–070 current — AE–MENU Tablet [LAYER–BY SCREEN–SET].

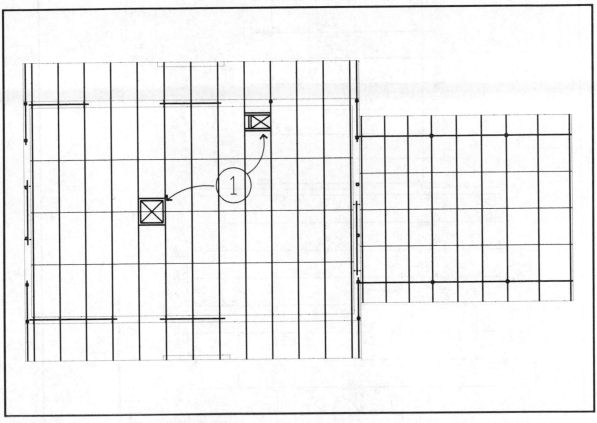

Framing for Roof Openings Illustration

① DEFINING OPENING FRAMING. Use LINE to place the openings that the design requires. The mechanical opening spans between joists and is five feet long. The roof hatch opening is four feet wide by three feet long.

Adding Text and Titles

Now that you have drawn in all the framing members, add the text and notes that describe them. Then, add the drawing title.

❑ LAYER. Set layer TXT-010 current — AE-MENU Tablet [LAYER–BY SCREEN–SET].

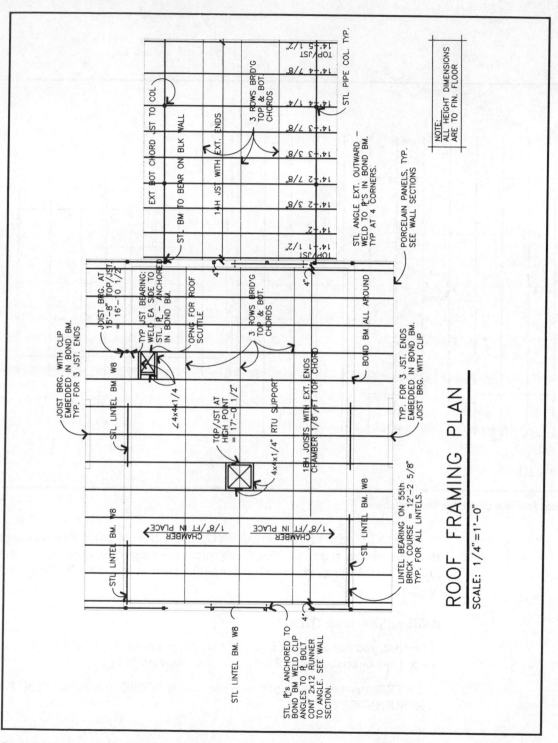

ROOF FRAMING PLAN

SCALE: 1/4" = 1'-0"

Completed Roof Framing Plan Illustration

❏ ADD TEXT AND NOTES. Use DTEXT to add all text and notes as illustrated above, but first use STYLE to set the current style to ROMANS with a six inch height.

❏ LAYER. Set layer TLE-010 current — AE-MENU Tablet [LAYER–BY SCREEN–SET].

❏ ADD TITLE. To add the title, use the STYLE option of the TEXT command to set ROMAND. Use a height of 18 inches.

❏ END DRAWING. You are now finished with the roof framing plan. Save the drawing by using END.

Let's move on to the mechanical plan.

Mechanical Plan Project Sequence

To create the mechanical equipment and duct system layout, you can use the enhanced BASEPLAN drawing (that you WBLOCKed earlier) as a foundation. First, begin a new drawing and insert the BASEPLAN. Unlike the roof framing plan, you won't need to change this base drawing.

After the BASEPLAN is in place, create two different grille symbols and insert them into the drawing. Then, place the equipment, add the ducts, and create a grille and exhaust fan for the restrooms. You complete the drawing by adding the text and titles.

Adding Supply and Return Grilles

Start by creating the mechanical grilles required to supply the building with conditioned air.

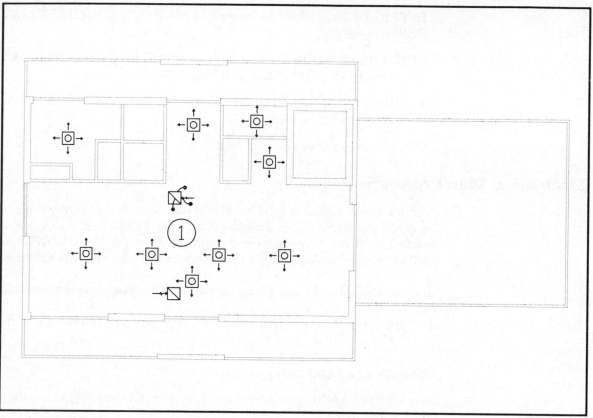

Supply and Return Grilles Illustration

① ADDING THE SUPPLY AND RETURN AIR GRILLES. Use the following technique to create and add the grilles.

Technique for Supply and Return Grilles

There are two types of grilles for the supply and return mechanical system. Create these grilles at full scale with the supply being 2'x 2', and the return 1'x 1' in size. Use INSERT to place the grilles in their proper locations.

Setting Up Supply and Return Drawing

 Enter selection: 2   Edit an existing drawing named 8910ME.

Select the [VIEW–RESTORE–SCRATCH] command from AE-MENU.

Enter selection: 1   Begin a NEW drawing named AE-SCR9.

Verify the settings shown in this chapter's AE-SCRATCH Setup Table.

To get started, set layer 0 as current. Begin by creating the Supply Grille symbol.

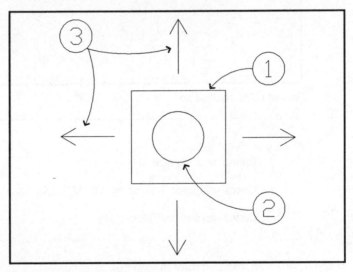

Supply Grille Illustration

Supply Grille

Command: **LINE**   Create a 2' square ①.

Command: **CIRCLE**   Place a 18" diameter circle in center of square ②.

Command: **LINE**   Create and repeat for each arrow and leader ③.

Command: **BLOCK**

Convert to block named SUPPLY with insertion point in center.

Now that you've created the Supply Grille symbol, use a similar method to make the Return Grille symbol. Draw a 1'x 1' square and add a line passing through it diagonally. Then, draw an arrow to indicate air direction. The sequence below shows how to create the Return Grille.

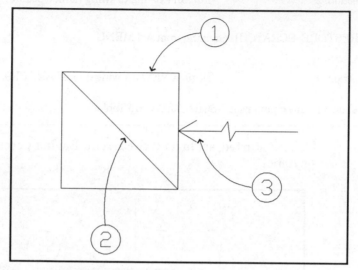

Return Grille Illustration

Return Grille

Command: **LINE**   Create a 1' square ①.

Command: **LINE**   Create diagonal line using OSNAP INT ②.

Command: **LINE**   Create arrow and leader ③.

Command: **BLOCK**

Convert to block named RETURN with insertion point in center.

Now that you've finished the blocks for the supplies and returns, use INSERT to put them in the appropriate locations within the plan. (See the illustration at the beginning of this technique).

Inserting Supplies and Returns

Command: **INSERT**

Place all supplies. Default scale and rotation angle.

Command: **INSERT**

Place all returns. Default scale and rotation angle

By taking the time to create symbols for the supply and return grilles, you are able to insert each type multiple times instead of recreating them over and over again.

Benefits

If you are going to have multiple occurrences of an item in a drawing, try to take the time to create a block and use INSERT to place them. Don't redraw each one. If you take the time, it won't be long before you have an extensive symbol library that will help you produce projects in the future.

Defining Equipment and Ducts

Mechanical engineers make their catalog selections of roof top equipment for buildings based upon their own calculations. We won't give you the details for our selection. If you want, you can do your own calculations, or simply approximate our illustration. After you have the equipment in place, add the ducts to tie the grilles together.

❑ LAYER. Set layer EQU-150 current — AE-MENU Tablet [LAYER–BY SCREEN–SET].

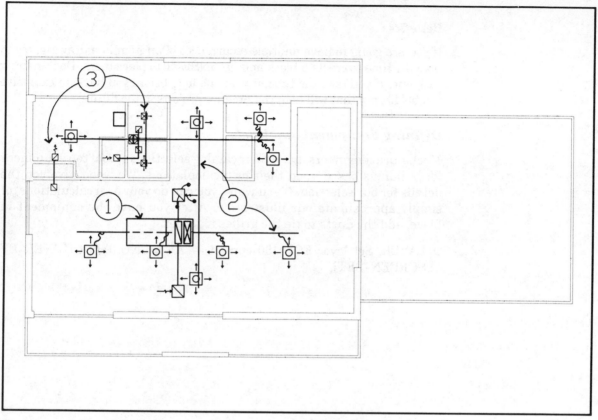

Equipment and Ducts Illustration

① DRAWING IN EQUIPMENT. Use LINE to add the equipment, making use of the roof framing plan mechanical opening to define the location.

② ADDING DUCTS. Now, use the LINE and PLINE commands to add all the supply and return ducts.

③ ADDING SPECIAL GRILLES AND EXHAUST FANS. To add the grilles and exhaust fan in the restrooms, simply use LINE and CIRCLE.

Adding Text and Titles

Add the text and notes to label the members you've just defined. Finally, add the drawing title.

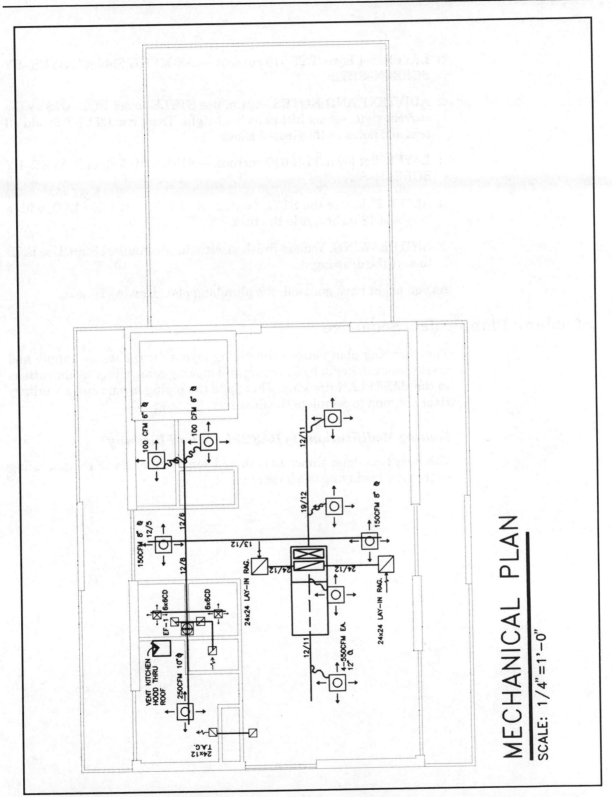

Completed Mechanical Plan Illustration

❏ LAYER. Set layer TXT-010 current — AE-MENU Tablet [LAYER–BY SCREEN–SET].

❏ ADD TEXT AND NOTES. Again, use STYLE to set ROMANS as the current style, set six inches as the height. Then, use DTEXT to add all text and notes as illustrated above.

❏ LAYER. Set layer TLE-010 current — AE-MENU Tablet [LAYER–BY SCREEN–SET].

❏ ADD TITLE. Use the STYLE option of TEXT to set ROMAND, with a height of 18 inches. Add the title.

❏ END DRAWING. You are finished with the Mechanical Plan. Use END to save the drawing.

As you might have guessed, the plumbing plan drawing is next.

Plumbing Plan Project Sequence

The plumbing plan defines the piping layout for the water supply and waste removal. Begin by inserting and making some minor modifications to the BASEPLAN drawing. Then, add the piping layout and a sanitary riser diagram to complete the drawing.

Making Modifications to BASEPLAN for Plumbing

The next technique shows how to add some emphasis to the plumbing fixtures by modifying the base plan.

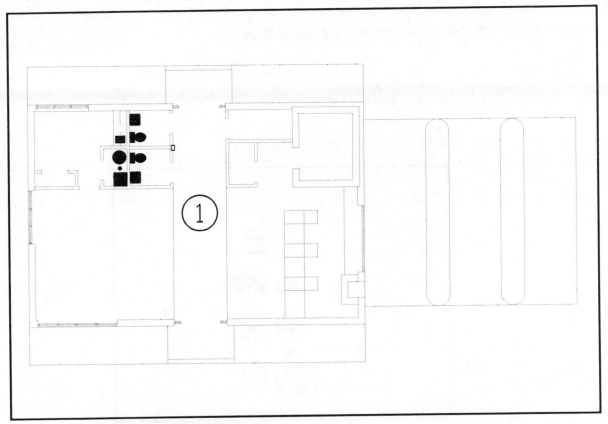

Modifying BASEPLAN for Plumbing Illustration

① MODIFYING THE BASEPLAN FOR THE PLUMBING LAYOUT.
Try the following technique to modify the BASEPLAN drawing.

Technique for BASEPLAN Plumbing Modifications

In this technique, we show you how to use HATCH to emphasize the plumbing fixtures. Hatching the fixtures adds to the readability of the drawing. Start by inserting the BASEPLAN. Next, ZOOM in on the fixtures of the plan and erase all the internal lines. Finally, add a hatch pattern to the fixtures.

Setting Up BASEPLAN Plumbing Drawing

    Enter selection: 2   Edit an existing drawing named 8910PL.

Select the [VIEW–RESTORE–SCRATCH] command from AE-MENU.

    Enter selection: 1   Begin a NEW drawing named AE-SCR9.

Verify the settings shown in this chapter's AE-SCRATCH Setup Table.

Set layer PLU-150 as current.

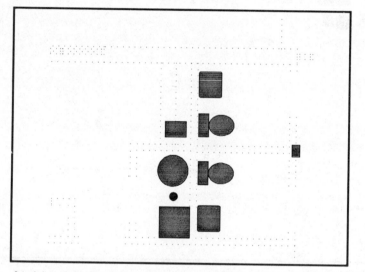

Applying BASEPLAN Plumbing Modifications Technique

BASEPLAN Plumbing Modifications

Command: **INSERT**   Insert *BASEPLAN drawing.

Command: **ERASE**   Select the internal fixture lines.

Command: **HATCH**

Use the LINE pattern, scale 3, and repeat for all fixtures.

The closely-placed LINE hatch pattern enhances the readability of the plumbing drawing.

Benefits

Even though this technique is simple, it can be helpful for the engineer. It means that the data he adds to the architect's BASEPLAN will standout, and won't confuse or mislead the contractor when he bids on, or constructs the project.

➡ *NOTE! As you know, making drawings clear and readable is just as important as the data you add to them. If contractors have a hard time understanding the drawings, the bids tend to come in high, and there are often a lot of change orders.*

Drawing Piping Layout

Since the plumbing fixtures are already in place, the first additions you will make are the piping and risers for the vent system. The first step is to add three-inch diameter circles for the vent and waste piping at each fixture. Then it's only a matter of connecting these circles to create the piping. Finally, you'll add lines to define the supply piping and add the text to the drawing.

❑ LAYER. Set layer VNT-150 current — AE-MENU Tablet [LAYER–BY SCREEN–SET].

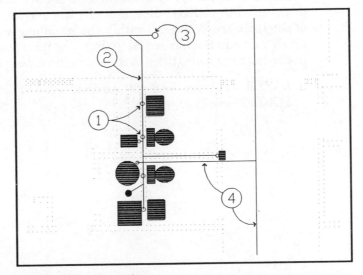

Piping Layout Illustration

① DEFINING VENTS. To define the vents, simply place a three-inch diameter circle at each fixture by using CIRCLE.

❑ LAYER. Set layer WAS-150 current — AE-MENU Tablet [LAYER–BY SCREEN–SET].

② DEFINING WASTE LINES. You can use the LINE command, or our AutoLISP routine [RXHAIR] to draw lines to indicate the piping for the waste lines.

③ DEFINING CLEAN OUT. Now use CIRCLE and place a six-inch diameter circle at the waste line to the street.

❑ LAYER. Set layer SUP-150 current — AE-MENU Tablet [LAYER–BY SCREEN–SET].

④ DEFINING SUPPLY LINES. Use LINE again to draw a line representing the supply to the hot water tank and to the hose bibs.

Creating the Sanitary Riser

Sanitary riser diagrams are visual representations of the piping for the sanitary lines. They are mainly for design purposes and are not true-to-scale working drawings as the other drawings in the set are.

To create the sanitary riser diagram for the project, use the SNAP function with the STYLE option and specify ISOMETRIC. Start this drawing on the RIGHT side, defining the vertical and horizontal piping of the main trunk located within the plumbing wall. Then, move to the LEFT and add the horizontal piping. The final steps are adding the text to the riser diagram, titles to it, and to the piping layout.

❑ LAYER. Set layer WAS-150 current — AE-MENU Tablet [LAYER–BY SCREEN–SET].

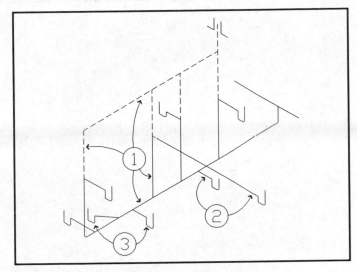

Sanitary Riser Illustration

❑ DEFINE DRAWING MODE. Begin by using the STYLE option of the SNAP command. Toggle to the RIGHT.

① DEFINING VERTICAL RISERS. Use LINE to draw vertical and horizontal lines representing the main trunk on the waste piping.

❑ REPOSITION ISOPLANE. Use the toggle Control E <^E> to flip to the LEFT.

② DEFINING BRANCH PIPING. Now use LINE to define the piping and traps for each fixture.

③ FILLETING THE INTERSECTIONS. You need to use FILLET with a radius of two inches to add curves to all pipe intersections.

Adding Remaining Text and Titles

The last task is to add the titles to the two separate drawings of the plumbing plan. Change to the correct layer and add the titles as defined in the illustration below.

❑ LAYER. Set layer TXT-010 current — AE-MENU Tablet [LAYER–BY SCREEN–SET].

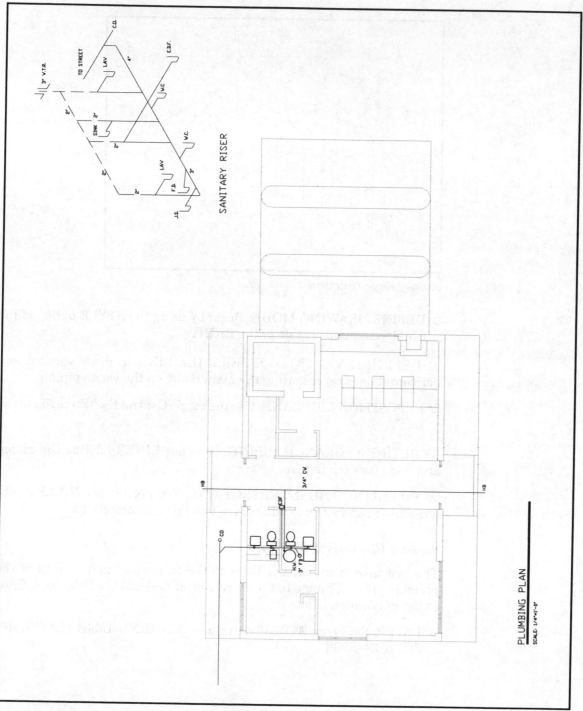

SANITARY RISER

PLUMBING PLAN
SCALE: 1/4"=1'-0"

Completed Plumbing Plan Illustration

□ TEXT STYLE. Use STYLE to specify ROMANS as the text style; specify six inches as the text height.

□ ADD REMAINING TEXT AND NOTES. Use DTEXT, once again with six-inch ROMANS style, to add the remaining text and notes shown above.

□ LAYER. Set layer TLE-010 current — AE-MENU Tablet [LAYER–BY SCREEN–SET].

□ ADD TITLES. Use the STYLE option of TEXT to set ROMAND, with a text height of 18 inches. Add the titles to the drawing.

□ END DRAWING. You are now finished with the plumbing plan. Save the drawing with END.

It's time to move on to the last engineering drawing, the electrical lighting plan.

Electrical Lighting Plan Project Sequence

For the electrical lighting plan, we will use the architectural reflected ceiling plan drawing, 8910CP, as a base. This existing ceiling plan drawing contains all of the fluorescent fixtures required for the interior spaces and the drive through area. You need to add incandescent flood lights to the roof overhang, and a letter indicator for each fixture type. Then, add three- point arcs to represent the wiring for these fixtures.

➡ *NOTE! If you are completing the entire drawing, use 8910CP as the prototype. Load AutoCAD and select Option 1, Begin a New drawing. Enter the drawing name as 8910EE=8910CP. This will create the working copy of 8910CP and name it 8910EE.*

Modifying Reflected Ceiling Plan 8910CP

If the base ceiling plan is in place, you are ready to modify it for the electrical plan. Start by changing the linetype to dotted with CHPROP. After you redefine the linetype, erase the unrelated text, and set the electrical layer as current. We will step you through the setup using the AE-MENU.

TURN LAYER OFF. Use the [LAYER–BY ENTITY–OFF] command and select the light fixtures.

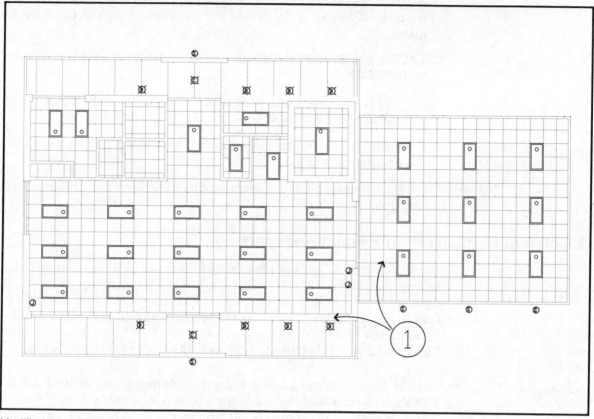

Modified 8910CP Illustration

① CHANGING PLAN AND GRID LINETYPE. Use the CHPROP command, select the entire plan with a Window, and change the LINETYPE to DOT.

❑ TURN LAYER ON. Turn layer LIG-160 ON — AE-MENU Tablet [LAYER–BY SCREEN–ON].

❑ SET LAYER. Set layer ELE-160 current — AE-MENU Tablet [LAYER–BY SCREEN–SET].

❑ MAKE FINAL MODIFICATION. Use ERASE to remove all text that relates to the ceiling plan.

Now that the base is complete, you can make the fixture modifications and additions.

Defining and Adding Fixtures

The next task is to add circles representing the junction boxes, add the roof overhang fixtures, and the letters that indicate the types of fixture you want to use.

❑ LAYER. Set layer LIG-160 current — AE-MENU Tablet [LAYER–BY SCREEN–SET].

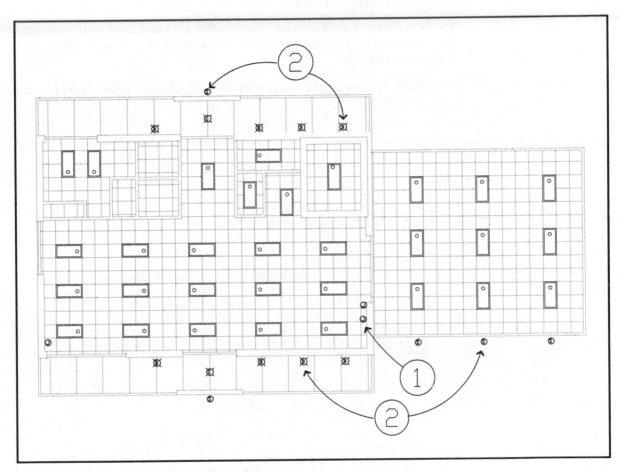

Fixture Modifications and Additions Illustration

① CREATING JUNCTION BOX AND INDICATOR. Use CIRCLE and specify a radius of three inches for the junction box. Using TEXT, place a six-inch high letter within the symbol. Now, use COPY with the MULTI option, select the circle and the letter (be sure to turn ORTHO off) and copy the boxes to the other light fixtures as required. Use CHPROP if the letter needs to be revised.

② ADDING ROOF OVERHANG FIXTURES. Use CIRCLE and TEXT, with a height of six inches, to create a flood light fixture and letter indicator. Just use COPY, as you did before, to place the fixtures.

Now that you have defined the fixtures and have the junction boxes in place, the last drawing task is to add the wiring.

Creating Wiring Layout

The last drawing task is to add three-point arcs and slashes to represent the wire runs and the number of wires required in each run. Also, add a switch symbol.

❑ LAYER. Set layer ELE-160 current — AE-MENU Tablet [LAYER–BY SCREEN–SET].

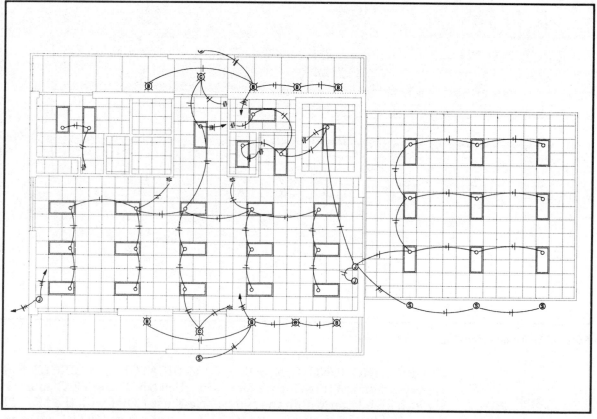

Wiring Layout Illustration

□ CREATING AND PLACING SWITCH SYMBOLS. This step is straightforward. Start by simply using TEXT, with a height of six inches, type in an "S". Then use LINE to draw a vertical line through the center of the letter to finish the symbol. Using the BLOCK command, insert the SWITCH into the drawing where required.

□ DEFINING WIRE RUNS. For this step, use ARC with the three-point option to create arcs between all fixtures as shown in the Wiring Layout drawing above.

□ DEFINING WIRE NUMBER INDICATOR. Use LINE to create a slash, and copy it as required to each wiring run.

Finally, add the text.

Adding Text and Titles

The last task is to add the text and titles to the plan.

□ LAYER. Set layer TXT-010 current — AE-MENU Tablet [LAYER–BY SCREEN–SET].

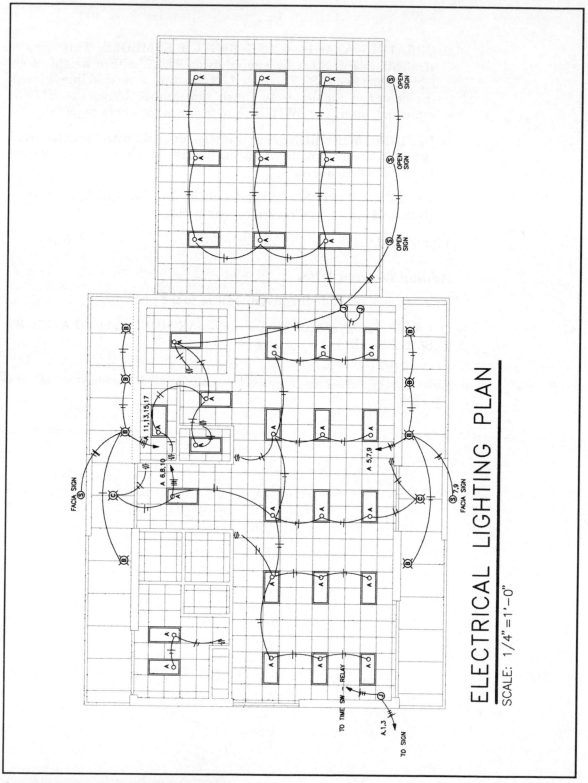

ELECTRICAL LIGHTING PLAN

SCALE: 1/4" = 1'–0"

Complete Electrical Lighting Plan Illustration

❑ ADD TEXT AND NOTES. Use DTEXT to add all text and notes as illustrated above, again setting the style to ROMANS and the height to six inches.

❑ LAYER. Set layer TLE-010 current — AE-MENU Tablet [LAYER–BY SCREEN–SET].

❑ ADD TITLE. Use the STYLE option of TEXT to set ROMAND as the current text font. Add the title.

❑ END DRAWING. That's it for the electrical plan. Be sure to save the drawing with an END.

Summary

In this chapter you made your drawing tasks immensely simpler by using a block that was created from the floor plan as a base drawing. With a few modifications and enhancements for the individual engineering plan types, you had all the background data available for the new plans required.

You are finished creating the major architectural and engineering drawings for the simulated project. We hope that by following us through this process you have learned more about AutoCAD and how to manage it. We also hope that you've experienced firsthand how making a few enhancements and following some procedural techniques can increase your productivity and make your production life a lot easier.

Just because you've finished the drawings on the screen doesn't mean that you are finished using AutoCAD. Next, let's take a look at how to manage plotting drawings. Then you'll see how to use the drawing data you've created to help write the specifications for the project.

DRAWING PLOT LOG

DRAWING NUMBER	DRAWING TITLE	DATE	SHEET SIZE	QUANTITY PLOTTED	PLOTTED BY
8910SP	SITE PLAN	4/4/89	24x36	1	JMA
8910FP	FLOOR PLAN	4/4/89	24x36	1	JMA
8910CP	CEILING PLAN	4/4/89	24x36	1	JMA
8910EL	EXTERIOR ELEV.	4/4/89	24x36	1	SW
8910BS	BUILDING SECT.	4/4/89	24x36	1	DL
8910ST	STRUCTURAL PLAN	4/4/89	24x36	1	SW
8910ME	MECHANICAL PLAN	4/4/89	24x36	1	JMA
8910PL	PLUMBING PLAN	4/4/89	24x36	1	JMA
8910EE	ELECTRICAL PLAN	4/4/89	24x36	1	JMA

Plot Log Illustration

CHAPTER 10

Project Completion and Revisions

Introduction

In the previous chapters you created standards, symbols, and drawings for the branch bank project. At this point, in the production of most projects, you would generate hardcopies of the drawings for bidding, revisions, and construction. Actually, the hardcopy drawings have already been shown for this project. They are contained in the previous development chapters. In this chapter, we will discuss some plotting features of these branch bank drawings, but we also want to take a broader view of plotting techniques and devices to help you plot your own projects.

Using the standards for line weights and layer colors that were developed earlier (see the section on prototype drawings in Chapter One) and the drawings developed for the project, we will show you how to customize your plotting procedures to give you consistency in plotting output. Line weights will be consistent from one drawing to the next and text heights will be the same size on your plotted sheets, even though they may be at various scales. If you combine the methods of system setup and the drawing standards we have presented so far with the methods of project completion that we will present in this chapter, you will see improvements in your final output.

Let's do a quick overview of what plotting devices are available and discuss some problems you might have run into had this been a real *live production* of a full working drawing set for the simulated project.

Plotter History

Plotters have come along way in a short period of time. The first plotters that we used in our practice were giant six to seven hundred pound units on dual-roller systems. They were not only expensive to buy, but very awkward to use. Operating methods for these *high-tech* monsters were similar to the way we run our pen plotters today, but the machines themselves were more difficult to maintain.

Back then, the market for plotters was small and the competition for sales was scarce. Today, we seem to be having a *gold rush* in plotters for the CAD market. Companies are busy creating and trying to sell us every kind of device imaginable. We have found that the most important features to investigate before purchasing a plotter today are the quality and availability of service, the quality of the plots produced, and the compatibility of the device with our current (and future) systems. We try not to let cost be the major deciding factor. You don't have to buy the most expensive device on the market, but in general, you get what you pay for.

Plotter Technologies

There are currently two types of imaging technologies used to produce drawings in the A/E environment. The first is vector, which is used in pen plotters. The second is raster imaging, used in dot-matrix, laser, and electrostatic plotters.

Vector Technology

Vector technology is a method of direct communication from the host computer to the plotter. The computer tells the plotter to draw a line from A to B to C, and while the plotter is processing the information and drawing the image, the next group of commands is forwarded. With the first couple of versions of AutoCAD, you could actually see how the operator created the drawing because the plotter would reproduce the drawing in the same sequence that the operator used. The plotter would select a pen from the carousel and draw a few lines; it would select another pen and draw a few lines; then the plotter would return to the first pen to draw some more lines. This was not very efficient. AutoCAD has solved this problem — the software now directs the plotter to draw everything required by a single pen before it selects another.

Pen plotters outnumber all the other printers or plotter types used today because of their low cost and ability to handle media up to E size sheets.

Raster Technology

While basic raster technology is not new — it has been used for quite some time in dot-matrix printers. Raster plotting is the newest technology in plotters. The difference between vector and raster plotters is simple. Both receive vector file transmissions in the form of AutoCAD output. But, unlike the vector plotter, the raster plotter waits to start drawing until it receives all of the vector file. It then *rasterizes* the file (converting it to a dot image), and begins *painting* a portion of the drawing contained in a specific width of the paper. It then repeats the process in a back-and-forth manner until the image covers the medium.

You might think that a vector plotter would be faster, since it starts drawing almost immediately after it begins receiving the vector file. However, all other things being equal, a raster plotter actually takes about one twentieth of the time that a vector plotter takes to complete the same drawing.

Without going into each raster device type in detail, we'll touch on the major features of each.

- Dot-matrix Printers — A dot-matrix printer is the basic word processing printer that you find in most offices. Some of them have graphic capabilities that allow you to create drawing images along with typed text. The most widely available resolutions on these printers range from a low end of 70 dpi (dots per inch) to a top end of around 300 dpi. The higher the dpi, the better the resolution of the plotted image. A drawing produced on a 70 dpi printer comes out with a rough and jagged appearance. However, one produced on a 300 dpi printer is close enough to pen plotting that you can't tell the difference. Most dot-matrix printers are modestly priced and easy to maintain. Like typewriters, you just replace the ribbon when it runs out of ink.

- Laser Printers— Laser printers are raster printers which generate drawings in a manner similar to the way office copiers work. The raster image is produced electronically on a drum. Toner adheres to the positive electrons of the image and is transferred onto the medium. Laser printers cost more than dot-matrix printers and are more expensive to maintain (we went through several, rather costly, toner packages in producing this book), but are faster than dot-matrix printers. The resolution of these printers is commonly 300 dpi, while the manufacturers of some newer models are claiming resolutions of 600 dpi.

  One drawback of both dot-matrix and laser printers is that they have only one line thickness. To overcome this problem, you can use thick polylines to achieve line weight. But you pay a price, plines add to the size of a drawing file. Most pen plotters can use up to eight pens, of various widths and colors, to give the drawing a more professional appearance.

- Electrostatic Plotters — The last raster device we will talk about is the electrostatic plotter (EP). The EP is a cross between its sisters, the dot-matrix and laser printers. The *head* of an EP contains anywhere from 1,000 to 16,000 printing elements or points, depending on the resolution (and width). These points apply an electrostatic charge to the paper for the image of the drawing. Next the paper passes over a pool of toner, where particles adhere only to the charged areas.

The cost of an EP system ranges from $20,000 to $60,000 depending on whether you want 200 dpi, or 400 dpi, in either monochrome, or color. The yearly cost for special EP media, toners, and maintenance can run as high as $15,000 when the device is used in a moderately sized A/E office. While EP plotters are definitely the most expensive of the devices we've described, their productivity to cost ratios is far better than those of pen plotters.

Plotter Bottlenecks

Production bottlenecks are the main causes of costly project overruns. As an example of how bottlenecks occur when you plot drawings, let's create the following scenario (which occurs all too often in most A/E firms):

It's two days before the final working drawings for your project are to be released for bids, and your CAD department is finishing up some final changes. Your production manager is from the old traditional drafting school and thinks his team can pull together all of the project's drawings before he goes to lunch; and they'll have a *check set* waiting for him when he gets back.

Unless you've got an armada of plotting devices and a large crew to operate them, your production manager is going to have a major case of stress-induced indigestion when he returns. Not only is the check set incomplete, but the prospects of finishing the job in the next two days are dismal.

The production manager's expectations are unrealistic, of course, but not uncommon — the overall production benefits of computers and their peripherals are frequently overestimated.

To a large degree, you can avoid the kind of plotter bottleneck that is bound to be precipitated by this scenario. This is where good production methods can make all the difference. Let's look at a few questions you should ask yourselves, as players in this scenario, to help understand the problem and formulate some solutions.

- How long does it take to plot a single drawing on your current system?

- How much time did you devote to doing check plots earlier in the project?

- How much does it cost in personnel time to take an operator completely away from his station to devote the next two days to a plotting marathon?

Bottleneck Solutions

First, let's look into the question of how much time it takes to produce a single drawing on a pen plotter. Let's assume in our example that you *are* using a pen plotter, since most A/E firms have at least one. A realistic production time for a pen-plotted drawing, whether a single plot or a series of plots for pin registration, is one hour. Let's also assume that the working drawings you're trying to get out consist of 40 sheets. If it takes an hour to plot one sheet, and you have 80 sheets to plot (40 for the check plots and 40 for the final documents), there is no way you can fit 80 hours of plotting into your two day time frame. And, that doesn't even include the time you'll have to spend *picking up redlines* from the review of the check plots.

Waiting until the end to pull out all the drawings was the first mistake. The second mistake was expecting the staff to *rise to the occasion* and stay all night to work on a project that was impossible to complete anyway. Does this sound familiar?

Solution One — Buy an Electrostatic Plotter

One solution to this problem would be to buy an electrostatic plotter. An EP could plot the identical single drawing in five minutes. Following simple arithmetic, you could plot the same 80-sheet job in less than seven hours and the operator could be off and running on another project producing revenue for the firm.

Although we agree that the cost of operation is high with an EP, if bottlenecks occur very often, the cost is equally as great or greater to operate without one. It all depends on where you want to spend your money — on the loss of productive employee time, on lost revenue you incur because you have a reputation for missing your deadlines; or on a new piece of equipment that will contribute to the firm's production even at nonpeak times.

An interesting observation that we've made in our office since we bought an electrostatic plotter (and one that we think serves as a good indicator of a lowered stress level among the staff members) is that we hear a lot fewer expletives muttered during the plotting process.

Solution Two — Use a Check Plot Management Technique

In spite of all the advantages that electrostatic plotters offer, not everyone is ready to take the plunge and buy one. You can manage drawing production and increase throughput with a pen plotter. It is not unrealistic to assume that the 40-drawing set could have been check plotted three times. That would represent three weeks of work (not including any forced labor over the weekends)! If you are charging the client by the hour, that's one thing, but how do you recapture some of this lost operator/machine time on set fee projects?

With last minute changes, there really isn't a way to get around doing last minute check plots. Our approach is not to avoid doing check plots, but to plot them at half scale. While not every drawing lends itself to being drawn at half scale, the time savings gained is astounding. Not only do half scale drawings take less time to plot, about one quarter of the time that full scale drawings take. Their smaller size makes them easier for checking and red-lining.

If you use this half-scale technique, check plots that used to take an hour to generate would only take 15 minutes. To plot check the first half of the 80 drawings, you would only need 10 hours. This doesn't mean that you'll hit the two-day deadline, since the remaining 40 drawings will still take 40 hours to plot at full-scale, but it does mean that you won't overrun the deadline by nearly as much.

Another way to improve your production time on check plots is to purchase a C size dot-matrix printer. This will also help speed your check plots.

Standards for Plotting

Your AutoCAD system should be a strong selling point for your services to potential clients. One striking feature of a well-managed AutoCAD system is the greater consistency of plot definitions that you can obtain compared to a manual system. One draftsman uses text and symbols differently than another. This causes discrepancies between, and even within, manually drafted projects. Setting up standards within AutoCAD helps solve this problem, and takes very little monitoring once everyone is familiar with the system.

If you have followed along, you have completed (or at least read through) nine drawings for the branch bank project, each with its own layering system related to the type of drawing. Each layer within each drawing has a color that relates to a pen width for plotting. If you have forgotten the layer setup, you may want to review the layering system that we described in Chapter One. Here is a list of the drawings.

```
                      BRANCH BANK DRAWINGS
        8910SP    =       Site Plan                 A—1
        8910FP    =       Floor Plan                A—2
        8910CP    =       Ceiling Plan              A—3
        8910EL    =       Exterior Elevation        A—4
        8910BS    =       Building Section          A—5
        8910ST    =       Structural Plan           S—1
        8910ME    =       Mechanical Plan           M—1
        8910PL    =       Plumbing Plan             P—1
        8910EE    =       Electrical Plan           E—1
```

Project Working Drawing List

Cross-Referencing Drawings

We recommend that you cross-reference the final drawings for any project with the drawing number log in the project work book to see if any omissions have been made, or if any additions are required.

Let's begin our discussion with some standardization techniques that we have available to plot the Branch Bank drawings. We also will show you how to insert title block information.

Title Blocks

Some A/E firms using AutoCAD create their title blocks and borders with conventional methods. They do this because of the time it can take to plot these areas, which often contain solid borders, intricate company logos, information on jobs and project owners, dates, sheet numbers, and revisions. A single title block becomes a sizable drawing file which can require a large amount of time to run on a pen plotter. On the other hand, plotting a title block with a electrostatic plotter requires little additional time and yields more consistent and professional looking results.

We've left adding this information to the branch bank drawings until we were ready to plot to help keep the drawing sizes down.

Creating a Title Block and Completing a Border Drawing

As you may recall, in Chapter Three an area was created for a custom title block as part of the custom border drawing, called AEBORDER. Now this block, called AETITLE, needs to be created to define the sheet title areas within the border.

We'll show you how to fill in the titles and use these areas to complete the drawings by using the [EDITATTR] command from the AE-MENU, answering the prompts with the project titles and sheet information.

You gain many benefits by doing the sheet titles as a separate block, but the most important is that the title information doesn't have to add to the drawing size until you are ready to plot. Some firms have found using a plotted border so desirable that they have even reduced their sheet drawing area to incorporate this procedure. Another benefit is that you can extract the relevant attribute data from these areas to print out for the project work book. This is an excellent way to let AutoCAD help you keep track of this data. For the method to operate at its best, you'll still need a plotter capable of producing the title blocks in a reasonable amount of time.

Technique for Creating the Title Block

Create the AETITLE at one AutoCAD drawing unit so that you'll be able to insert the block into any other drawing at any scale factor. This will keep you from having to create a different title block for every drawing scale. Create this drawing on layer 0, using information in the AETITLE Setup Table below and the default settings that AutoCAD initiates when you begin a new drawing.

➡ *NOTE! The following example is based on a 24" x 36" sheet of tracing paper and will have to be adjusted for other sizes.*

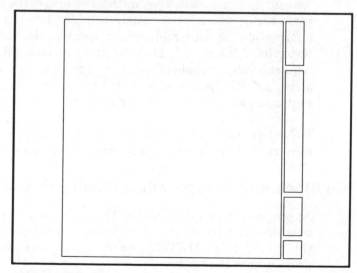

AETITLE Illustration

Drawing Setup for Chapter Ten

Since this chapter has only one drawing exercise, there is no need to create a scratch drawing. Follow the table and commands below if you want to create the AETITLE drawing. If you have the AE DISK, you have a copy of the AETITLE drawing.

COORDS	GRID	SNAP	ORTHO	UCSICON
ON	0	0	ON	WORLD

UNITS	Default all UNIT settings.
LIMITS	0.0000,0.0000 to 9.0000,12.0000
ZOOM	Zoom All.

Layer name	State	Color	Linetype
0	On	7 (white)	CONTINUOUS

AETITLE Setup Table

Begin by defining the drawing area, or main area, of the AETITLE drawing.

Setting Up AETITLE Drawing

 Enter selection: 2   Edit an existing drawing named AETITLE.

Enter selection: 1   Begin a NEW drawing named AETITLE.

Verify the settings shown in the AETITLE Setup Table.

When you are ready, use ZOOM with the Window option, specifying coordinates of 0,0 for the lower left, and 1.5,1.5 for the upper right. Create the Main Area.

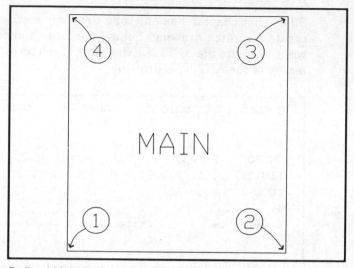

Defined Main Drawing Area Illustration

Main Area

```
Command: LINE
From point: 0.025,0.020    Start point ①.
To point: @0.425<0         ②.
To point: @0.460<90        ③.
To point: @0.425<180       ④.
To point: C
```

Now define the sheet title area, including spaces for the sheet number, firm name, project title, and revisions.

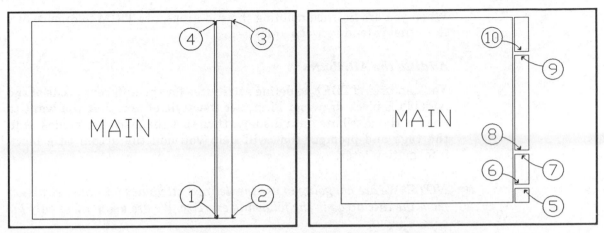

Sheet Title Area Illustration   Sheet Title Sub-Areas Illustration

Use LINE to outline the entire sheet title area. Then use OFFSET and TRIM to define the subareas listed in the table below.

➡ *NOTE!! Since you will create your AETITLE drawing at the correct coordinates for the AEBORDER drawing, use the default insertion point of 0,0.*

Sheet Title Areas

```
Command: LINE
From point: 0.455,0.020    Start point ①.
To point: @0.035<0         ②.
To point: @0.460<90        ③.
To point: @0.035<180       ④.
To point: C
```

Now that you have the sheet title area defined, use OFFSET to divide this area into the sub-areas shown in the Sheet Title Sub-areas illustration above. The offset distances are shown in the following table.

OFFSET DISTANCE	ANNOTATION
.035	5
.01	6
.075	7
.01	8
.235	9
.01	10

Sheet Title Sub-areas Table

When you are finished defining the sub-areas, use TRIM to remove any lines that extend between areas.

Adding the Attributes

You can use ATTDEF to define attributes for the different areas of the AETITLE block drawing. First, set the style of text that you want to use for the attribute data display; then, set the attribute modes. Add the tags and prompts that will help you add the actual title block information later.

➡ *NOTE! We are not going to use the defined attributes for data extraction from the title areas in the following exercise. We are using them only for ease of input.*

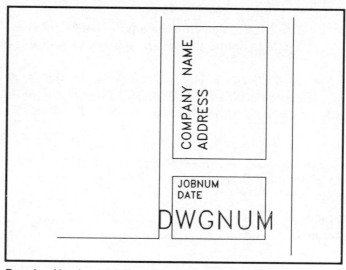

Drawing Number and Company Attributes Illustration

Attributes

```
Command: STYLE              Specify ROMAND for default style.

Command: ATTDEF
Attribute modes  --  Invisible:N  Constant:N  Verify:N  Preset:N
Enter (ICVP) to change, RETURN when done: <RETURN>
Attribute tag: DWGNUM
Attribute prompt: DRAWING NUMBER:
Default attribute value: <RETURN>
Start point or Align/Center/Fit/Middle/Right/Style: M
Middle point:              Select point in middle of area.
Height <0.2000>: .4
```

```
Rotation angle <0>: <RETURN>

Command: ATTDEF
Attribute modes  --  Invisible:N  Constant:N  Verify:N  Preset:N
Enter (ICVP) to change, RETURN when done:
Attribute tag: JOBNUM
Attribute prompt: JOB NUMBER:
Default attribute value: <RETURN>
Start point or Align/Center/Fit/Middle/Right/Style:       Select a point in upper left corner.
Height <0.4000>: .2
Rotation angle <0>: <RETURN>

Command: ATTDEF
Attribute modes  --  Invisible:N  Constant:N  Verify:N  Preset:N
Enter (ICVP) to change, RETURN when done: <RETURN>
Attribute tag: DATE
Attribute prompt: DRAWING COMPLETION DATE:
Default attribute value: <RETURN>
Start point or Align/Center/Fit/Middle/Right/Style:       Select a point in upper left corner.
Height <0.2000>: <RETURN>
Rotation angle <0>: <RETURN>

Command: TEXT
Start point or Align/Center/Fit/Middle/Right/Style:
Height <0.2000>: .3
Rotation angle <0>: 90
Text: COMPANY NAME                                         Enter your company name.

Command: TEXT
Start point or Align/Center/Fit/Middle/Right/Style:
Height <0.3000>: <RETURN>
Rotation angle <90>: <RETURN>
Text: ADDRESS                                              Enter your company address.
```

Now that you have set up the attributes for the drawing number, job number, and date, and have personalized the AETITLE drawing with your company name and address, add the attributes for the project title area.

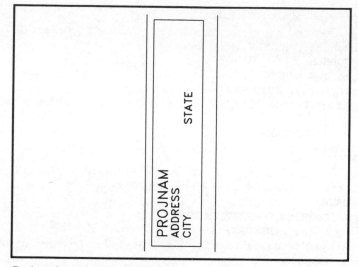

Project Area Attributes Illustration

Project Attributes

```
Command: ATTDEF
Attribute modes -- Invisible:N Constant:N Verify:N Preset:N
Enter (ICVP) to change, RETURN when done: <RETURN>
Attribute tag: PROJNAM
Attribute prompt: PROJECT NAME:
Default attribute value: <RETURN>
Start point or Align/Center/Fit/Middle/Right/Style:      Select a point in the bottom left corner.
Height <0.3000>: .4
Rotation angle Z: <RETURN>

Command: ATTDEF
Attribute modes -- Invisible:N Constant:N Verify:N Preset:N
Enter (ICVP) to change, RETURN when done: <RETURN>
Attribute tag: ADDRESS
Attribute prompt: PROJECT STREET ADDRESS:
Default attribute value: <RETURN>
Start point or Align/Center/Fit/Middle/Right/Style:      Select a point below the last.
Height <0.4000>: .3
Rotation angle <90>: <RETURN>

Command: ATTDEF
Attribute modes -- Invisible:N Constant:N Verify:N Preset:N
Enter (ICVP) to change, RETURN when done:
Attribute tag: CITY
Attribute prompt: PROJECT CITY:
Default attribute value: <RETURN>
Start point or Align/Center/Fit/Middle/Right/Style:      Select a point below the last.
```

```
Height <0.3000>: <RETURN>
Rotation angle <90>: <RETURN>

Command: ATTDEF
Attribute modes -- Invisible:N Constant:N Verify:N Preset:N
Enter (ICVP) to change, RETURN when done: <RETURN>
Attribute tag: STATE
Attribute prompt: PROJECT STATE:
Default attribute value: <RETURN>
Start point or Align/Center/Fit/Middle/Right/Style:    Select a point below the last.
Height <0.3000>: <RETURN>
Rotation angle <90>: <RETURN>
```

With your project area complete, add the attributes for recording the drawing revision numbers and dates.

➡ *TIP! You can preset your attributes and make them invisible so that the insertion of the title block doesn't require that all the prompts be answered.*

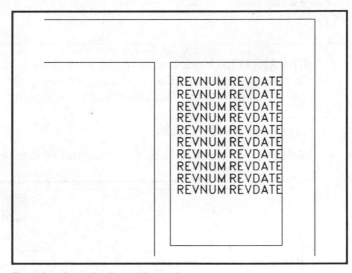

Revision Area Attribute Illustration

Revision Attribute

```
Command: ATTDEF
Attribute modes -- Invisible:N Constant:N Verify:N Preset:N
Enter (ICVP) to change, RETURN when done: <RETURN>
Attribute tag: REVNUM
Attribute prompt: REVISION NUMBER:
Default attribute value: <RETURN>
Start point or Align/Center/Fit/Middle/Right/Style:    Select a point in upper left corner.
Height <0.3000>: .2
```

```
Rotation angle <90>: 0

Command: ATTDEF
Attribute modes -- Invisible:N Constant:N Verify:N Preset:N
Enter (ICVP) to change, RETURN when done: <RETURN>
Attribute tag: REVDATE
Attribute prompt: REVISION DATE:
Default attribute value: <RETURN>
Start point or Align/Center/Fit/Middle/Right/Style:      Select a point adjacent to last.
Height <0.2000>: <RETURN>
Rotation angle <0>: <RETURN>

Command: COPY
Select objects: 1 selected, 1 found.                     Select the revision number attribute.
Select objects: 1 selected, 1 found.                     Select the date attribute.
Select objects: <RETURN>
<Base point or displacement>/Multiple: M
Base point:
Second point of
```

Copy as many attributes as you may need.

```
Command: END
```

The AETITLE block is now ready to be inserted into the final drawings. Before you plot the final working drawing set, you can add specific attribute data to the block by using the following sequence.

Defining Border Titles

Add the title block data by following the sequence below.

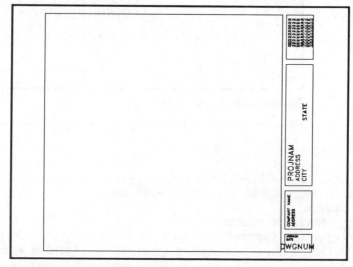

Completed AETITLE Illustration

❑ LAYER. Set layer TLE-010 current — AE-MENU Tablet [LAYER–BY SCREEN–SET].

❑ REDEFINING BLOCK ATTRIBUTE DATA. Use either the [EDITATTR] command from the AE-MENU, or use the DDATTE command of AutoCAD and answer all the prompts for drawing number, project number, date, project name, address, city, and state. Refer to the branch bank drawing list in the first part of this chapter for some of this data.

❑ SAVE. Save the drawing data to disk.

Now that you have completed the border and title areas, let's take a look at how line weights, views, and scale impact plotting.

Line Weights, Views and Scale

Line weights are important to plotting. They enhance the drawing by making it easier to read and understand. Dimension lines should be light weight, exterior wall lines heavy, match lines the heaviest, and so on. These are all standard. But, many times when you look at a final plot this line differentiation is not there. This usually happens because the right weights were not specified when the drawing was set up.

In Chapter One, we proposed the following line weights to go along with the layering system set up for the prototype drawings. We suggest that you use a similar set of standards to make your own drawings more readable and consistent.

LAYER COLOR	COLOR NUMBER	LINE WEIGHT	PEN SIZE
Red	1	x-fine	.18mm
Yellow	2	x-fine	.18mm
Green	3	fine	.25mm
Cyan	4	fine	.25mm
Blue	5	medium	.35mm
Magenta	6	medium	.35mm
White	7	bold	.50mm

Layer Colors and Line Weights

Plotting Views

Views are extremely helpful in drawing large projects since they let you move from one part of the drawing to another using ZOOM, without having to regenerate the drawing all the time. You can also plot your drawings by using *saved views*. You saved a view of the drawing area if you ran the AE-SETUP routine and inserted the AEBORDER drawing in Chapter Four. The view's name is PLOT.

Because you named and saved this view, you could printer plot your drawings by simply using PRPLOT, answering the prompt for what to plot with V (for view), and specifying the view name as PLOT.

Scaling Plots

The scale at which you plot your drawings is based on the limits or size of your sheet. It is better to decide what scale you want to use for each drawing at the beginning of a project rather than to change scales once you've created the drawing(s). If you do change scales, you'll probably also have to change all your text heights, some of the blocks that you inserted based on the original scale of the drawing, and the linetype scale.

The following table shows the overall sheet sizes of media in inches, and how sheet size coordinates with the scale you want for the final drawing.

SCALE	"A" 8.5x11	"B" 11x17	"C" 18x24	"D" 24x36	"E" 30x42
1"=10'	85x110	110x170	180x240	240x360	300x420
1"=20'	170x220	220x340	360x480	480x720	600x840
1"=30'	255x330	330x510	540x720	720x1080	900x1260
1"=40'	340x440	440x680	720x960	960x1440	1200x1680
1"=50'	425x550	558x850	900x1200	1200x1800	1500x2100
1"=1'	8.5x11	11x17	18x24	24x36	30x42
1/2"=1'	17x22	22x34	36x48	48x72	60x84
1/4"=1'	34x44	44x68	72x96	96x144	120x168
1/8"=1'	68x88	88x136	136x192	192x288	240x336

Plotter Sheet Sizes in Inches

Technique for Using Macros and Script Files for Plotting

AutoCAD provides a couple of ways to automate plotting commands. Creating script files is a good way to standardize your plotting procedures. From a plotting standpoint, a script file is merely a series of AutoCAD commands which you use to *batch* process your plotting jobs.

You also can use macros, which you looked at in the menu systems of Chapter Three, to develop some simple routines for controlling plotting.

Macros

Though macros aren't as powerful or flexible as AutoLISP routines, they are an easy way to automate and standardize for plotting. You can select the following macro, [PLOT], from the tablet menu of the AE-MENU. It automates the start up of the PLOT command with your saved view, called PLOT. Macros will perform some of the same functions as script files (you will see an example in a minute), but you still have to manually enter the majority of the answers to the PLOT command prompts. Here is a listing of our [PLOT] macro.

```
[PLOT]^c^cPLOT;V;PLOT;
```

Script File

Using a script file, you can automatically issue commands to change parameters to match the requirements of specific job types. These commands can start the plotter, set the origin and perform many of the functions that you are probably now specify from the screen.

You create script files with a text editor or word processing program. Each separate file has an extension of .SCR. By using script files, you not only create standards, but save yourself a lot of interaction with the keyboard.

The following files are simple script files which will perform the operations required to set up and plot the branch bank drawings, using the standards we set up for the project drawings. With the exception of the site plan, the scale for the other drawings is 1/4"=1'-0".

➡ *NOTE! These script and macro plot routines assume that you have assigned the pen numbers to the correct layer color.*

AEPLOT Script File

Check to see that you have the AEPLOT.SCR file in your AE-ACAD directory.

Read along, or create the script file

Here is a listing of the script file.

AEPLOT

PLOT	The Plot Command.
V	Plot a View.
PLOT	View Name.
Y	Yes, change parameters.
N	No, do not change pen/color/linetype.
N	No, do not plot to a file.
I	Inches.
0,0	Plot origin.
D	Size of sheet.
0.01	Pen width.
N	No, do not adjust for pen width.
N	No, do not Hide lines.
1/4"=1'-0"	Scale of plotted drawing.
	Add a blank space to start plot function.
	Add a blank space to return to drawing editor.

To run this script file from the command prompt, follow the command sequence below.

Invoking AEPLOT Script

```
Command: SCRIPT
Script file: AEPLOT
```

AutoCAD will retain the plot settings after the first plot using AEPLOT. If you wanted to plot another drawing of the same scale, you would use this next script file. (This file is not on the AE DISK.)

AEPLOT–2

PLOT	The Plot Command.
V	Plot a View.
PLOT	View Name.
N	No, do not change parameters.
	Add a blank space to start plot function.
	Add a blank space to return to drawing editor.

➡ *TIP! It is a good idea to create different script files with different parameters for your standard plots, and use AutoCAD's PLOT command only for plots that are out of the ordinary.*

For more information on script files and their operation, refer to INSIDE AUTOCAD or CUSTOMIZING AUTOCAD.

Plotting the Project Drawings

The site plan drawing is a different scale than the rest of the drawings for the project. If you plot the drawings in a sequence so that they are grouped according to scale, you'll only have to change the plotting scale once. This will save time because AutoCAD retains the previous plotting scale as a default until you change it.

If you plotted the site plan first, the following sequence shows the actual plot routine that the AEPLOT script file would generate for the floor plan. The listing shows how the scale setting in the routine (which would have defaulted to the value from the site plan) would be changed.

Plotting 8910FP

```
Command: PLOT
What to plot -- Display, Extents, Limits, View or Window : V
View name MAIN: PLOT

Plot will not be written to selected file
Sizes are in Inches
Plot origin is at (0.00,0.00)
Plotting area is 33.00 wide by 21.00 high (D size)
Plot is NOT rotated 90 degrees
Area fill will NOT be adjusted for pen width
Hidden lines will NOT be removed
Scale is 1=240            Default scale factor for 8910SP.

Do you want to change anything? <N> Y      Display of default settings.
```

Entity Color	Pen No.	Line Type	Pen Speed	Entity Color	Pen No.	Line Type	Pen Speed
1 (red)	1	0	36	9	3	0	36
2 (yellow)	2	0	36	10	4	0	36
3 (green)	3	0	36	11	5	0	36
4 (cyan)	4	0	36	12	6	0	36
5 (blue)	5	0	36	13	1	0	36
6 (magenta)	6	0	36	14	2	0	36
7 (white)	1	0	36	15	3	0	36
8	2	0	36				

```
Line types   0 = continuous line
             1 = ..................................
             2 = ----    ----    ----    ----
             3 = -----   -----   -----   -----
             4 = ------.  ------.  ------.  ------.
             5 = ---- -   ---- -   ---- -   ---- -
             6 = --- - -  --- - -  --- - -  --- - -
```

Do you want to change any of the above parameters?   <RETURN>
Write the plot to a file?   <RETURN>
Size units (Inches or Millimeters) : <RETURN>
Plot origin in Inches ,0.00: <RETURN>
Do you want to change any of the above parameters? <N> <RETURN>
Write the plot to a file? <N> <RETURN>
Size units (Inches or Millimeters) <I>: <RETURN>
Plot origin in Inches <0.00,0.00>: <RETURN>

Standard values for plotting size

Size	Width	Height
A	10.50	8.00
B	16.00	10.00
C	21.00	16.00
D	33.00	21.00
E	43.00	33.00
MAX	44.72	35.31

Enter the Size or Width,Height (in Inches) <D>: <RETURN>
Rotate 2D plots 90 degrees clockwise? <N> <RETURN>
Pen width <0.010>: <RETURN>
Adjust area fill boundaries for pen width? <N> <RETURN>
Remove hidden lines? <N> <RETURN>

Specify scale by entering:
Plotted Inches=Drawing Units or Fit or ? <1=20'>: **1/4"=1'**
Effective plotting area: 33.00 wide by 21.00 high
Position paper in plotter.
Press RETURN to continue or S to Stop for hardware setup **S**
Do hardware setup now   Use this option with pen plotters for setup.
Press RETURN to continue: <RETURN>
Processing vector: XXX   Vectors are processed and plotting begins.

Plot complete.
Press RETURN to continue: <RETURN>   Return to drawing editor.

Command: **END**

If you want to continue plotting the project drawings, use the AEPLOT–2 script file. You won't have to change the scale setting again.

Now that you have some tools for helping to plot your drawings, let's talk about why you should create a plot log chart for the project, adding it to the project work book.

Plot Logs

Plot logs are a vital part of the project manager's job book. A project plot log lets you see what drawings are currently being worked on and the status of the latest drawing. Records like this are not just more paperwork, but can be real life-savers if whoever was responsible for plotting the project is not able to finish the job. The operator who inherits the project can keep it rolling if he or she can become familiar with the project. The plot log should contain all the information needed to pick up the project.

DRAWING PLOT LOG

DRAWING NUMBER	DRAWING TITLE	DATE	SHEET SIZE	QUANTITY PLOTTED	PLOTTED BY
8910SP	SITE PLAN	4/4/89	24x36	1	JMA
8910FP	FLOOR PLAN	4/4/89	24x36	1	JMA
8910CP	CEILING PLAN	4/4/89	24x36	1	JMA
8910EL	EXTERIOR ELEV.	4/4/89	24x36	1	SW
8910BS	BUILDING SECT.	4/4/89	24x36	1	DL
8910ST	STRUCTURAL PLAN	4/4/89	24x36	1	SW
8910ME	MECHANICAL PLAN	4/4/89	24x36	1	JMA
8910PL	PLUMBING PLAN	4/4/89	24x36	1	JMA
8910EE	ELECTRICAL PLAN	4/4/89	24x36	1	JMA

Project Plot Log Illustration

Adding Revisions

If you have to make changes in the design or discover omissions after you have plotted the drawings and sent them out for bid, you will have to revise them, note the changes, and reissue the drawings to the contractor.

Technique for Adding Revisions

You already know how easy it is to make changes to drawings with AutoCAD. Clearly noting the revisions on the drawings is the second step. The following AutoLISP routine makes it simple to show which areas of the drawings were revised by placing a *cloud* shape around them.

Use the [CLOUD] command to invoke the routine if you are using the AE-MENU System. Use the [EDITATTR] command to note the date and revision number in the title block area.

The [CLOUD] menu item is supported by the CLOUD.LSP routine.

CLOUD.LSP AutoLISP Utility

Check that you have a copy of CLOUD.LSP file in your AE-ACAD directory.

Read along, or create the AutoLISP file.

This utility gives you a quick and easy way to demarcate revisions in the drawings.

Testing the CLOUD.LSP Routine

Now that CLOUD.LSP is installed, load AutoCAD and test the routine. If you are using the AE-MENU, test the [CLOUD] menu item.

CLOUD.LSP Operation

```
Command: (load "CLOUD")
C:CLOUD
Command: CLOUD
Polyline <Clockwise>...First Pt:          Pick first point.
To:                                        Repeat picking points until complete.
To: <RETURN>                               Will close Pline and form cloud shape.
```

By marking revision areas with the cloud shape and noting the revision in the title block area, you have a good way of clearly showing changes and of keeping track of revision dates and sequences.

Here is the CLOUD.LSP listing:

```
(prompt "Loading CLOUD utility...")
(defun C:CLOUD ( / mark p1 e1 e2 bulge vx data)
   (setvar "cmdecho" 0)
   (setq p1 (getpoint "Polyline <Clockwise>... First Pt:  ")
   )                                                              ;setq calculations
    (command "PLINE" p1 "W" "0" "")                               ;draw polyline
    (while p1
      (command p1)
      (setq p1 (getpoint p1 "To:  ")))
    )                                                             ;while
   (command "C")
   (setq e1 (entlast)                                             ;save it
          e2 (entget e1)                                          ;get it
          bulge (list (cons 42 -0.5))                             ;build cloud
          vx (cdr (assoc -1 e2))                                  ;set for lookup
          vx (entnext vx)                                         ;get next one
   )                                                              ;setq calculations
    (while vx                                                     ;if something there
       (setq data (entget vx)                                     ;get what's there
             data (append data bulge)                             ;tag to bulge
       );setq calculations
       (entmod data)                                              ;change entity
        (setq vx (entnext vx))                                    ;get next one
    )                                                             ;while
    (entupd e1)                                                   ;update entity

    (setvar "cmdecho" 1)
    (gc)  (princ)
 )                                                                ;defun
```

AutoLISP Routine for [CLOUD]

Alternate Output Methods

There are alternatives to going through the effort of plotting each drawing. These alternatives are useful for enhancing presentations. 3D color renderings, like the branch bank presentation drawing created with AutoShade in Chapter Five, frequently do not translate well to the printed, or plotted page. (Sometimes it seems that you have to muddy the screen image to get it to look better on paper!) To do your 3D renderings justice, you may want to look into using some of the devices mentioned below.

Photographic Output

Many companies now supply means of creating photographic slides of what is displayed on your AutoCAD screen. These hardware add-ons come equipped with their own high resolution color monitors and 35mm cameras to produce high quality slides. This is a good way to share the impact of AutoShade renderings, without having to plot or print the drawings. Slides are also convenient for presentations to large groups.

Video

Making a video recording of your AutoCAD output is a extremely effective way of presenting projects to large groups. It's the best way to take an AutoFlix movie to your client, since almost every office has access to a VCR.

Color Printers

Several additional options are available for AutoCAD output via printers, instead of plotters. Color printers do a better job at producing AutoShade renderings than monochrome printers. Output from these printers and other desktop devices (like some laser and simple dot matrix printers) are especially useful for exploring alternatives in the design phase, doing check plots, and for producing preliminary designs in a format that you can use in presentation booklets.

- Color ink jet printers. These printers have a wide price range, but still tend to provide inadequate resolution.

- High resolution color dot matrix printers. In general, these printers offer better resolution than ink jet printers, and can also handle paper widths up to size C.

You may find these printers useful additions to your current hardware. They can help you in the design and presentation phases of project production. In addition, high quality output from these can replace plotted drawings in many instances, helping you avoid plotting bottlenecks.

Summary

Creating consistent drawings is an important part of an A/E firm's presentation technique. Your drawings *will only become* consistent through careful management and an adherence to plotting standards. While every firm tries to implement a consistent style or standard in their work, individual draftsmen have their own writing styles and ways of doing things which tend to impede efforts to standardize. AutoCAD gives you a way to get a handle on most areas where inconsistency is a problem, such as lettering, dimensioning styles and symbol usage. The plotting standards we've discussed in this chapter can get you started towards producing consistent drawings.

We hope that we've also given you some ideas on how to avoid plotter bottlenecks. Getting your drawings out on time is just as important as what your drawings contain.

```
Apps Disk Create Edit Locate Frames Words Numbers Graph Print 12 50 pm
[....▼.....▼.........▼.........▼.........▼.........▼.........]....▼.....
4100

PART 1 GENERAL

1.01    RELATED WORK

A.    Section 04300 - Unit Masonry System:  Installation of mortar.

1.02    REFERENCES

A.    ASTM C91  - Masonry Cement.
B.    ASTM C94  - Ready-Mixed Concrete.
C.    ASTM C144 - Aggregate for Masonry Mortar.
D.    ASTM C150 - Portland Cement.
E.    ASTM C207 - Hydrated Lime for Masonry Purposes.
F.    ASTM C270 - Mortar for Unit Masonry.
G.    ASTM C476 - Grout for Reinforced and Non-reinforced Masonry.

1.02    ENVIRONMENTAL REQUIREMENTS

A.    Maintain materials and surrounding air temperature to minimum
      50 degree F prior to, during, and 48 hours after completion of
      masonry work.
```

CSI Specification Division Report

CHAPTER 11

Data Extraction, Specifications and Reports

Introduction

In this chapter we will present a *case study* on how to extract data from AutoCAD files and drawings. Our purpose is not to provide an exhaustive survey of data extraction, but to show you a single, *real life* example, using drawing data to generate specification documents and reports.

This is reporting system that we use every day. The key to this system is an AutoLISP routine which creates an ASCII text file containing project data, such as the names of drawing layers, drawing numbers, dates, revision dates, and job numbers. Since the layer names are a combination of descriptive abbreviations and the CSI numbering format, we will show you how to relate layer names contained in a file to existing specification divisions created on a word processing program. Once these relationships are established, we will show you how we generate reports for the project work book that we described in Chapter One.

We will also describe how to set up a *semi-automatic* specification and report generation system on a Local Area Network (LAN). We use our LAN every day, having tied together 12 AutoCAD workstations, 30+ employees, and (via modem) a branch office 1000 miles away.

Concepts of the Reporting System

When we assembled the software and hardware components of our system, the idea was to create a totally integrated environment where different types of data, generated by different types of software applications, including AutoCAD, could be collected at a central location, called a File Server. The file server provides a central data source that all the workstations have access to, whether they are individually dedicated to AutoCAD production, or to word processing, or run a combination of software applications.

We wanted to merge drawing data and written documentation so that we could track project drawing files according to their names and revisions, and consequently be able to generate specifications and reports from the data. Since we implemented the system, we have seen tremendous productivity gains, especially more frequent reuse of entities and existing drawings. We also have seen time saved by extracting report data that we used to have to keep track of manually. As a by-product of networking our workstations and defining the *communications loop* between the system data and the users, information has become more accessible, and easier to track.

Let's review some of the basic characteristics of the kinds of data that come into and out of the our system, and that you may want to report on with your own system.

Drawing Data

When you create a drawing with AutoCAD, every entity that you add and save with the drawing has a unique location in the drawing database. As you might imagine, the database for even a single drawing contains a tremendous amount of data, including information on:

- Every line and its associated data

- Every circle and its associated data

- Every arc and its associated data

- The location of each entity relative to the origin

- The layers used for each entity

- All attributes and their values

- The text entered and its characteristics

What you may not know, is that you can *export* this data with either standard AutoCAD commands or with AutoLISP routines. A simple application using this data, for example, might be to put together a report to verify that entities reside on the correct layers. Or you might try a more complicated application, like creating a detailed cost estimate of a project. AutoCAD makes it relatively easy for you to get to the data. the tricky part is figuring out how to put it in more useful report forms.

Specifications

Writing specifications is the most time-consuming and thought-intensive operation in most A/E firms. It's a dirty job and usually *nobody* wants to do it. However, neither we nor the client would consider any project viable without a good set of specification documents stipulating the specifics of our solution.

Writing specifications is an artistic process. The writer has to have good organizational ability and an intimate knowledge of construction materials and processes. He or she has to be able to write clear descriptions of every aspect of the project, and relate them to industry standard divisions and exacting requirements, including:

- Legal Requirements

- Insurance Requirements

- Bidding Procedures

- Materials or products, their source and quality

- Construction Workmanship

- Alternates or Substitutions

- Options

- Change Order and Payment Procedures

- Owner's Responsibilities and Limitations

- Contractor and Subcontractor Responsibilities and Limitations

- Inspections

- Testing Requirements and Reports

- Project Close-Out

Usually, we spend the most time generating the specification sections that deal with Materials and Products, and Construction Workmanship. These two areas relate directly to the working drawings and can involve a large amount of data. The specification writer has to examine and evaluate all this before he or she can even begin to write.

Most A/E firms use a *cut and paste* approach to these documents. They find specifications or divisions of specifications that have been previously written, copy the parts they can reuse, and reorganize the data to match the new project. Then the results have to be retyped. Some firms use a specification system such as MasterSpec to extract the divisions they need for specific projects.

But, what if you could take data from your drawings (where all the division data is generated), associate that data directly with a specification division database, sort the resulting comparison data and produce your specification document that way?

You can do it with the help of what we call *relationships*.

Relationships

The relationships we are talking about are associations made between AutoCAD drawing data and nondrawing data. For example, you can define a relationship between exported data about entities on layer CON-030 and a specification division, say ***03XXX, that is created on a word processor. The key to the relationship is to have common components in the names of the files, or data you want to associate.

We described our layer-naming convention in Chapter One. We use names that are combinations of three character abbreviations, describing the material or construction mode of entities to be drawn from those layers, and three digit numerics that relate to the CSI number format for each division. Having used these layer names and created specification divisions that conform to the CSI numbering format, we can directly define *relationships* between the layer drawing data and the nondrawing specification data.

Project Reports

In Chapter Ten, we discussed using the client's border and logo in the working drawings in place of the AEBORDER drawing. We also added sheet titles, completion dates, revision dates and drawing numbers as attributes. We will describe you how you can automate creating the project work book reports we described in Chapter One by using a system for data extraction for the sheet information and layers in each drawing.

System Software Components

The key to every productive computer system is the software that it runs. To create our report generation system, we used several software packages since there was no single software package that could manipulate every situation or type of data that we wanted to use. We also built our system over time.

Selecting and implementing multiple software packages can become very difficult, since not every package is compatible with every other package. Until individual software development firms create totally integrated systems to do the tasks we want, we'll probably be stuck with figuring out how to make different packages *talk* to each other. Even though most micro-based software packages can be tricked into communicating with each other, you have to decide whether it's worth the time and expense to find a *patch*, or some other work around, to fix compatibility problems.

We'll tell you what packages we used to create our system, but first let's go over the steps we went through to make our selections:

First, we had to confirm that the various packages would support one of the export file formats that AutoCAD makes available. If you can find a link between package features that already exist, you will have cleared your first hurdle.

Second, we had to find a way to program the packages to sort the different types of data and form the relationships we've discussed.

Networking

Before we describe the components of our system, we'd like to mention just why we chose to set up a network and offer you some alternatives, if *networking* is not something you are ready to do. What drove us into networking were the numbers of workstations and users in our office, and the quantities of data they produced.

If your office only uses two or three workstations, you can usually get along quite well *not* tying your systems to a common database. To use a system without a network, you have to do your communicating and data extracting via *floppies*, copying the data to a single workstation to extract the data and generate the reports. You have to be careful about maintaining the most current data while going through your reports and specifications processes. Revisions running concurrently with these processes can affect the accuracy of the end results.

You'll be ready to network your workstations when the numbers of workstations and users make it too difficult to track project data, or when your database grows so large that making the data more easily accessible becomes an important time factor in producing your projects. You can extract accurate, up-to-date data from a file server's database without the tedium of exchanging floppies, or the problems associated with concurrent revisions.

AutoCAD

AutoCAD software is, of course, the first component of any reporting system. It is where the data that we use to generate our specification document originates. If we create the drawings for a project using poor organization and methods, then the degree of misinformation contained in the drawings will translate directly to misinformation in the specification divisions of our document. The usability of the final output of our report generation system depends directly on whether we have followed good, sound standards of setup and operation when we created the drawings. The old truism "Garbage In, Garbage Out" definitely applies here.

Word Processor

After experimenting with different word processing packages, we found that MicroSoft Word served the operations of our firm the best. It is easy to use and learn, offers a wealth of tutorials on all facets of the software, and is well-supported by the supplier. It was important to us that it had the ability to communicate with other software packages. We used it to create almost all of the CSI specification divisions required for our system, and so far haven't run into any incompatibility problems with our other software packages.

Database and Spreadsheet Software

We choose to combine our spreadsheet and database function in one package. A lot of integrated software packages, defined as packages which provide more than one function (most incorporate three: spreadsheet, database and word processing), are currently available. Some functions are quite limited.

We chose Framework III from Ashton-Tate. We use the package's word processing program only for writing programs to run against its database and spread sheet components. Framework III accepts as many as 20 other formats from different word processing packages. We import our files into Framework as ASCII files.

Local Area Network

AutoCAD and the other packages we selected are not specifically designed to work within networks. We also didn't want to go to the expense of dedicating one of our workstations exclusively as a file server (where all drawing files are maintained). So, we had to select a networking software package that would let us run in nondedicated mode (in other words, still use the file server workstation to create drawings, etc.) and be small enough not to degrade AutoCAD's operation or performance on the other workstations.

We chose NOVELL's ELS II Local Area Networking Software to provide this function. We selected this software using the same criteria: it is easy to install, easy to use, and offers good support and documentation. We have found that ELS II meets all our requirements and more.

Since we implemented the system, we have grown past the packages' original eight node limit, and have installed a second nondedicated file server tied to the first. This gives us more storage capacity as well as a backup system should one of the file servers go down.

Other Options

We recognize that our software choices, and our solutions fit our situation. We don't expect our choices, particularly our software choices (not everyone has chosen Framework) to act as a general solution for everyone. If you don't have our mix of software, please read along for ideas on how to extract data from AutoCAD to get specifications and reports.

The Key to Extracting AutoCAD Data

Earlier we said that the key to our operation is an AutoLISP routine that let's us export data from the working drawings. AutoCAD generates a tremendous amount of drawing data, which is not all applicable (or required) for individual specification reports or applications. The benefit of our EXPORT AutoLISP routine is that it filters out unnecessary data. AutoCAD's ATTEXT command will filter out data by the design of the template file, but it is easier to achieve custom data extraction using AutoLISP.

Technique for Data Extraction

You will find our EXPORT.LSP among your AE DISK files. It supports the AE-MENU item called [EXPORT].

This routine has two parts: the first exports a list of all the layer names that have at least one entity drawn on them. The second exports all text entities on layer TLE-010, which we used for the title block attribute data of Chapter Ten.

EXPORT.LSP AutoLISP Routine

 Ckeck to see that you have a copy the EXPORT.LSP file to the AE-ACAD directory.

 Read along, or create the AutoLISP file.

Testing the EXPORT.LSP Routine

Now that EXPORT.LSP is installed, load AutoCAD and test its operation with one of the working drawings. Use one of the working drawing as a test drawing. If you added title attribute data to a working drawing (in Chapter Ten), use that drawing as a test drawing. Test the routine by loading the EXPORT.LSP file, or selecting the [EXPORT] item from the AE-MENU.

EXPORT.LSP Operation

```
Command: (load "EXPORT")
C:EXPORT
Command: EXPORT
Enter Name of Layer File for Export:      Enter name.
Enter Name of Title File for Export:      Enter name.
```

➡ *NOTE! If you wish, you can use the same file name more than once. The program will add or append new data to bottom of each file type, layer or title.*

Here is the listing for EXPORT.LSP:

```
(prompt "\nLoading EXPORT utility...")
(defun C:EXPORT (/ lyrlst nxtenty entdata lyrnm fl lyrout txtout)

  (setq lyrlst nil                              ;initialize
        nxtenty (entnext)                       ;get the first one
  )                                             ;setq calculations

  (while nxtenty
    (setq entdata (entget nxtenty)              ;get the next one
          lyrnm (cdr (assoc 8 entdata))
    )                                           ;setq calculations
    (if (not (member lyrnm lyrlst))             ;see if not on list
          (setq lyrlst (cons lyrnm lyrst))
    )                                           ;if
```

```
            (setq nxtenty (entnext nxtenty))        ;get next entity
       )                                            ;while
;Write out the list to disk file
       (setq fl (open "LAYER.DAT" "a"))             ;output to LAYER.DAT
       (if fl                                       ;only if file opened
          (progn
           (setq lyrout (car lyrlst))               ;get first from list
            (while lyrout
               (write-line lyrout fl)               ;write to disk file
                (setq lyrlst (cdr lyrlst))          ;drop from layer list
                (setq lyrout (car lyrlst))          ;get next one
            )                                       ;while
            (setq fl (close fl)
          )                                         ;progn
       )                                            ;if
                              ;Program to pull all text from layer = TLE-010
       (setq fl (open "TLE-010.dat" "a"))
       (if fl
          (progn
             (setq nxtenty (entnext))               ;start at beginning
              (while nxtenty
                (setq entdata (entget nxtenty))     ;retrieve entity data
                (setq lyrnm (cdr (assoc 8 entdata))) ;determine layer name
                (if (= lyrnm "TLE-010")             ;see if one we want
                   (progn
                     (setq txtout                   ;get actual text
                          (cdr (assoc 1 entdata)))
                     (write-line txtout fl)         ;write to disk
                 )                                  ;progn
               )                                    ;if
                (setq nxtenty (entnext nxtenty))    ;get next entity
             )                                      ;while
          (close fl)
        )                                           ;progn
      )                                             ;if
)                                                   ;defun
```

AutoLISP Routine for [EXPORT]

Now that we have shown you the key, let's talk about how to tie together a set up accept data and form the relationships necessary to generate the specifications and reports.

Setting Up the Software Components (Optional)

In the sequences that follow, we will show you the majority of the steps required to set up the software components of our system. If you are not going to set up a system, just read along to see how it is put together and how it functions.

Our first system task was to create the text files representing specification sections, named according to the CSI numbering format. Next, we needed to create Framework III database files containing CSI division numbers and titles for each section, named according to the section numbers. We will show the final piece of the system that we put together to export title block information and generate the project number and drawing number logs for the project work book.

Microsoft Word

Like a lot of firms, we began running AutoCAD on an IBM-AT with an EGA monitor. Also like a lot of firms, we eventually moved these workstations to perform bookkeeping and secretarial functions as we updated the AutoCAD workstations. This is when we added word processing to our operation. As we said, we chose MS Word for this function. We have used it to compile an extensive database of specification section documents.

Using MS Word's window feature has made it easy to create these specification documents. We can display more than one document on the screen at a time, letting us copy and insert pieces of our entire documents together to create complete specification sections.

Framework III

As we said earlier, Framework is the software package we selected to merge our different files and generate our specification documents and reports. We import our Word files into Framework as *unformatted* (ASCII) files.

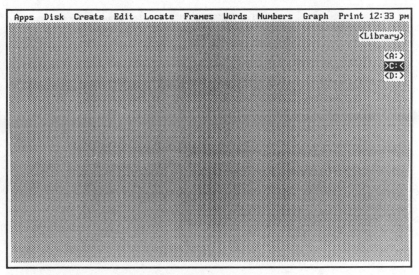

Framework III Desk Top Screen Illustration

We found the database feature of Framework useful, letting us create a very large number of records under each field name.

Framework Database Setup

In the following sequences, we want to give you the general steps that we went through to set up our database files. To gain a full understanding of how the package really works, you'll have to install the software yourself and experiment with it. If you do not have Framework, you can use this section as a means of *stretching your imagination* on how you might use AutoCAD drawing data to set up your own reporting system.

Just as we integrated a custom tablet menu to enhance the operation of AutoCAD, Framework must also be tailored or programmed to perform specific operations. Our setup operation involves tailoring two different areas of the program:

- Database — for sorting data for specifications and reports.

- Spreadsheet — for creating templates for report forms.

Once these areas are configured, we will discuss how Framework operates using these areas. We will begin by loading Framework and show you how we created the specification data bases.

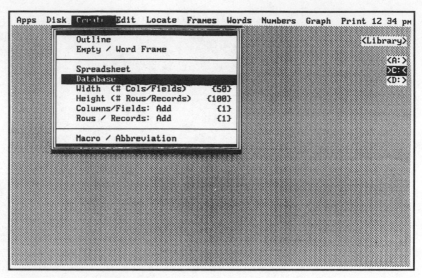

Database Setup Illustration

Specification Database

The specification databases for our system consist of MS Word specification section text files and Framework CSI division number files.

We created the text files ourselves, naming each file by the section number (for example, 02000, 04000) that corresponded to the files' contents.

To set up your own system, you would have to create these files, using whatever word processing package you have access to. With our MS Word text files in place, we can begin the following sequence of selections and data input by loading Framework.

➥ *NOTE! Framework supports cursor keys and/or a mouse as selection devices. We found that a mouse was well worth the investment.*

Select **Create**   The create menu will appear.
Select **Database**   An unnamed frame will appear.

You will notice that the border that defines the database frame is displayed in *bolded* format. When you move the cursor into this frame and press the left mouse button, the border changes to a narrow display. This indicates that the frame is active and ready for data input. In the bottom right corner of the display, you will see an area that contains two brackets [   ] with eight spaces between them. When you add the frame of a database file, the name appears between the brackets.

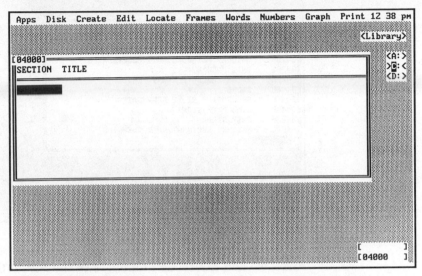

Database Frame Illustration

The database frame has two functional areas. The top area is for field names and the bottom area is for record names. Each area is defined by *cells*, much like a spreadsheet program.

First we specify the database name as the CSI Division Number. Then we specify two field names: SECTION, which we use for the CSI Section Numbers, and TITLE, will be used for the related title of each section number. A nice feature of Framework is that you can SIZE each cell so that all the data you enter is displayed.

Enter	**04000**	CSI Division File database name.
Pick		Use mouse to select first cell in top area.
Enter	**SECTION**	Use as reference for CSI section numbers.
Pick		Use mouse to select second cell in top area.
Enter	**TITLE**	Use as reference for CSI section titles.

Press the F4 function key for the SIZE command and use the cursor keys to increase the size of the TITLE cell.

Now that we have defined the SECTION and TITLE fields, we can input the section numbers for division 04000 with their associated titles.

```
 Apps  Disk  Create  Edit  Locate  Frames  Words  Numbers  Graph  Print  12 44 pm


                                                             <Library>

                                                                  <A:>
 [04000]                                                          >C:<
 SECTION   TITLE                                                  <D:>

 04100    MORTAR & SURFACE BONDING CEMENT
 04150    MASONRY REINFORCING & ACCESSORIES
 04200    UNIT MASONRY
 04235    PREASSEMBLED MASONRY PANELS
 04400    STONE
 04415    DECORATIVE STONE AGGREGATES
 04430    SIMULATED STONE
 04435    CAST STONE
 04500    MASONRY RESTORATION & CLEANING




                                                             [04000   ]
```

CSI Division Database Illustration

First, we add the section numbers to the SECTION field, then add the
section number titles to the TITLE field. After we have added all the
section numbers and titles for this division, we will save this frame and
move on to creating a frame for the next division.

➡ *NOTE! When you enter numbers in Framework, you need a leading spacebar
entry so that the program will not treat the number as a formula.*

Pick		Select cell in the SECTION Field.
Enter	**04100**	Section Number.
Pick		Select a cell in the TITLE Field.
Enter	**MORTAR & SURFACE BONDING CEMENT**	Section Number Title.

That's the basic format of our specification databases. To create the entire
specification database, you would continue on and open a new frame for
each division name and add all the section numbers and titles.

```
Apps  Disk  Create  Edit  Locate  Frames  Words  Numbers  Graph  Print 12 47 pm
                                                              <Library>
                                                                 <A:>
                                                                 >B:<
                                                                 <D:>



                                                         [07000        ]
                                                         [08000        ]
                                                         [09000        ]
                                                         [10000        ]
                                                         [11000        ]
                                        [02000        ][12000        ]
                                        [03000        ][13000        ]
                                        [04000        ][14000        ]
                                        [05000        ][15000        ]
                                        [06000        ][16000        ]
```

Completed CSI Division Specification Database Illustration

Once you have completed the last division, Framework will display a list of frames in the bottom right corner of the screen in the desk top area.

Importing Text Files

As we said earlier, Framework has an elaborate import function that can accept most data formats. For our purposes, using the ASCII text function was the easiest for importing both the specification and drawing exported text files.

First, let's talk about how to import the MS Word specification text files into Framework.

```
 Apps  Disk  Create  Edit  Locate  Frames  Words  Numbers  Graph  Print 12 48 pm

      Get File by Name                                              <Library>
      Save and Continue        A. Framework              {}
      Put Away                 B. dBASE                  {}         <A:>
      Clean Up Desktop         C. IBM DCA/DisplayWrite   {}         >C:<
                               D. WordStar               {}         <D:>
   -- Network File: Ask        E. MultiMate              {}
                               F. WordPerfect            {}
    ▶ Import                   G. ASCII Text             (4100)
    ▶ Export                   H. Lotus 1-2-3            {}
      DOS Access               I. Multiplan SYLK         {}
                               J. VisiCalc DIF           {}
      QUIT Framework III
                               1. Import Add-in          {}
                               2. Import Add-in          {}          [07000  ]
                               3. Import Add-in          {}          [08000  ]
                                                                     [09000  ]
                                                                     [10000  ]
                                                                     [11000  ]
                                                  [02000  ] [12000  ]
                                                  [03000  ] [13000  ]
                                                  [04000  ] [14000  ]
                                                  [05000  ] [15000  ]
                                                  [06000  ] [16000  ]
```

Framework III Import Function Illustration

First, we select DISK from the Framework Menu Area, then select IMPORT and G. ASCII Text. Then we enter the name of the Text File, 4100, and press <RETURN>.

```
 Apps  Disk  Create  Edit  Locate  Frames  Words  Numbers  Graph  Print 12:49 pm
[4100]
[.........................................................]   <Library>
4100
                                                               <A:>
PART 1 GENERAL                                                 >█:<
                                                               <D:>
1.01    RELATED WORK

   A.  Section 04300 - Unit Masonry System:  Installation of morta

1.02    REFERENCES

   A.  ASTM C91 - Masonry Cement.
   B.  ASTM C94 - Ready-Mixed Concrete.
   C.  ASTM C144 - Aggregate for Masonry Mortar.               [08000  ]
   D.  ASTM C150 - Portland Cement.                            [09000  ]
   E.  ASTM C207 - Hydrated Lime for Masonry Purposes.         [10000  ]
   F.  ASTM C270 - Mortar for Unit Masonry.                    [11000  ]
   G.  ASTM C476 - Grout for Reinforced and Non-reinforced Masonry [12000  ]
                                                               [13000  ]
1.02    ENVIRONMENTAL REQUIREMENTS                             [14000  ]
                                                               [15000  ]
   A.  Maintain  materials and surrounding air temperature to  min [16000  ]
                                                               [ 4100  ]
```

Imported Specification Text File Illustration

Framework will find our Word file, import it, and convert it to match its own format. It then displays the imported file in a word processor frame, naming it with our file name, and adding the file name to the directory of files shown in the bottom right corner of the desk top area (lower right). If any errors occur during the import function, Framework will display these areas. The same method can be used to import the drawing export files.

Framework Report Forms

We designed some report databases within Framework to automate inputting drawing information into forms that are part of the project work book, and that can be used later for production evaluations.

Project Number and Drawing Number Log Reports

Having already created specification databases, let's see how to create two different types of databases for report forms. These forms will be similar to the project number log, and the drawing number log that were illustrated in Chapter One.

We will name the database files PROJECT and DRAWING. The data exported to create the databases comes from the drawings and from keyboard entries of information suppling the records data.

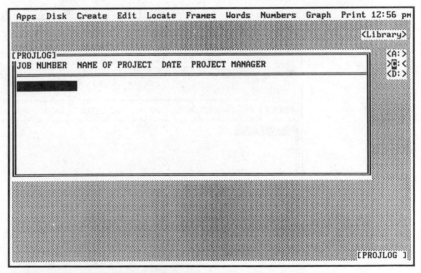

Project Number Database Illustration

As a refresher, recall that the project number logs are designed for tracking the assignment of job numbers and names, dates, and project managers on specific projects.

The project number log in the illustration (above) shows four Framework fields that will contain record data. Data for three of these records is gained from the exported data of layer TLE-010. Data for the project manager record is from keyboard entry.

We begin, as with the specification database, by first opening a frame. We enter a file name of PROJLOG and field names of Job Number, Name of Project, Date, and Project Manager.

Enter	**PROJLOG**	Database file name.
Pick		Use mouse to select first cell in top area.
Enter	**JOB NUMBER**	Use as reference for Job Number.
Pick		Use mouse to select second cell in top area.
Enter	**NAME OF PROJECT**	Use as reference for Name of Project.
Pick		Use mouse to select third cell in top area.
Enter	**DATE**	Use as reference for Date.
Pick		Use mouse to select fourth cell in top area.
Enter	**PROJECT MANAGER**	Use as reference for Project Manager.

Press the F4 function key for the SIZE command and use the cursor keys to increase the size of each cell as required.

Now that we have created the project number log report form, we will create a similar form for the drawing number log.

Drawing Number Database Illustration

The drawing number logs are designed to track single projects by documenting the project number and name, drawing number and name, the date the drawing was made, whom it was drawn by, the dates of any revisions, who checked them and any special remarks about the drawing.

In most practices today, we no longer include the names of the staff involved in creating or checking on the drawings themselves. We do this to stop contractors from calling the architect's office and asking for the person who created the drawing. This person may not be aware of all the conditions of the project, and consequently the contractor may get incorrect information. This is where a drawing number log comes in handy in keeping track of this information internally because both the creator and the checker of the drawing are listed with date and project information.

The drawing number log report form in the illustration (above) shows eight Framework fields. Data for all but two of these fields is from the exported data of layer TLE-010. Data for the remaining fields, DWG BY and CHK BY, is from keyboard entries.

We begin as before by opening a frame. We specify the name of the database as DWGLOG and add the Field names of Project No., Project Name, DWG No., DWG Title, Date, DWG BY, REV DATE, and CHK BY.

Enter	**DWGLOG**	Database file name.
Pick		Use mouse to select first cell in top area.
Enter	**PROJECT No.**	Use as reference for Job Number.
Pick		Use mouse to select second cell in top area.
Enter	**PROJECT NAME**	Use as reference for Name of Project.
Pick		Use mouse to select third cell in top area.
Enter	**DWG No.**	Use as reference for drawing number.
Pick		Use mouse to select fourth cell in top area.
Enter	**DWG TITLE**	Use as reference for individual drawing titles.
Pick		Use mouse to select fifth cell in top area.
Enter	**DATE**	Use as reference for individual drawing dates.
Pick		Use mouse to select sixth cell in top area.
Enter	**DWG BY**	Use as reference for drawn by name.
Pick		Use mouse to select seventh cell in top area.
Enter	**REV DATE**	Use as reference for individual drawing revision dates.
Pick		Use mouse to select eight cell in top area.
Enter	**CHK BY**	Use as reference for checked by name.

Press the F4 function key for the SIZE command and use the cursor keys to increase the size of each cell as required.

Now that we have each report form in place, the next task is to create spreadsheet report templates.

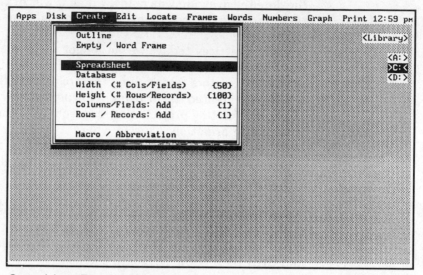

Spreadsheet Function Illustration

Spread Sheet Templates

To help generate the specifications, project number log, and drawing number log reports, we use the spreadsheet function of Framework. We use spreadsheet templates which contain cells that are labeled across the top with letters A, B, C , etc., and down the side with numbers 1, 2, 3, and so on. These labels are used as locations to transfer data from the project and drawing databases to the spreadsheet, and for creating a report.

To create the first template, we select CREATE from the Framework menu, then select SPREADSHEET and enter the template name of PROJECT.

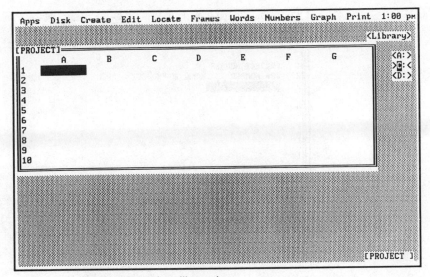

Project Spreadsheet Template Illustration

Framework will open a spreadsheet frame with letters and numbers ready for naming and data entry.

Enter	**PROJECT**	Template file name.
Pick		Use mouse to select first cell at (A) (1).
Enter	**PROJECT NUMBER LOG**	Use as template title.
Pick		Use mouse to select second cell at (A) (2).
Enter	**JOB NUMBER**	Use as reference for job number.
Pick		Use mouse to select third cell at (B) (2).
Enter	**NAME OF PROJECT**	Use as reference for project name.
Pick		Use mouse to select fourth cell at (C) (2).
Enter	**DATE**	Use as reference for individual project dates.
Pick		Use mouse to select fifth cell at (D) (2).
Enter	**PROJECT MANAGER**	Use as reference for project manager's name.

Press the F4 function key for the SIZE command and use the cursor keys to increase the size of each cell as required.

Now that we have set the titles for each column, it is a simple matter to relate the cells to corresponding areas in the databases. For instance, we relate cell (A) (3) for the project number to the PROJECT database field titled PROJECT NUMBER; and copy the data into the cells.

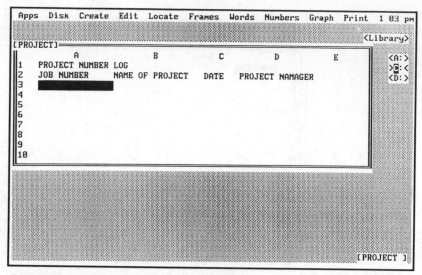

Project Spreadsheet Template Illustration

Next, we repeat the same process and create the DRAWING extraction spreadsheet template.

```
Apps  Disk  Create  Edit  Locate  Frames  Words  Numbers  Graph  Print  1:03 pm
                                                                    <Library>
[DRAWING]                                                           <A:>
          A              B          C      D        E        F     >C:<
1  DRAWING NUMBER LOG                                               <D:>
2  PROJECT NUMBER:                PROJECT NAME:
3  DRAWING NUMBER  DRAWING TITLE  DATE  DRW. BY  REV. DATE  CHECK BY
4
5
6
7
8
9
10
                                                            [PROJECT ]
                                                            [DRAWING ]
```

Drawing Spreadsheet Template Illustration

As before, we select CREATE, SPREADSHEET, and enter DRAWING as the template name. Then add the cells as defined below.

Enter	**DRAWING**	Template file name.
Pick		Use mouse to select first cell at (A) (1).
Enter	**DRAWING NUMBER LOG**	Use as template title.
Pick		Use mouse to select second cell at (A) (2).
Enter	**PROJECT NUMBER**	Use as reference for job number.
Pick		Use mouse to select third cell at (C) (2).
Enter	**PROJECT NAME**	Use as reference for project name.
Pick		Use mouse to select fourth cell at (A) (3).
Enter	**DRAWING NUMBER**	Use as reference for individual drawing numbers.
Pick		Use mouse to select fifth cell at (B) (3).
Enter	**DRAWING TITLE**	Use as reference for drawing titles.
Pick		Use mouse to select fifth cell at (C) (3).
Enter	**DATE**	Use as reference for project date.
Pick		Use mouse to select fifth cell at (D) (3).
Enter	**DRW. BY**	Use as reference for drawn by.
Pick		Use mouse to select fifth cell at (E) (3).
Enter	**REV. DATE**	Use as reference for project revisions date.
Pick		Use mouse to select fifth cell at (F) (3).
Enter	**CHECK BY**	Use as reference for checked by.

Press the F4 function key for the SIZE command and use the cursor keys to increase the size of each cell as required.

As you may have noticed, the titles of these two templates appear in the bottom right corner of Framework's desk top area.

Specification Reference Report Illustration

The final spreadsheet template that we need to create is one that we generate a report displaying the specification section text files available in the database that relate to the exported layer names from individual drawings.

As before, we select CREATE, SPREADSHEET, and enter SPECIF as the template name. Then we add the cells as defined below.

Enter	**SPECIF**	Template file name.
Pick		Use mouse to select first cell (A) (1).
Enter	**SPECIFICATION REFERENCE REPORT**	Use as template title.
Pick		Use mouse to select next cell (A) (2).
Enter	**DATA BASE FILES**	Use as reference for spec. database files.
Pick		Use mouse to select third cell (D) (2).
Enter	**EXPORTED SPEC. SECTIONS**	Use to reference for export layer names.

Press the F4 function key for the SIZE command and use the cursor keys to increase the size of each cell as required.

We have now completed all the basic setup operations required to implement a system for data extraction, export, and report generation.

Methods of System Operation

We have designed this section with two ways for you to practice the functions and generate reports from our system.

- The first uses Framework to create and generate the reports.

- The second uses standard DOS (Disk Operating System) commands to create and print similar reports.

If you want to follow along and actually create a project specification, you will need a word processing software package and a library of files named with the CSI Numbering Format.

We expect you may just want to read along the first time through, looking at the DOS commands. We hope that you take the concepts and operations that we show you and use them to help you design your own system, using either your existing software, or purchasing the software that seems right for you.

Technique for Generating Base Data from AutoCAD

Let's see how to generate the reports we have been talking about. To get started we'll assume that you have completed the drawings for the branch bank project, or have used the layer protocol defined in Chapter One to create another project. We'll also assume that, at this point, that you have already either used the [EXPORT] command from the AE-MENU System Tablet Menu or have typed in the EXPORT.LSP AutoLISP Routine, listed in the first section of this chapter. You need one or the other to create the base data to generate the following reports, and to help compile the project specifications.

➡ *NOTE! In the following exercises, we use the name 8910-EXP as our base data file.*

Generating a Specification Reference Report

The specification reference report uses the base data from an AutoCAD drawing, to create a report that contains the names of all the layers that have entities present. The [EXPORT] command will create a file that contains only the layers that actually have entities drawn on them. Even though a layer may be listed in the drawing, if it does not have at least one entity present at program run time, it will not be included in the exported file.

Once this report is created, you can relate the three numerical character portion of the layer name to your specification library, determining which CSI section numbers apply to the project's specifications.

Specification Reference Report Illustration

Method One (Requires Framework)

To get started, load Framework and open a "file cabinet", a frame displaying the directory that contains the export file, 8910-EXP. Then, select the exported file and bring it to the desk top area. Select from the main menu DISK, then IMPORT and G. ASCII TEXT, and select the export file from the file menu or desk top at the bottom right of the display. A frame will appear displaying the layer names in the order that they were extracted from the drawing. Now with your exported data frame displayed, you can reopen the file cabinet and load the specification reference report spreadsheet template, SPECIF.

Before you merge 8910-EXP with SPECIF and print your report, you must first use the "sort in descending order" function of Framework to correct the order in which the layer names appear. This step is necessary because you didn't create drawings one layer at a time. The type of ordering problem is easy to fix with the sort functions that database and spreadsheet programs provide.

Once you have sorted your data, you can use the CUT and PASTE functions of Framework to move the 8910-EXP data to the SPECIF spreadsheet template. Then you can hand enter the project name, drawing number, and date to complete the report.

The last task is to select PRINT from the Main Menu and print the SPECIF report. Then you can use this report to compare the exported specification data with the library of CSI numbered specification word processing files listed in the report.

To complete the task of generating the project specifications, select the listed files from your word processing library and print them. While the report might need some editing, you have semi-automated a very difficult and time consuming A/E task. If you are fortunate enough to have all of the system components that we described, you can also sort the exported file against the specification database and then print the specification sections automatically.

Method Two

This method is far simpler than Method One, but the results are limited. With 8910-EXP in place, use the DOS print command and print the contents. The printed file will appear as a single text line containing the layer names as they were extracted from the drawing. You can use this listing to compare to your word processing library to select the files to print for the project.

Using this simple method you will achieve the same end results, but will spend much more time tagging the proper files to print. Also, if you want to keep a hardcopy record of the exported data, you will have to either print it by hand or type in the print job name, description, and date.

Generating a Drawing Number Log

Of all of the reports you may want to generate, the drawing number log is one of the most important. This is true whether you use AutoCAD or conventional methods. The data contained in this log is important for management purposes, but unfortunately, the document often never gets created, or isn't well maintained. There always seems to be something more important to take care of than filling out a form, even if it would help locate a project and tell how many drawings were associated with that project.

When project drawings are created by conventional drafting methods, you can always look in the flat files to find the tracings. But with AutoCAD, you don't necessarily have these files or need them. The drawing can be stored on disk or tape, and placed in a vault. Now, how will you know which project has which drawings associated with it? By looking in the drawing number log, of course.

Automating this report form may help you in keeping the data on hand. While you're at it, you can store one copy with the project and another in the project work book archive files.

The EXPORT program extracts drawing title information. The file is named 8910-INF, and it contains all of the attribute data defined in each drawing when an AETITLE drawing is inserted just before plotting the project drawings.

```
┌─────────────────────────────────────────────────────────────────────┐
│ Apps  Disk  Create  Edit  Locate  Frames  Words  Numbers  Graph  Print  1 20 PM │
│                                                                       │
│                                                          <Library>    │
│                                                            <A:>       │
│                                                          >▓:<         │
│[DWGLOG]═════════════════════════════════════════════════════════     │
│║PROJECT NO. PROJECT NAME  DWG No. DWG TITLE    DATE    DWG BY  REV DATE  CHK BY │
│║                                                                      │
│║8910SP      BRANCH BANK   A-1     SITEPLAN     4/4/89  JMA             JWS │
│║8910FP      BRANCH BANK   A-2     FLOOR PLAN   4/4/89  JMA             JWS │
│║8910CP      BRANCH BANK   A-3     CEILING PL   4/4/89  JMA             JWS │
│║8910BS      BRANCH BANK   A-4     BLD SECTION  4/4/89  DL              JWS │
│║8910ST      BRANCH BANK   S-1     STRUCTURAL   4/4/89  DL              JWS │
│║8910ME      BRANCH BANK   M-1     MECHANICAL   4/4/89  SW              JWS │
│║8910PL      BRANCH BANK   P-1     PLUMBING     4/4/89  SW              JWS │
│║8910EE      BRANCH BANK   E-1     ELECTRICAL   4/4/89  JMA             JWS │
│║███████████                                                           │
│                                                                       │
│                                                          [PROJECT ]   │
│                                                          [DRAWING ]   │
│                                                          [SPECIF  ]   │
│                                                          [DWGLOG  ]   │
└─────────────────────────────────────────────────────────────────────┘
```

Drawing Number Log Illustration

Method One (Requires Framework)

Use the same sequence of commands to automate the drawing number log as you did the specification reference report. First open a file cabinet and bring 8910-INF to the desk top. Then you can load the DWGLOG database and use the IMPORT function to load 8910-INF. You'll be able to CUT and PASTE the data from 8910-INF to DWGLOG.

To get a printed report, you need to first load the DRAWING spreadsheet template and again CUT and PASTE the two documents. Then use the PRINT function of Framework to generate a printed drawing number log. If you have a database for a project on file, it is a simple matter to add data from other projects to the database to keep an accurate log of all drawings.

Method Two

If you extract the drawing title data and put it in a file or group of files, you can use the DOS PRINT command to create a hardcopy. Then you can file the data and use it as a reference for future requirements. You can enhance this DOS operation by using the COPY command, copying any small extracted files from the same project into one large file, then use the PRINT command. Or you might try using the DOS SORT command to sort a directory and arrange the files in descending project number order. Then, you can archive the sorted data to disk.

Generating the Project Number Log

The last report that we will discuss is the project number log. This log has a dual purpose: you can use it both to track single projects; and as the master project log for the firm.

You can key in to the database information for the Job Number and Name Of Project fields when the contract between the firm and the client is signed, then add the Date and Project Manager when the project is completed. Or you can extract the Job Number, Name Of Project, and Date from the drawing, and then key in the Project Manager data.

```
 Apps  Disk  Create  Edit  Locate  Frames  Words  Numbers  Graph  Print  1:22 PM
                                                                  <Library>
[PROJLOG]
JOB NUMBER  NAME OF PROJECT  DATE        PROJECT MANAGER              <A:>
                                                                     >C:<
8910        BRANCH BANK      4/2/89      JOHN ALBRIGHT               <D:>




                                                                  [PROJECT ]
                                                                  [DRAWING ]
                                                                  [SPECIF  ]
                                                                  [DWGLOG  ]
                                                                  [PROJLOG ]
```

Project Number Log illustration

Method One (Requires Framework)

If you chose to extract the data for this log from the drawing, you would follow the same sequence of commands as for the other log: bring the exported data file 8910-INF, the database file PROJLOG, and the spreadsheet template PROJECT to the desk top, and CUT and PASTE the data as necessary. Then you would use the PRINT command to generate the hardcopy. You can update this database, adding new projects any time you want.

Method Two

If you used the DOS COPY, SORT, and PRINT commands to get the drawing number log data, you can use the same commands in the same sequence to generate a project number log.

Summary of System Operations

Let's briefly summarize the overall extraction and reporting operations.

- After you have completed your drawings (or during specified phases of drawing completion) you first extract attribute data using either an AutoLISP routine (like our EXPORT routine), or using standard AutoCAD commands to extract the drawing data.

- You create a file of this data, then either print the data in the file with the DOS PRINT command, or import the data into an integrated system similar to our Framework system to manipulate, store and print the data.

- If you, or someone in your firm, has the ability, you can even write a program in the macro language furnished by your software to automate the report generation.

Summary

Using programs for data extraction and report generation gives us access to AutoCAD data and commands for our own purposes, letting us expand beyond just producing drawings. We can actually create specifications and reports. For a small A/E firm, this practically eliminates the need for hiring a specification writer. For a larger firm, it will assist the specification writer in assembling a more accurate set of specifications which are compatible with the drawing documents.

AutoCAD drawing data is a resource that we can *tap* for integrating graphic and nongraphic information, a process that is an essential, but time-consuming, part of our professional lives. Starting with simply controlling how we name our drawings, we can easily extract and convert the drawing data into specifications and project management reports that more accurately match our drawings.

Now that you have seen what we have done in our own office to prepare reports using AutoCAD data, we hope that you stretch your imagination and try a few ideas yourself. For every discovery you make for increasing your AutoCAD productivity, there are usually ten other ways to accomplish the same task. You have to explore the options yourself, and select the ones that suit your operation the best. Once you get started, we think you'll find it both interesting and rewarding to explore the ways you can use AutoCAD drawing data to perform functions that you are doing by hand today.

CHAPTER 12

Systems Management

Introduction

The key components required to successfully implement a systems management strategy are simple: use good *common sense*, have a working knowledge of the system, and plan carefully. If you make the plan detailed, easy to understand and follow, your efforts have a excellent chance of succeeding. The complexities of systems management arise when you deviate from your plan.

The most unpredictable variable in maintaining good systems is the *human factor*, involving not only the people who actually use the system, but the managers as well. Architects and engineers are often not *naturals* at managing other people. Some have to expend a great deal of time and effort to be effective in management roles. To effectively manage an AutoCAD system, the best approach is to implement procedures that carefully integrate people and machines into a predetermined course. In this chapter, we will put forward solutions to some of the systems management issues that we raised in the preceding chapters.

Getting the Best from AutoCAD and the People Who Use It

As we have said, using AutoCAD involves a continual process of educating yourself. You need to constantly research changes in hardware and software if you want to remain informed. One way to stay abreast of changes is to get involved with a user group. Depending upon the size of your firm, you may want to have one or more people from your office take part in these local forums to exchange ideas. Many times, you will discover that other people have run into problems that you thought were unique. User groups, particularly if their membership is made up of people in the same profession, can provide solutions to your problems.

User Groups

We recommend that you organize a user's task group within your firm, assigning responsibility for attending group meetings armed with hit lists of issues or problems to resolve. User groups are a good resource for information on techniques and procedures for enhancing your system production.

Conventions

Each year national conventions, like the one called A/E SYSTEMS, are held for education and marketing to architects and engineers. These conventions are held in major cities throughout the country and showcase the latest in hardware and software. The organizers hold seminars and offer tutorials, including systems management. These seminars are often general. However, if the speakers don't address the issues where you need assistance, you usually can find *someone* at the convention who can answer your questions.

Large displays are usually prepared by hardware manufacturers and software firms, giving you a chance to try out not only new computers and software, but output devices like plotters, printers, and scanners. We suggest that you and your firm attend these events. They will provide you with an opportunity to exchange ideas with fellow practitioners from other parts of the country who may, in fact, have already solved your current problem or concern. Conversely, you may have a solution that others may be seeking.

Project Planning

Project planning means different things to different people. Very simply, project planning is an orderly road map that you create to define a project's design and production process from the beginning to the end. How thoroughly you prepare this map and follow its course determines how successful the end result is. As architects and engineers, one of our greatest assets is our ability to plan. Most often however, we are only concerned with the end result — the building itself. Doing a thorough job of defining the steps necessary to reach that end is generally where we falter.

Time constraints set by our clients and premature promises on our part, more often than not, make it impossible to keep on a schedule without expending exhausting hours just before the due date. As a result, almost all architects and engineers have suffered through *the eleventh hour crisis.*

Design Phase

The design phase is always the most difficult portion of any job in terms of planning our time and resources. Because it is the most creative phase of our practices, it is the most difficult to *forecast* a completion. As a result, we usually continue to design right on through the subsequent phases of the work until the documents are finally released for bidding.

Every architect encounters difficult design problems that require more time than he or she first realized for sorting data and organizing it into a solution. There are no *pat* answers or solutions for a system of better management in the design phase. AutoCAD, however, does let you assemble portions of your work with prototype drawings and blocks, which certainly helps to create more complete and well-thought-out design solutions.

In the case of the simulated Branch Bank project, much of the drawing data for the equipment, toilet and partition enclosures, and furnishings can be stored in the system's database and retrieved for quickly solving a design problem in another project, whether it is a bank, an office building, or a hotel.

Critical Phases

Where project planning becomes most critical is during later phases of the work, like the design development and construction document, or working drawing phases. These phases of your practice involve the most time and resources, and therefore require you to rely even more on good management skills. To help illustrate some of these management issues, let's focus on construction documents and expand on the management tools we talked about in earlier chapters.

AutoCAD vs Manual and AutoCAD with Manual

New users of AutoCAD have the tendency to try to put everything on the system. We don't fault their enthusiasm, but question how practical this endeavor is. A newly installed system is virtually *ignorant* until you can teach it something about what you do in your practice. Its library is limited, but it is eager to gain from your knowledge and experiences. You should take a realistic, controlled approach to your initial integration of AutoCAD into your office.

We suggested in Chapter One that you consider primary drawings, such as floor plans, for your first ventures into AutoCAD. You and your engineering consultant will benefit by having these *backgrounds* prepared for creating the mechanical, electrical, and plumbing documents by manual methods, either by drafting directly on the plots, or by using them as base sheets with the popular *overlay* pin bar system.

How to structure your procedures is a management consideration. Mixing AutoCAD and manual drafting techniques in this way is a *natural*. Most of the time, modifications and revisions will require a tedious review process to insure that you and the engineer are using the same plan.

We suggest that you place annotations on the binding side of the drawing indicating the drawing name and the date the plan was prepared and/or revised. This gives you a way to track which plans are the most current. We also think it's a good idea to keep records of all your drawings by name and issue dates in a drawing specification log (which we discussed in Chapters One and Eleven).

Deciding which portions of your work to create on the computer system and which portions to do manually depends on how soon you intend to convert certain portions of your current manual efforts into the AutoCAD system. Set short, intermediate, and long term *sights* on your achievement of this goal. We offer the following as guidelines to help you increase your integration of AutoCAD into your practice while using manual drafting methods for other tasks:

Short Term Production Goals:
Prepare all plan and related plan work on AutoCAD
Build a library of frequently used blocks and symbols
Use AutoLISP routines for routine *short-cuts*"

Intermediate Term Production Goals:
Prepare all elevations with AutoCAD
Prepare all small scale building sections with AutoCAD
Continue to develop your symbol library
Develop other customized uses of AutoLISP routines

Long Term Production Goals:
Prepare all small scale detail and section work on AutoCAD
Integrate specification writing with AutoCAD
Develop a *customized* menu system based on the project type

Production Management

Production management is a *science*. The person who holds the position of project manager needs several key qualifications. He or she must maintain continuous contact with, and provide input to all phases of design and production, anticipate and plan all activities associated with the various phases of the work, have extensive experience in building technology, be a goal-oriented person and have the ability to *measure* accomplishments through the entire production process.

Integrating AutoCAD in a firm adds a new dimension to production management. Your production manager must be knowledgeable in using the system itself, determining the best balance between system use and manual tasks. His or her primary responsibility is generally considered to be authoring the contract documents. Many of the management tools we discuss are directed towards helping the project manager.

Project Work Book

We advocate creating a project work book for every project that you produce. Unless you organize responsibilities differently, we think the production manager should maintain this work book so that all the information and material relevant to a project's production is centralized. The larger the scale of the project, the more important its project work book becomes.

The following is an accounting of what we consider the essential contents of a good project work book. (See the illustrations at the start of this Chapter and in Chapter One for examples.)

Project Flow Chart

A project flow chart is simply a graphic representation of the production process. It describes the work in detail, the method in which it will be produced, and the *time-table* for production, and the events that must occur. This document helps you establish the order in which your data will be created. This is significant, because it helps make sure that adequate information is conveyed not only to personnel within the office, but to your client and outside consultants as well.

Drawing Specification Log

This log is intended to provide a record of the *vital statistics* of all the drawings to be produced for a project. It contains information about the drawing data as described by the computer system. With this record, the production manager can not only verify the characteristics of each drawing in the construction documents for the current project, but has detailed access as a *database* reference for planning future projects.

Specification Log Sheet

This document provides a *tickle file* for items to include in the project specifications and is based on the Construction Specifications Institute (CSI) format. The project manager can maintain a log of the items that must be included in this portion of the construction documents.

Project Drawing Layout

This document serves as a utility for producing a *mock up* of the actual drawings to be produced for the project. This allows the project manager to predetermine the extent of the work to be included in the contract documents. In addition, it gives you a way to see which drawings will be produced on the computer system and which will be produced manually. Finally, you can use the project drawing layout to determine which individuals in the firm will be responsible for producing the actual drawings.

Project Log Sheet

The project log sheet is simply a record of the project's assignment, including the personnel and critical dates involved.

Project Number Log

The project number log is a record of all the drawings produced for a project. It contains vital information like drawing descriptions, sheet numbers, dates of issue and revision, and distribution. An AutoCAD system running under MS DOS only provides for drawing names that contain eight characters/numbers or less. The number log document gives you the ability to expand drawing descriptions so that they are more easily recognized and retrieved.

System Standards Manual

The best tool to use for managing your AutoCAD system is a standards manual. If everyone would adhere to and maintain the standards, your management responsibilities would be minimized.

Your standards manual should contain information about all the elements of your system, from the format of your project work book, to how to manage the hardware. Each area should be clearly defined and presented in a way that is logical, easy to read, and easy to follow. All the symbols that you currently use should be illustrated, and all your custom AutoLISP routines documented.

Having the information in the standards manual available to everyone, and enforcing strict adherence, can forestall the majority of the problems that occur in operation of your system.

While any user of AutoCAD is a potential contributor to the standards manual, it should be the AutoCAD manager's responsibility to decide which items should or should not be incorporated. We recommend that you distribute a copy of the manual to every employee who uses the system. An up-to-date manual is a must. It pays to always revise and fine tune the manual so that it reflects your most current methods.

Managing Drawing Setup

The first step toward insuring your success with AutoCAD is to make certain that all the drawings of a new project are begun in the same manner.

If every drawing for a project is going to be created by a single operator, then you won't have a problem.

If the project is going to be comprised of drawings created by different operators, you will probably see a lot of variation when you plot the final drawing set. If the drawing formats vary greatly, you will have to make extensive changes to the drawings just to make them consistent with each other. This will require additional time and effort, usually when you can least afford to expend it.

Remember, the objective is to increase productivity. This is another good reason why a standards manual is important. The difficulties created by using different drawing formats should be problems of the past.

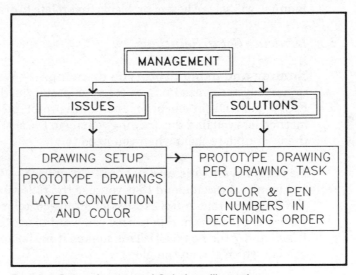

Drawing Setup, Issues and Solutions Illustration

Drawing Setup Issues

The major issues concerning drawing setup involve using prototype drawings and choosing layer names and associated colors. If you set up and use different prototype drawings (which you can select at setup time or specify with the drawing name) to match the drawing task you wish to perform, you can preset the variables required for that task, such as scale factors and layers. While prototype drawings are an additional resource to manage, they do help you keep your drawing sizes to a minimum and save unnecessary regeneration time.

We recommend that you tailor separate prototype drawings to drawing tasks as opposed to creating one large, generic prototype drawing. In a generic prototype drawing, every layer, variable, and scale factor has to be configured in, resulting in large drawing files that may contain data that is unnecessary for individual tasks. As we mentioned, common sense and good drafting practices can make a big difference in the performance of a workstation.

The following list shows management issues that need to be resolved before they become systems management problems.

❏ How to name and use prototype drawings

❏ What layer naming convention and associated colors to use

Your decisions on these two issues, when documented in the standards manual, will solve a host of problems that relate to drawing format and setup.

Drawing Setup Solutions

Naming and using prototype drawings — We recommend that you coin easily-recognized names for your prototype drawings, for example: ELEV and DET for elevation and detail prototype drawings. This simple method lets you find the drawings easily in the system so that you can use them, or update them when you need to.

Layer conventions and associated colors — When assigning colors to layers, we recommend that you use the following approach. Correlate the numbers that AutoCAD assigns to each color (1 = Red, 2 = Yellow, etc.) to the incremental pen sizes you want to use for plotting (1 being the finest and 7 the heaviest). This makes it easier to remember which pen goes in which slot on the plotter.

By using these simple techniques to manage your drawing setup procedure, you will find that your workstation and plotting performance will improve and your drawings will become more consistent.

Managing Symbol Libraries

Back in the days of conventional drafting methods, architects and engineers developed a symbol library, of sorts, by saving reusable details on 8 1/2" by 11" paper. These details were saved in a file where they could be retrieved for future projects. This library became the most valuable item the architect or engineer had for increasing productivity. The value of reusable entities has not changed over the years. In fact, AutoCAD let's you insert these reusable symbols and drawings into a new project just as if they were developed from scratch.

Symbols libraries are the main ingredient of a productive AutoCAD system. Without these libraries, every time you needed a lavatory in your drawing, you would have to draw it. If you consider how many lavatories you might have to individually draw for a large project, you can see that the process would be time-consuming, to say the least. But the catch is, as you continually build the system, your symbol library becomes more and more valuable.

Management issues arise when symbol libraries become too large and become out of date. Workstations will slow down because of the amount of time required to search the library. Even if you wanted to remove the older, out-dated symbols, it is only possible by inserting each symbol to figure out what it is. This is a management problem that will require many hours of tedious work to correct.

The solution is to continually monitor the library building process, purge the older, unused symbols or modify these symbols for your current and future use.

Symbol Library Issues

The best approach to eliminating problems with symbol libraries is to prevent the problems from occurring in the first place. In the days of conventional drafting methods, it was easy enough to thumb through the file cabinet, locate the detail you needed, copy it onto *sticky back* clear film with a copier, then paste it to the drawing.

With today's emphasis on speed and productivity, you can't afford to spend your time keeping track of your symbols this way. You now have access to a computer file cabinet, and can find and insert the symbols you need in a fraction of the time it used to take. Remember, AutoCAD is a lot faster at retrieving and editing existing library symbols than you are at revising hand-drafted, reusable details.

Speed won't make up for a lack of organization, however, and that is why the following issues must be resolved in the early stages of system integration so that they don't grow into major problems.

Here are the major issues:

❑ How do you document the symbols in the library?

❑ How do you name symbols?

❑ How and when do you update symbols?

❑ How do you use symbols with AutoLISP?

You may want to review Chapter Two where we discuss how to use symbols libraries.

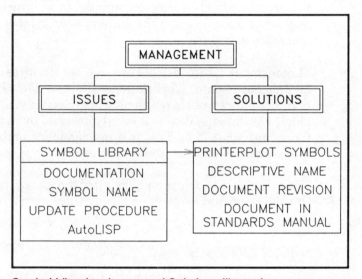

Symbol Libraries, Issues and Solutions Illustration

Symbols Library Solutions

Everyone has heard the old saying, "An ounce of prevention is worth a pound of cure." If you can realize that these symbol-libraries management issues are unavoidable, and step up to address them before they become major problems, you will get faster more effective solutions.

The following suggestions are ways to help manage your system's symbol library:

Document the symbols in the library — The best method for documenting the contents of your symbol library is: to run a printerplot of the symbol as you create it; and (before you do *anything* else) write its name on the plot, the date it was created, and the project and drawing where it was inserted. Then, you can file the drawing away in a three-ring binder labeled "Master Symbols Library" for future reference. Also, make a copy of the drawing and put it in the project workbook (or binder). That way all your symbols will be easy to find and will be documented as to the dates they were created.

Naming symbols — The best name for a symbol is a descriptive one that is easy to recognize for retrieval. Depending on the size of your library, symbol names may have to be a combination of letters and numbers to represent the CSI format. For example, you might want to name the first aluminum window that you create, of a certain type, W8300/1. The second window would be W8300/2, and so on. Using this method will help your specification writer identify those products that he or she must include in the specifications. Because of the limitations placed by MS DOS, you will be limited to eight characters for your naming convention.

Another way to name your symbols is to use letters to identify a symbol and numbers to define the order in which it was created. For example, if you needed to identify a bath tub you could name it TUB-001. The second tub would then be named TUB-002. This method will help you track symbols and their frequency of use. If you don't use TUB-004 often enough to justify keeping it within the library, you can delete it from the system. Never reuse a symbol name once it has been deleted from the system.

➡ *NOTE! Try to make it a practice to never reuse a retired symbol name (one that has been deleted from the system), since you might just overwrite a more current symbol when you reload a tape backup or archived floppy.*

Updating symbols — You should update symbols as changes are required and document the new symbol with a printerplot, just as you did when you originally created it. If you have a tape backup subsystem, it would be a good idea to first back up the symbol on a specially-designated tape for symbols, then do your update. This way you'll have a restorable record of the old symbol in case you ever need quick access to it. Again, to prevent overwriting a more current symbol, always avoid reusing a symbol name that has been deleted from the system.

Using symbols with AutoLISP — As you went through Chapters Two and Three, you may have noticed that for a majority of the book's project symbols, we used associated AutoLISP routines for inserting and assigning attributes or scale factors. If you are using AutoLISP routines, it is a good idea to include the associated AutoLISP routine in your project work book documentation of the symbol. Many times when a symbol is updated, variables associated with the symbol also change. In turn, these affect the AutoLISP routine. Having the symbol and AutoLISP routine together gives you a way to update both the symbol and the routine when you need to.

Managing Menu Systems

In Chapter Three, we discussed how to create or modify a menu system, and how to use that system as a tool for production. But the old adage that "nothing in life ever stays the same" holds true for menu systems. What gives you the best function today may not give you the best function tomorrow. That's because your entire AutoCAD system is built upon change. Each new release of AutoCAD changes some of the ways the system works and the functions it provides. A menu system that you (or someone else designs) to operate with Release 9 will be missing many of the key functions that a menu system designed to operate with Release 10 can provide.

A menu system for Release 9, for example, would lack the advanced functions for creating 3D entities, which were used in the presentation drawings. It also would lack the advanced AutoLISP functions that enhance the operation, function, and speed of AutoLISP routines. You wouldn't be able to define a User Coordinate System or use the Viewports command to show different views of the same drawing on one screen. You also couldn't make enhancements to 3D perspective views of a buildings.

Menu systems have to allow for change. For this reason, managing your menu system is vital to the long-term usability of your AutoCAD system.

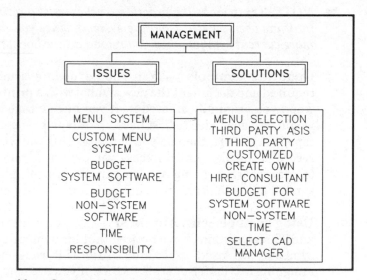

Menu Systems, Issues and Solutions Illustration

Menu Systems Issues

A good menu system is the backbone of any efficient A/E firm's AutoCAD operation. It can be based around screen menus, tablet menus, or a combination of both. But, if the menu system isn't budgeted for, maintained, and managed properly, problems can arise which either drastically reduce the menu system's effectiveness as a production tool, or disable its use altogether.

The following is a list of major issues and problems that surround menu systems. Carefully considering and resolving these issues will help you maintain optimum production performance.

❑ How do you obtain a custom menu system? Should you *roll your own*, purchase it from a third party, or hire a consultant to design it?

❑ How much do you budget for system software updates? What can you expect to pay for updates to a third party menu system, or for a consultant to update a custom menu system?

❑ How much do you budget for nonsystem software? How much does any additional software cost that your menu system may require for report generation or other advanced functions?

❑ How much time do you expect to spend managing the menu system? How much time can you allot for maintaining and updating it?

❑ Who's gets the job? Should you give one person the responsibility of maintaining the menu system?

If you don't decide on the best ways to resolve these issues, problems will arise to haunt you — if not immediately, sometime in the future.

Menu Systems Solutions

Obtaining a custom menu system. — Menu system designs range from the familiar and standard AutoCAD overlay template menu to some that are extremely advanced and complex (which may take a full time employee to maintain). The best menu system, however, is the one that incorporates all the functions that you need for your particular method of operation, plus gives you some special options or routines which let you customize features. Your decision on whether to purchase or create a custom menu system should be based on your staff's abilities and some common sense. Your major choices are:

■ Select a third party menu system and use it as is.

■ Select a third party menu system and customize it.

■ Design and create your own custom menu system.

■ Hire a consultant to create a custom menu system.

The first choice, purchasing a third party menu system, is the least expensive, but you will sacrifice some productivity since there is no single, purchasable, menu system that will meet all of your menu needs. If you purchase a menu system with the intention of using it *as is*, you will probably find one that is either very inexpensive, but does so little that the template only makes a good digitizer dust cover, or is very expensive and so complex that it actually slows down production.

The second choice may be your best bet depending on your staff's expertise and experience in customization. If you have someone in your office who has customization experience, then you can purchase the less expensive menu system, and add your own functions to it.

The viability of the third choice again depends on your staff. If there is someone in your office who has the expertise to create the whole menu system, then what better way to insure that your system is totally customized to your requirements? No one knows your operation and needs better than you and your own staff.

The fourth choice is a good one if you don't have the staff available to create or customize a menu system. Consultants, even though expensive to retain, can create your menu system based on your requirements, and may actually give you the largest productivity gains for the least amount of time invested.

Budgeting for system software updates — A whole set of problems can revolve around budgets (or the usual lack of budget). Just because you have purchased a menu system, or have paid a consultant to create it, doesn't mean you're finished paying. What you believed at the time to be the *latest and greatest* will have to change to keep up with progress in the underlying AutoCAD software. To keep production at its highest level, you have to make provisions for updating the menu system as you purchase updates from AutoCAD.

Third party developers generally offer *upgrades* at a cost. Once your system is customized, if no one in the office has the ability to update it, then you'll have to allocate funds for that purpose as well.

Budgeting for nonsystem software — There is the matter of adding new software or updating existing software that you use in conjunction with the menu system (such as Framework III or Microsoft Word that we use in our own system). Budget issues are important because any drop in production usually means a larger bite in operating funds. The bottom line is maintaining productivity. Even though you may think that putting, say, five thousand dollars, in the year's budget for new software and updates to existing software is a lot, the issue is whether you are buying some production gains with your budget.

Spending time managing the system. And who gets the job — The last two major issues involved in managing a menu system, are those of time and responsibility. These issues deal with the employees in your office who use the AutoCAD system. No matter how much software you purchase or how many consultants you hire, the people who use the menu system need to be directly involved in the design and operation of it. They are the ones who really know how the menu system needs to function. Without their input and commitment, your menu system will probably flop.

From our experience, we've found that once you put a third workstation into operation, you'll need to select an AutoCAD manager, who may be an operator, or at least have significant experience using the system. Your AutoCAD manager will need to oversee the system's operation, research the uses of third-party software, and solve the hundred or so little problems that other managers in the firm just don't have time for. For most small firms, this doesn't have to be a full time position — the AutoCAD manager can take care of his or her management responsibilities and produce drawings as well.

The most important thing is that he or she be permitted enough time to maintain the highest level of production possible. A major mistake that management seems to make concerning the production and operation of a system is in not giving the AutoCAD manager enough time or support to do his or her job right.

Menu System Options

No matter what menu selection route you chose to take, you should include consideration of the building types that your firm designs. For example, your firm might be involved in multifamily and single family residential, and hospital type projects. These building types have doors, windows, toilets, and millwork, all of which are candidates for library symbols.

However, the actual items are distinctively different for the two building types. In an apartment project, we generally use 2'-6" and 3'-0" hollow core doors with wood frames, while for hospitals we use 4'-0" laminate-coated doors in aluminum or hollow metal frames.

The window types show another dissimilarity. You wouldn't want to have both a 3'-0", single hung, aluminum window symbol for an apartment project and a 1 3/4" x 4" flush glazed window symbol for a hospital project on the same menu because of the physical space limitations on the tablet. Neither of these two symbols is applicable to the other project type.

If the symbols you use for different project types differ significantly, like these examples, you can probably justify using separate menu systems. When you select or create a menu system, consider basing your customization on the building types you generally work with. You can increase your productivity and produce more accurate results.

So far, we have discussed issues and problems that relate to system integration. Now let's move on to some issues and problems that can arise during the design development phase of AutoCAD drawing production.

Managing Design Development

AutoCAD is often overlooked by many A/E firms as a design tool. Don't be misled into thinking that today's AutoCAD is only good for producing 2D drawings, and it is not worth the time and effort for producing design drawings.

With Release 10 and its ability to create full 3D drawings (and by using AutoLISP to aid 3D operations), AutoCAD has become just as valuable a design tool as it has been a production tool. If you don't use AutoCAD in design, you're only using half of the functions already built in to the software. Why not use it fully? The added flexibility of being able to make quick changes for various design solutions and to reuse the design data for the base of the working drawings will definitely add to your productivity.

The management issues that arise when you use AutoCAD for design development are different than those we just discussed. They are less of a financial and employee burden, and more a matter of managing data. These issues have to be addressed if you want your use of AutoCAD for design to be worthwhile.

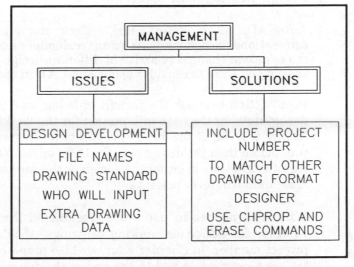

Design Development, Issue and Solutions Illustrations

Design Development Issues

When you integrate AutoCAD into the design development phase of project creation, a whole new area of production gains opens. Concurrently, a whole new group of management issues arise concerning how the data will be created; and how it will be incorporated into the working drawing phase of the project. Most of these questions can be resolved by just using logic and sound judgement, but others will take a little more time to deal with.

The following is a list of the major issues that come into play when you integrate AutoCAD in the project design phase. Try to address these issues before any integration takes place.

❑ What file names do you use with design drawings?

❑ What method or standard do you use to create the design drawings?

❑ Who will input the design data — the designer or the AutoCAD operator?

❑ How do you handle the extra drawing data created for design purposes?

Some of these issues may not seem *earth shattering* at first glance. However, the problems that arise from not resolving these issues are more difficult to fix later on. By the time they become a problem, you often have to learn to live with them.

Design Development Solutions

AutoCAD, as a design tool, offers numerous advantages over conventional design methods using trash paper and pencil. The ability to make design changes to arrive at different design solutions is the major reason you want to consider using AutoCAD in the design phase.

People often overlook the benefit of being able to reuse the AutoCAD design data for the base information for the working drawings, without having to re-input the data. This ability has a greater overall impact on production than the design benefits themselves. Your optimum use of an AutoCAD system in either the design *or* the working drawing phase is to *reuse* data wherever possible.

What file names to use —We recommend that drawing file names, whether for design or working drawings, should always include the project number. In Chapter Four, we also proposed the use of the letter "D" as a notation to add to the end of the file name to indicate that a drawing is a design type and not a working type. For example, we used the name 8910FP-D for our floor plan design drawing for the branch bank. This gives you a good way to keep track of all the individual drawings you produce, by project number and by phase

What methods and standards to use — Choosing methods and standards for creating design drawings may not seem that important when they are used only for presentation purposes. But you can reuse the design drawing data in the working drawing phase, so it's logical to use the same methods and standards in both phases.

If you create the design data using layers and symbols that you don't use for the working drawing data, you'll have to spend a lot of extra time changing the drawing entities to the right layers and replacing the symbols. This obviously reduces your productivity.

Who will input the design data — The issue of who will do what in producing design drawings is the one that always seems to cause turmoil. Should the AutoCAD operator, whose tight schedule demands full time use of a workstation, give up that workstation to the designer (who might not be able to operate AutoCAD as efficiently) so that he or she can input the design?

Or should the designer have access to an extra workstation that can be shared on a rotational basis? Or should the AutoCAD operator and the designer sit together with the operator drawing, and the designer guiding?

Usually, it's very ineffective for an AutoCAD operator to be involved in creating design drawings at all. We have found the best approach is to have an extra workstation, maybe not the best, for the designer to use. Then you don't have to pull the AutoCAD operator off his or her workstation (and away from what they are most productive at). The operator's time is worth a lot more than the cost of an extra workstation.

Handling extra drawing data — One issue that we have not addressed yet is workstation performance. When you create a design drawing, extra data in the form of 3D entities is produced. As a result, if the workstation you use is a standard IBM AT with an EGA monitor, for example, its performance using design data in the working drawings will gradually deteriorate.

If you recall in Chapter Six, we reused the branch bank design drawing as the base for our working drawing plans. We did not use any special techniques to clear out all the extra data because the project was small and performance degradation was not significant.

Think of how much 3D data would be carried over if you were using the design data of a twenty story office building for the base of the working drawings. In this case, the extra data generated for the design phase *would* affect the overall performance and production ability of a standard workstation.

To resolve the performance problem, you can begin the new working drawing as you did in Chapter Six by specifying the New Drawing Name=Design Drawing Name. Then, use CHPROP and change all thicknesses to 0 to eliminate a portion of the 3D data. You also can erase all other data that does not apply to the 2D working drawing.

If the project is complex and the 3D image will help you create the working drawing, then rotate the design drawing into a 3D view and create slides of it. Then use CHPROP and ERASE to correct the drawing. This will give you the data you need, and not cause the workstation's performance to suffer.

We have tried to address some major management issues related to integration and producing design drawings. Next, let's look at issues revolving around managing production drawings. This is the area where management issues can create the largest problems, resulting in the greatest loss of productivity.

Managing Project Production

AutoCAD became the number one micro-based CAD software package because of its flexibility and relatively low initial cost. Also, Autodesk has continued to support production needs by developing new releases that *handshake* with old, entered data, thereby preventing drawing obsolescence. In addition, third party support and attachments to AutoCAD have enhanced its flexibility and power even more.

AutoCAD's is a *user friendly* system that electronically draws lines similar to the way architects and engineers draw them manually. The *overlay* drafting system was a revolutionary production method that had similar procedures and techniques that let you make the transition into AutoCAD. Very simply compared, both systems allow you to *toggle* layers to get the desired results. Both systems also let you *cut and paste* old drawings and data to create new drawings.

But with all this flexibility and power, management issues become inevitable. The next question is "How do you manage this flexibility and power?".

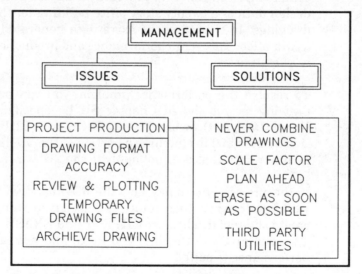

Project Production, Issues and Solutions Illustration

Project Production Issues

A/E firms using AutoCAD seem to fall into three categories.

There are those that fail in their attempts to integrate and use AutoCAD. Their efforts to get *up to speed* become such a management problem that they characterize the system as unusable and abandon it for the more familiar manual drafting methods.

There are those that succeed and go on to becoming experts. They work through and resolve management issues and are the leaders in AutoCAD use today.

And finally, there is a large group somewhere in the middle, struggling but still sticking with it. These *never say die* souls keep running up against a series of walls, solving the management problems that plague their operations.

The following list outlines some of these production issues, geared towards this last group of users.

❑ How do you format drawing data?

❑ How do you define the accuracy and tolerance of drawings?

❑ How do you review drawing files and plotting?

❑ Should you create temporary drawing files?

❑ How do you archive drawing data?

If you are using conventional production methods, equivalent issues have long since been settled. But with AutoCAD and computers, a whole new dimension has been added as a you need to resolve these management issues to reach a goal of gaining productivity in the production phase.

Project Production Solutions

In the days before AutoCAD, these management issues were relatively simple to solve. One reason was that every production effort was visible at all times. If you wanted to review a drawing's progress, it was generally taped to a draftsman's drawing board.

We have all experienced having the *boss* look over our shoulders to help with a tricky construction detail. In addition, check prints or progress prints were not as difficult to get because we didn't have to wait on the computer and plotter *to do its thing* before a print could be run.

If these issues are of concern, we hope to give you some ideas to help resolve them.

Formatting drawing data — Our main concern with formatting data is to input it in such a way that it doesn't hinder workstation performance. You want to minimize vector and plotting time when progress and final plots are needed. If you plan your data entry and organize your drawing files so that you minimize the file size, your final output and the general performance of your computer will be quicker.

For example, if you used AutoCAD to create a floor plan, ceiling plan, and roof plan on the same drawing, and you modified the layers for each, the amount of data you would have generated by the time you reached the roof plan would seriously impact the performance of your workstation. Ultimately it would impact the plot itself.

The larger the file, the slower the response time. This is the reason we created a very simple floor base plan drawing in Chapter Six, made a block of it, and used it as guide for the other plan drawings. Think ahead as you create a new drawing. Think about what data from your current drawing you will need for the next drawing. When you have completed that portion and that portion only, create a block of it so that you transfer the minimal amount of data to the next drawing.

Accuracy and tolerance of drawings — One of AutoCAD's assets is that it can maintain accuracy up to 16 decimal places. You might argue that since a contractor can't build a building to those tolerances, why bother taking the time to enter data to that degree? To a point you are right. But, you still must base your degree of accuracy on the *drawing task itself*.

For all practical purposes, floor plans for most building types can be worked to the closest quarter of an inch (0.25, two decimal places). Construction details and sections should generally be set to a closer tolerance of a sixteenth of an inch (0.0625, four decimal places), or even a thirty-second of an inch (0.03125, five decimal places).

Accuracy and tolerances are no less important in designing and producing the drawings for a building than they are in actually constructing the building. Consider that if you draw a column grid system without the appropriate degree of accuracy, other building component like the walls, partitions, and openings will also be inaccurate. Entering accurate data doesn't really take any longer than entering inaccurate data. One of AutoCAD's greatest tools let's us pin the intersections, or endpoints to place a single dimension or a series of dimensions.

Let's use a column grid example. Say that you completed a column grid layout, but because of a design change, you had to move a single column line 3 1/2". Say you just adjusted the dimensions involved. For all practical purposes, the floor plan, plotted at 1/4"=1'-0", is not going to show this *out of scale* error. But what happens to every other building line that is referenced to this single column line? Who is going to keep track of every inaccuracy that was overlooked during the production phase?

The consequences may only show up to haunt you during the construction phase. What happens? You guessed it, the contractor calls your office to tell you that the wall is going to have to be relocated because it misses the plumbing roughin by the 3 1/2" dimension. Now you have the onerous task of explaining the change order for this relocation to your client. Embarrassing? To say the least.

The moral of this story is — when AutoCAD provides you with the ability to maintain accuracy in your drawings, use it.

Reviewing drawing files and plotting — Reviewing drawing data before you plot the drawings is an important issue. Using a plotter is time-consuming and can actually hamper production compared to producing drawings by hand. Planning for this task becomes more critical with AutoCAD than under manual methods. With AutoCAD, you can create your working drawings with such speed that it's easy to miss mistakes. One undiscovered mistake in design or format can precipitate massive changes later.

Improving your technique will help, but nobody is perfect. There will always be mistakes. We've found that the best solution is to keep mistakes from tying up your plotter in the first place. Spend time reviewing the drawing on the screen instead of being frustrated at the unpunctual, plotted results.

To save computer processing time you can implement a third party plot spooling program, use an electrostatic plotter (instead of a pen plotter), or dedicate a computer or hardware device to plotting. You can do this on a standalone, or in a network. This will save you from having to tie up a workstation and an AutoCAD operator for check plots. These hardware and software devices will let you spend more time in the drawing phase, instead of consuming productive time with plotting.

Creating temporary drawing files — With conventional drafting methods, we frequently create or work out details using *trash* paper before the details are incorporated into the final working drawing. With AutoCAD, we can create and plot (or printer plot) temporary drawings for the same purpose.

Be careful if you do this.

All too frequently, we don't purge these temporary AutoCAD drawings from the system when we complete a project. Instead, they just take up valuable storage space on a workstation, or even worse, on a file server. They also slow the computer's processing time. It's easy to forget about them until storage space becomes limited, and by that time, no one seems to know if the drawings are worth keeping or not. Most often, you are stuck with either leaving worthless data on the system or dedicating an inordinate amount of someone's time to sort through all the data to determine what to keep and what to *pitch*.

Another danger is that if do you allow old, unused data to build up to such a *critical mass*, there is always the possibility of losing some current data during the cleanup process.

We suggest that at the end of each project's working drawing phase, before you archive the data, you plot and review each temporary drawing left on a workstation or file server to see if the drawing, or some portion of it, is worth keeping. Then, one person should take the responsibility for removing the unusable work.

Archiving drawing data. — The final production management issue that we want to address is determining how to archive the data once you've completed a project. In Chapter Two, we discussed using a four digit number (representing all or part the project number) as part of the drawing file name. This becomes even more important when you have completed a project and you want to archive the drawings. Labeling a diskette, tape, or optical disk with a project name and number is fine for today, or maybe even for a year from now. But how about three, or five, or ten years from now, when the client wants you to design an addition for the project?

Paper labels fade and are sometime inadvertently removed. If a label is unreadable or gets thrown away, how will you be able to distinguish what project the data belongs to, especially if you use common file names, like FLR1, FLR2, CL1, or EL1 between projects? In this case, your only recourse would be to load the drawing on the system and call the drawing up to look at the title block. If a number of labels are missing, this could take you a lot of time.

If you use our method of including the project number in the file name, you could simply perform a DOS DIR (directory) command and read the job number on the screen. Simple.

Your approach to resolving these issues, as they relate to AutoCAD, is using a standards manual for everyone to follow. Otherwise, you may be hitting your head against the wall in frustration for a long time.

Managing Communications

In the days before AutoCAD, communication between architects and engineers required the architect to produce a blue line print of his or her design and working drawings for the engineer to use. The only management requirement that existed then was to correctly fill out a letter of transmittal showing which drawings were sent, when they were sent, and to make sure the engineer received the latest revisions during the course of the project. This was a very simple and straightforward process.

The process, as you know, is no longer straight forward for firms using AutoCAD. Communication issues can cause a real management headache.

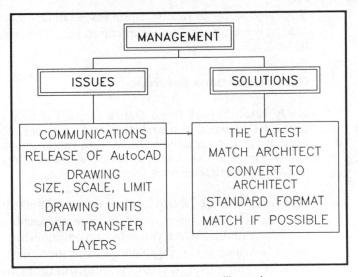

Communications, Issues and Solutions Illustration

Communications Issues

Transferring AutoCAD drawing data between architects and engineers is different than transmitting drawings created by conventional means. The ideal scenario would be for both the architect and engineer to use the exact same drawing format and data input methods. No extra time would be required by either party to resolve drawing incompatibilities.

Like architects, however, engineers have adopted their own formats and procedures, suited to their own needs. Conversely, engineers frequently provide services to various architectural firms who most likely have their own individual drawing formats and methods.

The issue is how to merge two different formats without causing either party a lot of work and a loss in productivity. To make such a merger possible, and to determine the best method of communicating drawing data, the following major issues need to be resolved with a preproduction conference between the architect and engineer.

❑ Which Release of AutoCAD do you use?

❑ Which drawing sheet size, scale and limits do you use?

❑ Which format of drawing units do you use?

❑ What medium do you use to transfer data?

❑ What layer names and colors do you use?

You need to try to resolve these issues before a project begins to insure that at least minimal communication requirements are met. Other issues do still exist.

Communications Solutions

Early resolution of these issues (above) should minimize any related problems between the architect and engineer. If left unresolved, data arriving at the architect's office from the engineer (and visa versa) will cause somebody to have to spend a lot of time needlessly adapting drawing data to the *house* format.

Which Release of AutoCAD to use. — This issue should really never come up when a project is begun. The answer is simple. Use the latest release available. When you select an engineering consultant, you should not only evaluate the engineer on experience with the project's building type, but manpower availability, and whether the engineer is operating with the latest release of AutoCAD.

Drawing sheet size, scale and limits. — The standards for sheet size, scale, and limits should be established during the preproduction conference. This conference should address and establish a plotting schedule for all check plot reviews as well as final plotting.

Determining what drawing units to use is probably the most frequently debated topic between architects and engineers, especially civil engineers. Civil engineers sometimes use the COGO language for their site data input. These third-party software programs can cause major problems for an architect who wants to use the site plan drawings.

These site drawings are created with AutoCAD, but with COGO used for locating the building on the site. The building would have been drawn in architectural units of feet and inches, and the site plan would have been draw in decimal units of 10. This means that one of the two parties will have to convert their drawing units and scale to match the other. Usually, the first step the architect takes when he or she receives this type of drawing is to convert it to architectural units. Depending on the complexity of the building and the site, allowances will have to be made for this conversion.

What medium to use to transfer data. — This issue is usually the simplest to resolve. Data can be transferred between the architect and engineer by means of either 5 1/4-inch or 3 1/2-inch floppies. You can determine the type you need during the preproduction conference. In the case of archiving, both parties have to be using the same type of archiving software to insure the data's readability.

Archiving procedures involve data compression, so to assure total transferability, determine before hand which software will be used.

The second way you can transfer data is via modems and telephone lines. This method is particularly useful if the architect and engineer reside in different cities or even different states. We suggest that if you make data transmission a part of your practice, you should install a separate, dedicated telephone line for this purpose. Never use a line that has the telephone company's call waiting feature because a second caller will interrupt your valuable transmission time.

There are a number of quality modems and software mediums on the market and most will do the job. However, since you will be transmitting drawing data (which tends to be large files), we advise that you purchase a modem that transmits at baud rates of 2400 bps or greater. This will minimize the transmission time, long distance phone charges, and, of course, the computer's and operator's time.

What layer names and colors to use. — The last communications issue is one dealing with a layer naming convention. Communications are necessary only if the architect and engineer have implemented a similar naming system. Again, there needs to be some discussion regarding this subject to avoid possible problems.

Architects and engineers can work together quite successfully even if their naming conventions are completely different. This should not be confused with the reasons for maintaining a system within the separate offices themselves. Since the engineer's work is to be overlaid or layered on the architect's work, it is only important that both parties understand what each of their respective layers represent. If you need to, refer to Chapter Eleven for the reasons why we prefer to use the CSI format for naming layers. An additional benefit that this format provides a starting point for a standard between the architect and engineer.

Choosing layer colors, however, does have an impact on the project when it is plotted. In this case, the architects and engineers should maintain the same layer colors so that the final plots will have a uniform and more professional appearance.

Managing the System

In this section we will discuss some issues that surround the system management of the computer as a shared resource. We'll include some tips on how to manage an AutoCAD Local Area Network (LAN). These issues also effect the management of non-network workstations, but the impact is not as severe for *standalone* workstations.

A networked workstation, from a configurational standpoint, looks just like a non-networked, or standalone workstation. Both should have high resolution graphic displays, 80286 or 80386 class CPUs with math coprocessors to match, keyboards, and digitizers. The real difference is in what size their hard disk drives should be.

The hard disk drive for a workstation attached to a network doesn't have to be especially large or fast because the file server stores the files and does most of the work for data access and storage. The attached workstation does need to have enough disk storage to maintain the applications software and the operating system.

The following issues deal with how to manage file storage for either workstation.

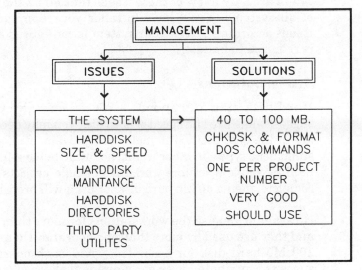

The System, Issues and Solutions Illustration

System Issues

System management issues will vary with the systems you use. There are a few major concerns that affect all systems, no matter how they are used. These issues all center on storage capacity and speed, or more precisely, the lack of it.

The worst situation you can experience in a network is to have your file server disk, where all your valuable data is stored, go down. Disk failure is not uncommon among micro-computers and there's not a whole lot you can do to prevent it. You can anticipate the eventuality, though, and make sure that backups are maintained.

The next worst situation, is to have the hard disk drives on your file server and your standalone workstation completely full of data while all files are current and active. You can avoid this all-too-common problem if you follow some good practices and think ahead about:

❑ Hard disk size and speed

❑ Hard disk maintenance practices and time periods between maintenance sessions

❑ Hard disk directories and file management

❑ Use of third party file maintenance and DOS menu programs for system management

These are only a few of these issues that effect the every day operations of all systems. You'll have to tailor your approach to resolving these issues according to how your system is configured and how much data it is required to handle.

System Solutions

Hard disk size and speed — We can only give you a general guide to suggest what size and speed of hard disk you may need for your system. You are in the best position to make that judgement. If your workstations are standalone types (in other words, not attached to a file server) and they are used by single operators working on single projects until the projects are complete, then a 40 Mb hard disk for each will probably suffice.

However, if you use the workstations for more than one project at a time, and they are used by more than one operator, then we generally prefer a 100 Mb hard disk for each. This gives each operator enough working space to contribute his or her individual share of *trash* data.

If you are using a file server, a good rule of thumb is to allow 40 Mb of storage per attached workstation. As far as speed is concerned, we suggest that you purchase a hard disk with at least a 28 millisecond-second access time.

If your system is networked, then disk access speed becomes even more important when the remote workstations need the data from the file server disk.

Disk storage prices have come down in recent years, so over-buying disk space doesn't impact the office budget as severely as it once did. Remember, the disk sizes that we've given you are just guidelines. Your individual requirements could vary substantially if your work load is heavy or if you have dedicated the system to a very large project with a large number of files that have to remain active.

Hard disk maintenance practices and time periods — One of the simplest ways to monitor a hard disk is to run the DOS utility CHKDSK at least once a week. You can even include the CHKDSK command in your AUTOEXEC.BAT file so that your disk's status gets checked every time you turn the computer on. This utility will find any lost clusters or open files that are just taking up space.

Another standard method of hard disk maintenance is to periodically back up all your data to diskettes or tape and then reformat the hard disk. We recommend you do this about every six months.

➥ *NOTE! If you use the DOS FORMAT command on your hard disk, all data on that disk will be erased. So, be sure you have backed up your data correctly and that it is stored before you FORMAT the hard disk.*

Hard disk directories and file management — There's really no difference in deciding how to name and use hard disk directories for standalone or network applications. We prefer to name all directories, like our drawing files, with the project number. This let's us make quick searches when we need to find drawing files. It also makes it easier for us to backup and archive project data when projects are complete. Again, the key to this method is being able to recognize directories and files at a glance.

Third party file maintenance and DOS menu programs — The last issue of this section is the use of third-party software DOS utility programs. There are two general types. The first are DOS menu types, for copying and deleting files; and the second are for hard disk maintenance and performance enhancements.

The DOS menu programs give you a more friendly *menu* approach for dealing with files. They display the directories and files within each disk and allow you to select files for erasing, coping, or backing up. These programs are well worth their minimal cost.

One of the most useful functions that hard disk utilities software can provide is a way to *unerase* files that you erased by mistake. This is a very attractive feature, especially since DOS doesn't respond to an OOPS command!

Hard disk utilities also provide special routines that will sort hard disk data and reformat it to optimize the disk's performance (much like reformatting the hard disk and restoring the backup data all in one step). These programs are also well worth the cost, and we recommend that you use them on each workstation to help your maintenance operations.

Summary

While we consider systems management a science, some people seem to have a sixth sense about how it is done. But, if you are not one of these fortunate few, you probably need a little help with the responsibilities like the rest of us. The size of your firm and the types of projects you design have a direct bearing on how intense your management effort may be.

If you typically design buildings with repeatable entities, such as apartments, hotels, and hospitals, you will have a somewhat easier time maintaining control over these repeatable blocks, drawings, and symbols. Following the simple guidelines we have provided will help you be even more successful using AutoCAD.

If you typically design buildings that contain fewer repeatable entities, you will have to establish a more detailed method of management to avoid spending an inordinate amount of time searching for reusable data and recreating symbols.

Systems management may well be the key to successfully integrating AutoCAD into your business. Hopefully, the issues and solutions that we have presented in this chapter should give you a *handle* on the problems that can arise from *poor* methods of planning, production, management, and maintenance.

Hopefully, we also have given you some ideas for tools to keep the issues from becoming problems in the first place.

Earlier in this chapter we described three types of AutoCAD users. If you are an AutoCAD user who has *run aground* with problems using AutoCAD, don't abandon the ship. It's never too late to get your system back on course. The guidelines we have provided in these pages will help you do that.

For those of you who have become experts at using AutoCAD, our hats are off to you. We hope that this book provided new ideas and techniques that you can use and develop even further.

And finally, for the *never say die* individuals for whom we wrote AUTOCAD FOR ARCHITECTS AND ENGINEERS, we hope our efforts, techniques, guidelines (and a little practice) will move you up in the ranks to join the experts.

Techniques and AutoLISP Routines

Techniques

Chapter Four

Technique for Property Line Input
Technique for Creating Setback Lines
Technique for Creating Setback Block
Technique for Calculating Areas
Technique for Floor Plan Area Layout
Technique for Adjusting the Areas
Technique for Defining Wall Thicknesses
Technique for Defining Mullions
Technique for Adding North Parking Area Striping

Chapter Five

Technique for Enhancing and Correcting Exterior Entities
Technique For Adding Columns
Technique for Adding 3D Curbs
Technique for Adding 3D Recessed Areas
Technique for Outlining 3D Roof Curves
Technique for Defining 3D Vertical Ruled Surfaces
Technique for Adding Remaining 3D Curved Surfaces
Technique for Defining 3D Horizontal Surfaces
Technique for Adding 3D Ceiling Surfaces

Chapter Six

Technique for Adding Concrete Walk Construction Joints
Technique for Drawing New Contours
Technique for Labeling Contour Lines
Technique for Adding Landscaping
Techniques for Dimensioning
   User-Defined Dimension Block
   Adding Special Notes to Dimensions

Technique for Adding Room, Door, and Window Annotation
Technique for Wall Pouching
Technique for Creating Ceiling Grid System
Technique for Inserting Light Fixtures

Chapter Seven

Technique for Defining Major Components
Technique for Defining Sloped Roof Lines
Technique for Adding the Grade Lines and Curbs
Technique for Creating and Placing Window
Technique for Creating Base Drawing for West Elevation
Technique for Creating Store Front and Doors
Technique for Placing Elevation and Annotation Symbols

Chapter Eight

Technique for Defining Foundation, Floor Slab, and Wall
Components
Technique for Creating Joist Web Members
Technique for Creating Other Half of Section
Technique for Creating Detail Base Drawing

Chapter Nine

Technique for BASEPLAN Framing Modifications
Technique for Supply and Return Grilles
Technique for BASEPLAN Plumbing Modifications

Chapter Ten

Technique for Creating the Title Block
Technique for Using Macros and Script Files for Plotting
Technique for Adding Revisions

Chapter Eleven

Technique for Data Extraction
 Exporting Drawing Data
 Framework for Specifications and Reports

AutoLISP Routines

Chapter Two

DOORSYM.LSP — Insert and Increment Door Annotation Symbols.
WINSYM.LSP — Insert and Increment Window Annotation Symbols.
ROOMSYM.LSP — Insert and Increment Room Annotation Symbols.
EELEV.LSP — Insert Detail and Arrow Annotations.
EDITATTR.LSP — Edit multiple attributes.

Chapter Three

LARBYENT.LSP — Turn OFF or FREEZE a layer by selecting an entity on the screen.
TOLAYER.LSP — Change an entity's layer to that of an entity selected on another layer.
LINEMEND.LSP — Prompts for two lines then replaces the two lines with one.
ACAD.LSP — Call standard AutoCAD commands with the keyboard entry of a few characters.

Chapter Four

XYZBOX.LSP — Lay out areas with polylines.

Chapter Five

CUT.LSP — Break or cut a line at a selected point with one operation.

Chapter Six

TREE.LSP — Add tree symbol with user-defined diameter.
FIXTURE.LSP — Insert light fixture blocks of user-defined size.

Chapter Seven

DULXEXTD.LSP — Select multiple lines to extend at one time.
DULXTRIM.LSP — Select multiple lines to trim at one time.

Chapter Eight

ANGLE.LSP — Create an angle shape to user-defined dimensions.
BEAM.LSP — Create a wide flange structural shape, beam or column, to user-specified dimensions.

Chapter Nine

> RXHAIR.LSP — Rotate the crosshairs of the display by either selecting an object already at the proper angle or by specifying a point and typing in an angle.

Chapter Ten

> CLOUD.LSP — Demarcate revisions to the drawings.

Chapter Eleven

> EXPORT.LSP — Export layers used and title block information.

Bonus Routines — Appendix B

> TXTHGT.LSP — Change text height of a "selection set" of text by automating the CHANGE command.
> TXTSTY.LSP — Change text style of a "selection set" of text by automating the CHANGE command.
> BRKLINE.LSP — Used to break a polyline.

AE-MENU System

Introduction

This Appendix contains a listing of all the AE-MENU files, AutoLISP files, and a listing of the drawing files provided on the AE DISK. In addition, this Appendix provides a listing of the ACAD.LSP file and three AutoLISP bonus routines.

The AE-MENU is designed to be used with the standard AutoCAD menu and to be placed in Tablet Area One of the AutoCAD menu. See Chapter Two for details on installing the AE-MENU..

AE DISK Installation

The AE DISK contains all the menu, support, and drawing files for the book. These files are compressed on the disk. The disk contains two install routines that will load these files onto your hard disk. These routines are called AE-LOAD and DR-LOAD.

AE-LOAD which is contained on the AE DISK, will copy all the symbols, AutoLISP routines and menus to the correct directory, AE-ACAD. After you have created the directory, change to the AE-ACAD directory, and insert the diskette into drive A: and type in **A:AE-LOAD** at the C:\AE-ACAD> prompt. The install program will decompress the disk files and load them into your AE-ACAD directory.

DR-LOAD will copy the book's completed drawing files to the correct directory, 8910. Follow the same procedure for the drawing files. Create the drawing file directory, called 8910, change to the directory, put the AE DISK in drive A:, and type A:DR-LOAD. The program will decompress the book's drawing files and load them into your drawing directory.

AE DISK Install

```
C:\> MD C:\AE-ACAD        The menu directory.
C:\> MD C:\8910           The project drawing directory.
C:\> CD \AE-ACAD          Change to menu directory.
C:\AE-ACAD> A:AE-LOAD     Loads menu and support files.
C:\AE-ACAD> CD \8910      Change to drawing directory.
C:\8910> A:DR-LOAD        Loads drawing files.
```

When the install operation is complete, you will have two directories on your hard disk. All the book's menu, symbol, AutoLISP, and prototype files will be in your AE-ACAD directory. All the book's drawing files will be in your 8910 drawing directory. Use the DOS DIR command to list the files.

Verifying AE-MENU and AutoLISP Files

Change to your AE-ACAD directory.

Verify that you have the following files.

 If you don's have the AE-DISK yet. Just read along.

```
C:\AE-ACAD> DIR *.*
```

AE-MENU.MNU	The book's menu program.
AEPLOT.SCR	Script file for plotting.
ACAD.LSP	The book's AutoLISP routines.
ANGLE.LSP	
BEAM.LSP	
CLOUD.LSP	
CUT.LSP	
DOORSYM.LSP	
DULXEXTD.LSP	
DULXTRIM.LSP	
EDITATTR.LSP	
EELEV.LSP	
EXTLINE.LSP	
FIXTURE.LSP	
LARBYENT.LSP	
LINEMEND.LSP	
ROOMSYM.LSP	
RXHAIR.LSP	
TOLAYER.LSP	
TREE.LSP	
TXTHGT.LSP	
TXTSTY.LSP	
WINSYM.LSP	
XYZBOX.LSP	
AEBORDER.DWG	The book's symbol library.
AETITLE.DWG	
AE-MENU.DWG	
ARROW.DWG	
DETAIL.DWG	
DRSYM.DWG	

```
EELEV.DWG
ELEV.DWG
EXIT.DWG
JB.DWG
LAVRT.DWG
LGT.DWG
LIGHT.DWG
NORTH.DWG
RECEPT.DWG
RETURN.DWG
RMSYM.DWG
SECT.DWG
SUPPLY.DWG
SWT.DWG
TELE.DWG
TOILET.DWG
TREE.DWG
WINSYM.DWG

ENGR-PT.DWG
PLAN-PT.DWG
SECT-PT.DWG
SITE-PT.DWG
```

The book's prototype drawings.

Verifying Drawing Files

 Change to your 8910 directory.

Verify that you have the following files.

 If you don's have the AE-DISK yet. Just read along.

```
C:\8910> DIR *.*
```

```
8910BS.DWG
8910CP.DWG
8910EE.DWG
8910EL.DWG
8910FP.DWG
8910FP-D.DWG
8910FP-P.DWG
8910ME.DWG
8910PL.DWG
8910SP.DWG
8910SP-D.DWG
```

The book's drawing files.

```
8910ST.DWG
BASEPLAN.DWG
```

ACAD.LSP Installation

The book's AE-MENU requires a modified ACAD.LSP file that comes with the AE DISK. The book's ACAD.LSP file plays an important role in how the AE-MENU system functions. Without the file you will miss the ability of being able to turn layers off or on, freeze or thaw or set layers by selecting them from the screen menu area. Since a major amount of time is spent dealing with layers, these routines become important from a production stand-point. Follow the instructions in Chapter Two to load and test the ACAD.LSP file from the AE DISK.

AutoCAD requires (and uses) only one ACAD.LSP file. Do not copy your standard ACAD.LISP file to the AE-ACAD directory, use the book's ACAD.LSP file in your AE-ACAD directory. To protect your old ACAD.LSP (which we assume is in your ACAD directory) use the DOS RENAME command and make a backup copy. Call the file something like ACAD.OLD.

ACAD.LSP Listing

Verify that the book's ACAD.LSP file is in your AE-ACAD directory. Use a text editor to examine it. Or look at the listing below.

Use the listing below, if you want to create the file with your text editor.

➡ *NOTE! If you do not make a backup copy of your standard ACAD.LSP file, and you copy the book's ACAD.LSP file to your ACAD directory, you will overwrite your standard ACAD.LSP file with the book's ACAD.LSP file. Make a back up copy of your standard ACAD.LSP file.*

AE-MENU ACAD.LSP Listing

If you wish to examine the AE-DISK's ACAD.LSP file, here is the listing:

```
    (vmon)
    (setvar "cmdecho" 0)
;
    (DEFUN C:BR () (COMMAND "BREAK"))
    (DEFUN C:ZE () (COMMAND "ZOOM" "E"))
    (DEFUN C:EN () (COMMAND "END"))
    (DEFUN C:ZW () (COMMAND "ZOOM" "W"))
    (DEFUN C:ZP () (COMMAND "ZOOM" "P"))
    (DEFUN C:ER () (COMMAND "ERASE"))
    (DEFUN C:OF () (COMMAND "OFFSET"))
    (DEFUN C:MO () (COMMAND "MOVE"))
    (DEFUN C:CO () (COMMAND "COPY"))
    (DEFUN C:FI () (COMMAND "FILLET"))
    (DEFUN C:CH () (COMMAND "CHANGE"))
    (DEFUN C:EL () (COMMAND "ERASE" "L" ""))
    (DEFUN C:CR () (COMMAND "CIRCLE" "2P"))
    (DEFUN C:LI () (COMMAND "LINE"))
    (DEFUN C:RE () (COMMAND "REDRAW"))
;This routine creates the layer list
  (defun C:LAYLOAD (/ tmp)
   (setq layers (cdr (assoc 2 (tblnext "LAYER" T))))
   (setq layers (list (cdr (assoc 2 (tblnext "LAYER"))) layers))
   (while (setq tmp (tblnext "LAYER"))
         (setq layers (cons (cdr (assoc 2 tmp)) layers))
   )
)
;This program loads the layer list into the screen area
  (defun C:LAY-UTIL (/ mo bw lay num lim valid)
   (if (not (boundp layers)) (setq layers (C:LAYLOAD)))
   (setq mo 0 lay nil)
   (setq len (1- (length layers)))
   (setq num (fix (/ len 18)))
   (setq lim (rem (1+ len) 18))
   (while (not lay)
     (if (< mo num) (setq tt1 18) (setq tt1 lim))
     (setq bw 0)
     (while (< bw tt1)
         (grtext bw (nth (+ bw (* mo 18)) layers))
         (setq bw (1+ bw))
     )
     (while (< bw 20)
         (grtext bw "          ")
         (setq bw (1+ bw))
     )
```

```
        (cond ((= mo 0) (grtext 18 "next"))
              (( mo num) (progn (grtext 18 "next") (grtext 19 "previous")))
              (T (grtext 19 "previous"))
        )
      (setq valid nil)
      (while (not valid)
        (setq temp (nth 1 (grread)))
        (cond
            ((and (/= mo num) (= temp 18))
              (progn
                (setq mo (1+ mo))
                (setq valid T)
              )
            )
            ((and (= temp 19) (/= mo 0))
              (progn
                (setq mo (1- mo))
                (setq valid T)
              )
            )
            (( temp tt1)
              (progn
                (setq lay (+ temp (* mo 18)))
                (setq valid T)
              )
            )
        )
      )
    )
  (grtext)
  (setq tt1 (nth lay layers))
)
;This routine sets the layer from a screen menu pick
(defun C:LAYERSET()
   (command "LAYER" "SET" tt1 "")
)
;This routine turns the layer off from a screen menu pick
(defun C:LAYEROFF()
   (command "LAYER" "OFF" tt1 "")
)
;This routine turns the layer on from a screen menu pick
 (defun C:LAYERON()
   (command "LAYER" "ON" tt1 "")
)
```

```
;This routine thaws the layer from a screen menu pick
  (defun C:LAYERTHW()
    (command "LAYER" "THAW" ttl "")
)
;This routine freezes the layer from a screen menu pick
  (defun C:LAYERFRZ()
    (command "LAYER" "FREEZE" ttl "")
)
```

ACAD.LSP Listing

AE-MENU Installation

When you load the book's menu and support files into your AE-ACAD directory, you will have a copy of the full AE-MENU system menu file. To install the menu, refer to the instructions in Chapter Two.

The AE-MENU.MNU File

We created the electrical and north arrow symbols at full scale. The electrical symbols are half size, six inches of full scale. If this size is too large or small when you insert them, add a scale factor of .5 for three inches, 2 for twelve inches, or whatever scale you need, to the macros that insert these symbols. See the following example for how to make a scale factor addition to the macro that inserts the electrical receptacle symbol.

```
[G-15] ^c^cINSERT;RECPT;scale factor;;\
```

By adding the scale, two semicolons and a slash, the scale factor will be accepted and you will be prompted for the rotation angle.

AE-MENU.MNU Listing

Verify that you have a copy of the AE-MENU.MNU file in your AE-ACAD directory. Use your text editor to examine it.

Read the descriptions in Chapters 2 and 3.

Bonus AutoLISP Routines

The following two AutoLISP routines are designed to be activated with a tablet menu selection. When you load the book's menu and AutoLISP files onto your hard disk, you will find these bonus files, TXTHGT.LSP and

TXTSTY.LSP. These routines support the AE-MENU menu items [CHANGE–TEXT–HEIGHT] and [CHANGE–TEXTYLE].

AutoLISP TXTHGT and TXTSTY.LSP

 Verify that you have the TXTHGT.LSP and TXTSTY.LSP files in your AE-ACAD directory. Examine the files with your text editor.

Look at the listings below, or create the AutoLISP files.

The TXTHGT.LSP routine lets you change either selected text or all the text to a different height.

```
(defun C:TXTHGT ( / mark th th1 ss1 s1 cnt mark e)
(setvar "CMDECHO" 0)
(setvar "EXPERT" 3)
(setq mark 0

(setq curr (getvar "TEXTSIZE"))                    ;select new and old text
(prompt "\nNew text height <")                     ;get new height
    (setq th (getdist)
    (setq th (if (null th) curr th)                ;check for null
    (setq th1 (getdist "\nOld text height <*>: "))  ;get old height
    (setq th1 (if null th1) "*" th1                ;check for null

(prompt "\nSelect objects or Return for all text:") ;select objects or text
    (setq ss1 (ssget))
       (if (null ss1) (setq ss1 (ssget "X" '((0 . "TEXT")))))

    (setq s1 (sslength ss1))                       ;determine loop count
    (setq cnt 0)

(prompt "\nUpdating text height...)                 ;change text
    (setvar "highlight" 0)
    (repeat s1
        (if (= "TEXT" (cdr (assoc 0 (setq e (entget (ssname ss1 cnt))))))
        (if (or (= (cdr (assoc 40 e)) th1) (= th1 "*"))
            (progn
                (entmod (subst (cons 40 th) (assoc 40 e) e))
                (setq mark (+ mark 1))
            )
        )
    )
)
```

```
(setq cnt (+ cnt 1))
)
(prompt "\n")
     (prin1 mark) (prompt " Lines of text updated")
(setvar "cmdecho" 1)
(setvar "highlight" 1)
(gc) (princ)
)
```

AutoLISP Routine for [CHANGE–TEXT–HEIGHT]

Testing the TXTHGT.LSP File

To test the TXTHGT.LSP, use the sequence below.

TXTHGT.LSP Operation

```
Command: (load "TXTHGT")
C:TXTHGT
Command: TXTHGT
Loading TXTHGT utility...              Loading indicator.
New text height <??>:                  Specify new height.
Old text height <*>:                   Specify which height to change or all.
Select objects or Return for all text: Select text to change.
Updating text height...                Working indicator.
?? Lines of text updated               Finished and displays the number of entries changed.
```

This AutoLISP routine changes either selected text, or all the text to a different style.

The next routine called TXTSTY.LSP, changes text styles. Here is the listing.

```
(defun C:TXTSTY (/ mark tst curr new old ssl sl tbl e cnt)
(setvar "CMDECHO" 0)
(setvar "EXPERT" 3)
(setq mark 0

(setq tst 0)                                              ;select and check for sty
(setq curr (getvar "TEXTSTYLE"))
(while (= tst 0)
   (prompt "\nNew text style <")
   (princ curr)
   (prompt ">: ")
   (setq new (strcase (getstring)))
   (setq new (if (= "" new) curr new))
   (if (null (tblsearch "STYLE" new))
     (prompt "\nThat style does not exist")
     (setq tst 1)
   )
)                                                         ;end while statement
 (setq old (strcase (getstring "\nOld text style <*>: ")))  ;select and check old
   (progn
       (prompt "That style does not exist")
       (setq old (strcase (getstring "\Old text style: <*>: ")))
   )
)                                                         ;end while statement
(setq old (if (= old "") "*" old))

(prompt "\nSlect objects or Retrun for all text:")       ;select objects or text
(setq ss1 (ssget))
(if (null ss1) (setq ss1 (ssget "X" '((0 . "TEXT")))))

(setq sl (sslength ss1))                                 ;determine loop count
(setq cnt 0)

(setq tb1 (tblsearch "STYLE" new))                       ;change text
(prompt "\nUpdating text style...")
(setvar "highlight" 0)
(repeat s1
   (if (= "TEXT" (cdr (assoc 0 (setq e (entget (ssname ss1 cnt))))))
      (progn
         (setq e (subst (cons 7 new) (assoc 7 e) e))
         (setq e (subst (assoc 41 tb1) (assoc 41 e) e))
         (setq e (subst (cons 51 (cdr (assoc 50 tb1))) (assoc 51 e) e))
         (setq e (subst (assoc 71 tb1) (assoc 71 e) e))
         (entmod e)
```

```
          (setq mark (+ mark 1))
      )
    )
)
 (setq cnt (+ cnt 1))
)
 (prompt "\n")
 (prin1 mark) (prompt " Lines of text updated")

(setvar "CMDECHO" 1)
(setvar "HIGHLIGHT" 1)
(gc) (princ)
)
```

Testing TXTSTY.LSP File

Try the routine.

TXTSTY.LSP Operation

```
Command: (load "TXTSTY")
C:TXTSTY
Command: TXTSTY
New text style <??>:                               Specify new style.
That style does not exist                          Error message.
Old text style <*>:                                Specify style of all.
Select objects or Return for all text:             Select text to change.
Updating text style...                             Working indicator.
<??> Lines of text updated                         Displays number of changes.
```

If you have not read CUSTOMIZING AUTOCAD and INSIDE AUTOLISP (New Riders Publishing) yet, you can use both books as references for creating custom menus and AutoLISP routines.

EXTLINE.LSP Routine

Here is one last AutoLISP routine on the AE-MENU that lets you extend lines. It supports the AE-MENU item [EXTEND–LINE].

```
AutoLISP Routine for [EXTEND-LINE]
(defun C:EXTEND (/ spnt epnt pnt dpnt egrp int npnt fpnt deltax deltay)
   (setq egrp (entget (car (entsel "\nPick the line to extend : "))))
   (setq pnt (getpoint "\nPick the point to extend to : "))
   (if (or (null egrp) (null pnt)) (prompt " Try again.")
      (progn
      (setq spnt (cdr (assoc 10 egrp))            ;select start point
            epnt (cdr (assoc 11 egrp)))           ;select end point
    (if (> (distance spnt pnt) (distance epnt pnt))   ;set distance
          (setq npnt epnt fpnt spnt agroup 11)
        (setq npnt spnt fpnt epnt agroup 10)
   )
     (setq deltax (- (car npnt) (car fpnt))
           deltay (- (cadr npnt) (cadr fpnt)))
     (setq dpnt (list (- (car pnt) deltay) (+ (cadr pnt) deltax)))
     (setq int (cons agroup (inters spnt epnt pnt dpnt nil)))
     (setq egrp (subst int (assoc agroup egrp) egrp))
     ))
   (princ)
)
```

We hope you will find these routines helpful in increasing your AutoCAD productivity.

Authors' Comments

These are our thoughts and comments on the practice of architecture, the computer industry, and why we wrote this book.

John M. Albright

I decided that I wanted to become an architect after my first drafting course in junior high school. My dream was to become the best. When I could not afford college, I attended a state junior college at night and worked at any drafting job I could get during the day in order to pay tuition and to practice my chosen profession. Then in 1983, AutoCAD came from out of the blue. Now, with some fourteen years of drafting by conventional methods and six years of using AutoCAD, I can no longer draw without my trusty computer by my side. Although, I have lost some of my manual drafting skills, I have learned that to be the best, you have to continue to learn and keep an open mind to progress.

The introduction of micro-computers and AutoCAD has probably affected the architectural profession more than any other development. Firms that resisted change either closed or were dissolved into other more progressive firms. Today it doesn't matter how well you letter, or that you can produce a drawing that has the proper line weights. AutoCAD and plotters have taken care of that. The key now is the ability to use different software packages, to think in a logical manner and to have some knowledge of the profession. To be the best today, you need to have a thorough knowledge of architectural practices, construction methods and computer technology, from the hardware to the software. As my mother says, "Experience is the best teacher, but also the hardest."

I hope this book will keep you from having to learn the same lessons that I did the hard way. To all you who read this book, GOOD LUCK with AutoCAD.

Elizabeth H. Schaeffer

I had no idea that I would eventually end up in the computer business when I got out of college. Back then, I didn't really know what I wanted to do with my life. I toyed with the idea of going to medical school, or of getting an advanced degree in microbiology, or, ironically, of becoming an architect. I enjoyed so many different subjects in college that I could never totally devote myself to a single discipline.

When I had the opportunity to join IBM in 1981, I found the job fit a combination of my interests. In my years at IBM, I learned a lot about applications, configuring hardware, different marketing techniques, customer satisfaction, how operating systems work, what it's like to work for a fine employer, and how to swim upstream against a flood of bureaucracy. I also learned that there is an absolute need for the developers of computer hardware and software to communicate better with the people who buy their systems, especially with the written word. *Techno-babble* is running rampant. Computer manuals are a joke (and a bad one, at that) to many customers. Frequently, the manual is the last place we go to look for something. It is extremely frustrating to have to wade through text that is unnecessarily convoluted and filled with cryptic acronyms. Creating straightforward documentation is a fairly recent endeavor in the computer industry. I am a firm believer that short of being *cute*, computer documentation should be at least as friendly as the systems it describes.

It was a real challenge for me to help create this book. Juggling the ideas of five different people, learning the fundamentals of AutoCAD were tasks that I found formidable but a lot of fun.

B. Dwayne Loftice, AIA

Having been in the profession for 23 years, I have witnessed enormous changes in our roles as architects. To some, we still just *draw plans* for buildings. To others, we play an important part in influencing people's lives by providing a better work, play and home environment. Our profession is changing rapidly with the spiraling technology of building products and procedures. Today, we are confronted with literally thousands if not hundreds of thousands of pieces of data that have to be sorted in order to design the simplest of building forms. We have found ourselves having to be versed in a myriad of other subjects such as finance, psychology, environmental issues, politics, and yet, we still design buildings.

Sometimes I wonder where the architectural profession will be in the next century. When I compare the role of the present day architect to that of the master builders of ancient Rome and Greece, I am amazed at the differences and yet, the similarities. The major difference is that the architect of those days conceived the design, engineered the solution and built it as well. The major similarity, on the other hand, is that our creative ideas are still conveyed by paper and pen. Unlike our predecessors of yesteryear, today, we have consultants of various disciplines to provide our engineering support and builders or contractors to put our concepts and ideas into a reality.

The one thing that has been changing over the past decade is the method in which we communicate our building design. CAD, as a partner and someday maybe as the replacement to the paper and pen, has revolutionized our ability to process and provide more information, in less time and with a greater degree of accuracy. This powerful tool has given us a new opportunity to be more than just building designers and production technicians. We now have the ability to truly view and analyze objects, spaces and buildings in three dimensions as we have only been able to visualize or draw them in the past.

This revelation can provide us with an opportunity to become better designers as well as better communicators. AutoCAD as a design and production tool is not merely a replacement of the pencil and paper medium. It provides you with an opportunity to effectively integrate the creative mind with a proven tool whose capabilities are almost unlimited. The changes in AutoCAD, since its inception, have opened doors for the architect's and engineer's design and production capabilities. The examples, suggestions, and techniques illustrated in this book, as a tutorial for a specific project, can provide you with guidelines on how you may develop further possibilities for your types of projects.

The End is just the Beginning.

Scot A. Woodard

I have seen micro-computers improve significantly since my first personal computer CAD station six years ago, and they do not seem to be leveling off in price any time soon. Speeds have increased from four megahertz to twenty-five megahertz, storage capacities have risen from dual-floppies to a gigabyte in size, floppy drives from 360 kilobytes to 1.44 megabytes, and software updates occur at least twice a year. If I were to maintain my hardware/software during this six years, a large sum of money would have been paid for the upgrades.

To keep an edge in today's market among other AEC firms, one must stay abreast of the technology. We as partners, associates, and managers need to realize the necessity for continuous funding to keep our CAD systems, and especially our firms, competitive with other firms.

The hardware/software are not the only costly items in CAD operations, the personnel to run the machines have also become a costly commodity. These personnel costs run parallel to the increasing prices workstations, and this will not change for at least five to seven years when the need for operators will lessen. Computers have given us great advancements in our fields, but along with advancements come high costs. Be prepared!

J. Wise Smith

I have been in MAIN line Architecture since 1979. By the term MAIN line, I mean designing and building structures. This 15 years spans the era from pre-microchip to today's technology. I have had to be in a constant state of change due to the technology explosion. Not least in this era of change is the nature of the technological support personnel. All of these factors drove me to seek a method of producing contract documents that could keep up with the changing environment. At first, I ordered an IBM system 60 which required conditioned space and a lot of it. Thanks to their time delay, they could not ship and I missed this *opportunity*. At this point, AutoCAD software had not been invented. Next came the *Mega* system, *Intergraph* — the do-all system. I now know enough not to move too quickly. Plotters at this time were plane tables with pen bars.

Then came AutoCAD, a software system compatible with individualized workstations and not tied to a main frame system. By now, educational institutions were beginning to recognize the need to train people in CAD technology — mostly to supply the market in engineering and other peripheral professions. Architecture was too aesthetically sensitive to allow such a precise drafting method to spoil the beauty of drawings. Consequently, no one in architecture was taught *drafting*.

Then came the revolution, and vocational schools and secondary colleges began to fill this void. Now they no longer taught drafting by conventional methods (pen, ink, pencil, etc.), but switched to computer drafting. Major accredited universities with architectural programs acquiesced to letting the secondary colleges and vocational schools teach CAD, while they stayed with the basics of History, Design, Math , etc. And in this 15-year time frame, most of the old line draftsmen either died or disappeared to a peripheral industry.

Now the profession consists of three broad categories of people:

1. Architects of my age who know both sides of the microchip era. Usually, this group consists of architects who were schooled in the conventional drafting methods and learned AutoCAD on their own. Generally, this group becomes bored with the actual operation of the computer and soon moves up to management positions.

2. Architects of recent vintage who have graduated in the last three to five years, but have little knowledge of how to produce a project.

3. A broad range of button pushers who know AutoCAD inside and out but can't tell a brick from a door handle.

The challenge today is to make these three groups function as a homogeneous unit. It is upon this challenge that I have built J. Wise Smith Associates, Inc. and it seems to be working well. Only through AutoCAD could this be possible.

As we have discussed, anyone reading this book should be made aware that after three workstations, a CAD manager is required. This could be one of the three users, but time must be spent in search of new programs, updates, new hardware and standardizing the use of the workstations. And lastly, the only thing guaranteed is change. There will always be an update, a faster machine, a new component, etc. Plan for it.

Our System's Hardware

Introduction

Here is a brief description of the hardware we use at J. Wise Smith Associates, Inc. We don't use elaborate high resolution 19-inch monitors or display list graphic cards. We use simple, but we think powerful, workstations that perform adequately for our projects.

Hardware

Selecting the right hardware to make up a system is just as important as selecting the software. We experimented with different hardware, and discovered quickly that we not only got what we paid for, but we paid in maintenance costs later for what we got. When you move one half to one meg files or make 20 or 30 plots a day, you won't be able to tolerate a slow, under-configured file server, especially if that file server is also used as a workstation. It's a waste of everyone's time when the demands put on your file server degrade the performance of the whole system. You don't have to buy super high-end systems to avoid the frustrations and missed deadlines that under-configuring causes — many of those systems require very high cost maintenance procedures.

We found that our networked system of IBM and Compaq 386 computers gave us the best performance per dollar investment over the long run.

Workstations

Keeping up with changes in technology is a problem that all firms eventually encounter. When workstations became too outdated, our original strategy was to rotate those workstations out of use with AutoCAD and into use for accounting or word processing functions. We would then replace these older models with the best new workstations we found available at the time.

When workstations got very old, but were still operational, we donated them to schools and other institutions. Because our network and the market for our services has grown so quickly, our donation plan has faltered. Our current workload won't allow us to part with our old workstations.

At present, all of our workstations are either IBM PS/2 Model 80-041's or Compaq 386/20-40's with Zenith Flat Screen Displays. Our systems are equipped with math coprocessors, Compaq VGA graphic cards in the Compaq computers and standard VGAs in the PS/2 computers.

While this configuration may not be the fastest, or the best on the market, we find the resolution, color, and speed to be quite adequate for our purposes. For input devices, we have 12 x 12 digitizer tablets attached to each workstation. To connect the workstations to the network, we selected IBM Base Band Adapters that we cabled together with twisted pair cables in a Ring fashion. This network configuration gives us the flexibility to add workstations without spending a large amount on each adapter card or cable attachment.

File Servers

The two file servers we use on our system are Compaq 386/25's with a 300 MB hard disk and a tape backup subsystem for each. We use the same displays and graphic cards for these devices as we do for the workstations. They have math coprocessors which match the processor's speed. One of the file servers also has a Hayes 2400 External Modem attached so that we can access systems at our satellite office and communicate with our outside engineering consultants. To allow each workstation and file server access to all our files, we added an IBM Base Band Extender which allows us to configure the network with a total of eight file servers, each of which can attach eight workstations. This arrangement gives us access to any file server from any workstation.

➡ *NOTE! We are not giving a broad endorsement to base band networks. They have some serious limitations that you will have to consider for your own firm. It so happens that base band fit our firm's needs.*

Printers and Plotters

We have attached an NEC LC890 laser postscript printer to the one file server which handles most of our printing needs. We use an IBM Proprinter II, attached to the other file server, for file server maintenance and reports only. We also have a Calcomp electrostatic E size plotter attached to this file server, which let's us plot drawings from any workstation without having to use switch boxes.

This configuration of workstations, file servers, and plotter has given us room for growth with the networking costs comparable to that of a minimally equipped 286 computer. By networking the total system, we can produce and track projects with less effort and fewer man-hours than we could before.

Index

A

ACAD.LSP, 3-27, B-4
AE-AutoLISP
    EXPORT.LSP, 11-7
    FIXTURE.LSP, 6-46
    LARBYENT.LSP, 3-26
    LINEMEND.LSP, 3-27
    Menu Routines and AutoLISP, 3-1
    Systems Standards Manual, 12-6
    TOLAYER.LSP, 3-26
    TREE.LSP, 6-16
    TXTHGT.LSP, B-8
    TXTSTY.LSP, B-8
    utilities, 3-27
    with symbols, 12-11
    XYZBOX.LSP, 4-16
AE-AutoLISP Routines, 2-2
    ANGLE, 8-4
    AutoLISP insertion, 2-2
    BEAM, 8-5
    CHANGE-TEXT-HEIGHT, B-8, B-9
    CHANGE-TEXT-STYLE, B-11
    CUT-LINE, 5-8
    Detail and Arrow, 2-35
    DOOR, 6-28
    Door Symbol, 2-29
    EXPORT, 11-25
    DOORSYM.LSP, 2-30
    Edit Attribute, 2-37
    EDITATTR.LSP, 2-37
    EELEV.LSP, 2-35
    EXTEND-MULTIPLE-LINE, 7-5
    EXTLINE.LSP, B-11
    FIXTURE, 6-47
    LAYER-BY SCREEN-ON, 9-5
    REVISION, 10-24
    ROOM, 6-28
    Room Annotation Symbol, 2-33
    ROOMSYM.LSP, 2-33
    RXHAIR, 9-5
    Systems Standards Manual, 12-6
    TREE, 6-17
    TRIM-MULTIPLE-LINES, 7-5

WINDOW, 6-28
Window Annotation Symbol, 2-31
WINSYM.LSP, 2-31
with symbols, 12-11
XYZBOX, 5-8, 5-10
XYZBOX command, 4-16
AE-AutoLISP Utilities, 3-25
    ACAD.LSP, 3-27
    ANGLE.LSP, 8-4
    BEAM.LSP, 8-4
    CHANGE-LAYER, 3-26
    CLOUD.LSP, 10-24
    CUT.LSP, 5-7
    DULXEXTD.LSP, 7-4
    DULXTRIM.LSP, 7-5
    LAYER-BY ENTITY-OFF, FREEZE or SET, 3-26
    LINE-MEND, 3-27
    RXHAIR.LSP, 9-5
AE-MENU
    A:AE-LOAD, B-1
    ACAD.LSP Listing, B-4
    AE Directory Table, 2-23
    AE-DISK installation, B-1
    Main Screen Menu, 3-2, 3-16
    System, 4-16, B-1, B-4
    Tablet Area One, 3-16
AE-MENU.MNU, 3-3, 3-16, 3-33, B-7
AE-SETUP, 1-28 ,2-6, 3-16
AEBORDER, 1-27, 3-21
AutoLISP & Productivity, 3-24
bonus AutoLISP routines, B-7
border drawing & setup, 3-22
budgeting updates, 12-14
choices and solutions, 12-13
creating prototype drawings, 1-22
Custom menus, 12-13
Customization Choices, 3-1
Customizing and managing, 3-1
Defined MAIN & SCRATCH Areas Diagram, 1-28
Dialog Boxes, 3-5
Drawing Scale Selection Menu, 3-20
Drawing Task Menu, 3-17
ENGR-PT Layers Table, 1-26
Implementation Prerequisites, 3-2
Installation, B-1

issues, 12-12
issues to solve, 12-13
MAIN, 1-27, 3-22
Menu Routines and AutoLISP, 3-1
Menu System Concepts, 3-2
options, 12-15
PLAN-PT Layers Table, 1-24
SCRATCH, 1-27, 1-28
SECT-PT Layers Table, 1-25
Sheet Size Selection Menu, 3-18
SITE-PT Layers Table, 1-23
tailoring, 3-1
Menu System Diagram, 3-1
UCS, 1-27, 1-28
Using a UCS for 2D Drawings, 1-27
Using and Creating Border Drawings, 1-27
WCS, 1-28
AutoCAD for A/E
    AE-MENU Tablet Area One Diagram, 3-16
    third-party menu system, 3-2
AutoCAD's Menu
    ***BUTTONS, 3-4
    ***ICON, 3-7
    ***POP1, 3-5
    ***SCREEN, 3-9
    ***TABLET1, 3-10
    **3DViews, 3-7
    A Portion of AutoCAD's Tablet Menu, 3-10
    ACAD.MNU, 3-4
    Advanced User Interface, 3-5
    associated ICON & Tablet Menu areas, 3-3
    Font Icon Menu, 3-7
    Macro Codes, 3-12
    Standard Button Menu, 3-4
    Standard SubMenu Syntax, 3-9
    Tools Pull Down Menu, 3-5
    Button, 3-3
    Macro Command Structure, 3-11
    Menu Areas, 3-3
    Pull Down and Icon Menu Areas, 3-3, 3-5
    Screen Menu area, 3-3, 3-8
    Tablet Menu Area, 3-10
AutoFLIX, 5-4
    operating, 5-48, 5-52
AutoLISP
    ACAD.LSP file, 3-25
    AutoLISP and Productivity, 3-24
    Extended AutoLISP, 3-25
    Invoking AutoLISP Routines, 3-24
    Memory, 3-25

AutoShade
    Operating, 5-48
    Options, 5-48
    Setup, 5-45

B

BasePlan
    Base Floor Plan Drawing, 6-24
    BASEPLAN Illustration, 9-6
    Modifications to BASEPLAN for Plumbing, 9-22
    Modifying the BASEPLAN, 9-6
Building Section Drawings
    Applying Wind Break Wall Outline, 8-11
    Adding Foundation, Floor Slab & Wall
Components, 8-10
    Adding the Text, Titles & Detail Annotation, 8-31
    Applying Joist Web Members, 8-25
    Brick Illustration, 8-18
    Building Section, 8-1, 8-7, 8-27
    Building Section 8910BS, 8-3
    Completed Building Section, 8-31
    Concrete Block and Structural Framing, 8-20
    Correcting the Layers, 8-14
    Creating Joist Web Members, 8-24
    Enhancing the Exterior Walls, 8-14
    Exterior Wall Enhancements Illustration, 8-14
    Exterior Wall Lines Illustration, 8-9
    Interior Features Illustration, 8-30
    Joist Web Members, 8-25
    Layer Corrections, 8-14
    Major Component Elevation Lines, 8-7
    Making a Mirror Image, 8-27
    Metal Mansard Panels, 8-22
    One Half of Completed Building Section, 8-28
    Porcelain Wall Panels & Store Front, 8-15
    Project Building Section, 8-1
    Project Sequence for Building Section, 8-2
    Roofing System, 8-22, 8-26
    Suspended Ceiling, 8-17
    Wind Break Wall, 8-11
Bonus AutoLISP Routines, B-7

C

Ceiling Plan Drawing
    Adding Suspended Ceiling Grid System, 6-39

Adding Text, 6-48
Applying Suspended Ceiling, 6-41
Ceiling Plan Project Sequence, 6-37
Completed Ceiling Plan, 6-48
Enhanced Base Plan, 6-38
Enhancing Walls & Roof Lines, 6-38
Porcelain Soffit Panels, 6-42
Insert Base Plan, 6-37
Light Fixtures, 6-43
Soffit Panels, 6-42
Chapter Setup Drawing
AE-SCR4, 4-6
AE-SCR5, 5-6
AE-SCR5 Setup Table, 5-7
AE-SCR6 Setup Table, 6-4
AE-SCR7, 7-3
AE-SCR7 Setup Table, 7-4
AE-SCR8, 8-3
AE-SCR8 Setup Table, 8-4
AE-SCR9, 9-4
AE-SCR9 Setup Table, 9-4
Coordinate Systems Icon, 5-2
Concepts
Data Extraction "System", 11-1
Menu System, 3-2
Screen or Tablet, 3-2

D

Data Extraction
AutoCAD, 11-6
concepts, 11-1
CSI number format, 11-4
Drawing Data, 11-2
EXPORT AutoLISP routine, 11-7
Framework III, 11-10, 11-17
networking, 11-5
Project Reports, 11-4
purpose, 11-1
Relationships, 11-4
software compatibility, 11-4
Specification Database, 11-12
Specifications, 11-2
Data Extraction Programs
CSI Division Database, 11-13
Database Setup, 11-12
Drawing Number Database, 11-18
Drawing spreadsheet template, 11-22

Framework III, 11-11, 11-15
Import Text Files, 11-15
PROJECT, 11-20
Project Number Database, 11-17
Project Number Log, 11-29
Project Spreadsheet Template, 11-20, 11-21
PROJLOG, 11-18
SPECIF spreadsheet template, 11-24
Specification Reference Report, 11-23
Spreadsheet Function, 11-19
Definition
Brick Surface, 8-18
Concrete Block & Structural Members, 8-20
Exterior Wall Lines, 8-9
Interior Wall Features, 8-30
Major Component Elevation Lines, 8-7
Mansard Framing, 8-19
Metal Mansard Panels, 8-22
Porcelain Wall Panels & Store Front, 8-15
Roofing System, 8-26
Structural Roof Framing, 8-22
Suspended Ceiling, 8-17
Design Development
file names, 12-18
issues, 12-17
management, 12-16
managing extra data, 12-19
methods and standards, 12-18
solutions, 12-18
Design Drawings
8910FP-D, 4-14
8910SP-D, 4-38
AE-MENU System, 4-2
BASEPLAN, 4-38
Procedures and AutoCAD, 4-1
Phase, 12-2
Flexibility with AutoCAD, 4-2
Floor Plan, 4-4
Increased Speed with AutoCAD, 4-2
Project Design, 4-3
Sequence for Site Plan, 4-4
Sequence for the Floor Plan, 4-4
Site and Floor Plan, 4-1
Site Plan 8910SP-D, 4-5
Transferability of Data, 4-3
Detail Drawings
Adding reinforcing, 8-36
Adding Text and Title, 8-37
Changing Scale, 8-37
Completed Detail, 8-37

Creating a Detail, 8-33
Door Section, 8-35
Enhanced Detail Drawing, 8-35
Enhancing Brick Surface, 8-35
Enhancing Door Section, 8-35
Foundation Detail, 8-33
Project Detail, 8-2
Reinforcing, 8-36
Drawing Setups
Ceiling Plan 8910CP, 6-37
Floor Plan 8910FP, 6-25
for Chapter Eight, 8-2
for Chapter Five, 5-5
for Chapter Four, 4-5
for Chapter Nine, 9-3
for Chapter Seven, 7-3
for Chapter Six, 6-3
for Chapter Two, 2-2
management, 12-7
Setup Table, 2-3
Drawing Specification Log
purpose, 12-5

E

Electrical Drawings
Adding Text, Fixture Table and Titles, 9-33
Complete Electrical Lighting Plan, 9-33
Creating Wiring Layout, 9-32
Defining and Adding Fixtures, 9-31
Lighting Plan, 9-3
Lighting Plan Sequence, 9-29
Modifications & Additions, 9-31
Modified 8910CP, 9-29
Modifying Reflected Ceiling Plan 8910CP, 9-29
Wiring Layout, 9-32
Engineering Drawings, 9-1
Engineering Drawing Table, 9-3
Exterior Elevation Drawings
Annotations, 7-41
Adding Text and Titles, 7-42
Adding Concrete Curbs & Grade Lines, 7-13
Base Drawing, 7-20
Base Information for North Elevation, 7-26
Building Exterior Limits, 7-9
Building on Existing Data, 7-1
Creating Base Drawing for South Elevation, 7-38
Creating Base Drawing for West Elevation, 7-19

Creating Store Front & Doors, 7-32
Creating Teller Window, Remote Tellers, 7-25
Placing Storefront, 7-35
Creating Windows, 7-16
Curbs and Grade Change, 7-29
Custom Lettering of Signage, 7-42
Defining Curb & Grade Elevations, 7-29
Defining Drive Through Mansard, 7-23
Defining Exterior Limits, 7-9, 7-28
Defining Hatched Surfaces, 7-39
Defining Major Component Elevations, 7-26
Defining Components of East Elevation, 7-7
Defining Mansard Curve, 7-10
Defining Mansard Over Drive Area, 7-31
Defining Specialty Equipment, 7-37
Defining Steel Columns, 7-24
Defining Slope of Roof, 7-11
Drive Through Columns, 7-24
East Elevation, 7-7
Exterior Elevation 8910EL, 7-3
Exterior Elevation Illustration, 7-1, 7-6
Exterior Limits Illustration, 7-28
Hatch Pattern, 7-40
Horizontal Frame Members, 7-34
Major Components Illustration, 7-7
North Elevation, 7-26
Partially Completed East Elevation, 7-18
Placing Window, 7-36
Signage Illustration, 7-42
Slope of Roof, 7-11
South Elevation, 7-38
South Elevation Completed, 7-42
Specialty Equipment Illustration, 7-25
Storefront and Window Illustration, 7-32
Storefront Placement Illustration, 7-35
Vertical Frame Members, 7-33
West Elevation, 7-21
Windows, 7-6
Window Placement, 7-36

F

Floor Plan Design Drawing
Adding Exterior & Interior Wall Thickness, 4-21
Adding Teller Work Positions, 4-29
Adding Text, 4-36
Adding Windows, 4-32
Area Layout, 4-15

BASEPLAN Block, 4-24
Checking Design Layout, 4-26
Completed Floor Plan Design, 4-38
Correcting Layering, 4-25
Correcting Layering Table, 4-26
Creating the Base Plan, 4-24
Defining Floor Plan Areas, 4-15
Defining Sub-Areas, 4-23
Doors Locations, 4-28
Design Drawing Project Sequence, 4-14
Inserting Doors, 4-27
Inserting Plumbing Fixtures & Cabinets, 4-28
Layer Designations, 4-25
Main Area Layout Table, 4-15
Moving Design Layout to the Main Area, 4-26
SubArea Layout, 4-23
Teller Counter Dimension Table, 4-30
Windows, 4-33
Floor Plan Drawing
Adding Annotations, 6-27
Adding Remote Tellers and Columns, 6-30
Adding Text, 6-35
Applying Wall Poucheing, 6-33
Base Floor Plan Drawing, 6-24
Completed Floor Plan, 6-35
Corridor Area, 6-29
Floor Plan Dimensions, 6-31
Enhancing Exterior Walls, 6-25
Enhancing Interior Walls, 6-26
Floor Plan Enhancements, 6-25
Floor Plan Project Sequence, 6-24
Poucheing Exterior Walls, 6-32
Remote Tellers & Columns, 6-30
Wood Paneling on Vault Walls, 6-26
Flow Chart
Communications, Issues & Solutions, 12-25
Design Development, Issue & Solutions, 12-16
Drawing Setup, 12-7
Project Flow Chart, 12-5
Project Production, Issues & Solutions, 12-20
Symbol Libraries, Issues & Solutions, 12-10
System Management Flow Chart, 12-1
System, Issues & Solutions, 12-28

H

Hardware, D-2
12 x 12 digitizer tablet, D-3

"third party" maintenance, 12-31
Calcomp electrostatic E size plotter, D-3
Compaq 386/20-40, D-2
directories & file management, 12-31
File Server, 11-1
hard disk, 12-30
IBM Proprinter II, D-3
IBM PS/2 Model 80-041, D-2
Local Area Network, 11-6
management, 12-29, 12-30
NEC LC890 laser postscript printer, D-3
networking, 11-5
Printers and Plotters, D-3
VGA, D-3
Workstations, D-2
Zenith Flat Screen, D-2

L

Log Forms
Drawing Number Log, 11-27
Drawing Specification Log, 12-5
Project Log Sheet, 12-6
Project Number Log, 12-6
Systems Standards Manual, 12-6

M

Management Issues
A/E SYSTEMS, 12-2
archiving files, 12-24
AutoCAD vs Manual, 12-3
budgeting updates, 12-14
communications, 12-25
critical phases, 12-3
Custom menus, 12-13
data input, 12-18
Design Development, 12-16
drawing accuracy & tolerances, 12-22
Drawing Data, 11-2
Drawing setup, 12-7
Drawing sheet size, scale and limits, 12-26
Drawing Specification Log, 12-5
file names, 12-18
formatting drawing data, 12-21
hardware, 12-30
layers and colors, 12-8

Management characteristics, 12-4
managing extra data, 12-19
menu management, 12-15
method and standards choices, 12-18
Naming symbols, 12-11
networking, 11-5, 12-28
Plotter Bottlenecks, 10-4
plotting, 12-23
production, 12-4, 12-20
Project Drawing Layout, 12-5
Project Flow Chart, 12-5
Project Log Sheet, 12-6
Project Number Log, 12-6
Project Planning, 12-2
Project Production issues, 12-20
Project Work Book, 12-5
reviewing drawing files, 12-23
software compatibility, 11-4
Specification Log Sheet, 12-5
Specifications, 11-2, 11-3
Standards for Plotting, 10-6
Symbol libraries, 12-8
Symbol library documentation, 12-10
System issues, 12-29
Systems Standards Manual, 12-6
temporary drawing files, 12-23
third-party menu system, 3-2
transferring data, 12-27
updating symbols, 12-11
User Groups, 12-1
which AutoCAD Release, 12-26
years and colors, 12-27
Mechanical Drawings
Adding Supply & Return Grilles, 9-15
Adding Text & Titles, 9-20
Completed Mechanical Plan, 9-20
Defining Equipment & Ducts, 9-19
Inserting Supplies & Returns, 9-18
Mechanical Plan, 9-2
Mechanical Plan Project Sequence, 9-15
Project Sequence for Mechanical Plan, 9-2
Return Grille, 9-18
Supply Grille, 9-17
Movie
generating, 5-4

N

Naming
files, 12-18
layers and colors, 12-8
prototype drawings, 12-8
Symbol Library, 11-4
Symbols, 12-11
Network
Compaq 386/25, D-3
criteria, 11-5
File Servers, D-3
Hayes 2400 External Modem, D-3
IBM Base Band Adapters, D-3
IBM Base Band Extender, D-3
Local Area Network, 11-6
NOVELL's EL II, 11-7

O

Operations
Design data input, 12-18
DOS, 11-24
Framework III, 11-24
Generating Specification Reference Report, 11-25
Generating Drawing Number Log, 11-27
Generating Base Data from AutoCAD, 11-24
Generating Project Number Log, 11-29
Method One, Drawing Number Log, 11-28
Method One, Project Number Log, 11-29
Method One, Specification Report, 11-26
Method Two, Drawing Number Log, 11-28
Method Two, Project Number Log, 11-30
Method Two, Specification Report, 11-26
Methods of System Operation, 11-24
Summary of "System" Operation, 11-30

P

Plan Drawings
Ceiling Plans, 6-3
Contour Labels, 6-12
Floor Plan, 6-2
Project Plan Drawings, 6-1
Project Sequence For Ceiling Plan, 6-3
Project Sequence For Floor Plan, 6-2

Project Sequence For Site Plan, 6-1
Site Plan 8910SP, 6-3
Plotting
    Adding Revisions, 10-23
    AEPLOT, 10-20
    AEPLOT Script File, 10-19
    AEPLOT-2, 10-20
    Alternate Output Methods, 10-25
    Bottleneck Solutions, 10-5
    Color ink jet printers, 10-26
    Color Printers, 10-26
    Dot-matrix Printers, 10-3
    Electrostatic Plotters, 10-3
    High resolution color dot-matrix, 10-26
    Invoking AEPLOT Script, 10-20
    Laser Printers, 10-3
    Layer Colors & Line Weights, 10-17
    Line Weights, Views and Scale, 10-17
    Macros, 10-19
    Photographic, 10-26
    Plot Log, 10-1
    Plot Logs, 10-23
    Plotter Bottlenecks, 10-4
    Plotter History, 10-1
    Plotter Technologies, 10-2
    Plotting 8910FP, 10-21
    Plotting Project Drawings, 10-21
    Plotting Views, 10-17
    Project Plot Log, 10-23
    Project Working Drawing List, 10-7
    Raster Technology, 10-2
    Scaling Plots, 10-18
    Script File, 10-19
    Standards for Plotting, 10-6
    Using A Macro or Script File for Plotting, 10-18
    Vector Technology, 10-2
    Video, 10-26
Plumbing Drawings
    Adding Text & Titles, 9-27
    BASEPLAN Plumbing Modifications, 9-22, 9-24
    Completed Plumbing Plan, 9-27
    Creating Sanitary Riser, 9-26
    Drawing Piping Layout, 9-25
    Piping Layout Illustration, 9-25
    Plumbing Plan, 9-2
    Plumbing Plan Project Sequence, 9-22
Presentation Design Drawing
    3D Ceiling Surfaces, 5-38
    3D Curved Mansard Surfaces, 5-33
    3D Entrance Walks, 5-21

3D Horizontal Surfaces, 5-35
3D Planting Area Surfaces, 5-43
3D Roof Base Plan, 5-26
3D Site Surfaces, 5-41
3D Surfaces, 5-28
3D Vertical Curve-edged Surfaces, 5-30
8910FP-P, 5-6
Adding 3D Curbs, 5-16
Adding 3D Horizontal Surfaces, 5-35
Adding 3D Recessed Areas to Roof, 5-22
Adding 3D Roof Curves, 5-26
Adding 3D Vertical Surfaces, 5-30
Adding Columns for Canopy, 5-14
Applying 3D Recessed Area to Drive Through, 5-24
AutoFlix Basics, 5-4
AutoFlix Options, 5-51
AutoFlix Setup Illustration, 5-49
AutoShade and AutoFlix, 5-1
AutoShade Basics, 5-3
AutoShade Options, 5-48
AutoShade Setup Illustration, 5-45
Basic Rules of Ruled & Edged Surfaces, 5-29
Black Holes & Hollow Areas, 5-29
Color Enhancements for AutoShade, 5-13
Columns at Canopy, 5-14
Completed Wire Frame Model, 5-45
Completed Wire Frame Roof System, 5-40
Coordinate System Icon, 5-2
Correcting Windows & Frames, 5-12
Curb Surfaces, 5-42
Defining 3D Curb Surfaces, 5-42
Defining 3D Entrance Walks, 5-21
Defining 3D Island Surfaces, 5-43
Defining 3D Planting Area Surfaces, 5-43
Defining 3D Site Surfaces, 5-41
Drawings and Movie, 5-4
Generating A Movie, 5-54
Horizontal Voids, 5-29
Island Surfaces, 5-43
Making Color Enhancements, 5-13
Operating AutoFlix, 5-52
Preparing the Presentation Drawing Base Plan, 5-8
Presentation Drawing Base Plan Illustration, 5-9
Project Presentation Drawings, 5-5
Roof Base Plan, 5-22
RULESURF and EDGESURF Surfaces, 5-29
Setting Up AutoShade Drawing, 5-47
SURFTAB1 and SURFTAB2 control, 5-30

Three Dimensional AutoCAD Drawing, 5-1
Using Release 10's 3D Features, 5-2
Production
    accuracy and tolerances, 12-22
    archiving files, 12-24
    AutoCAD Release, 12-26
    communications, 12-25
    Design Development issues, 12-17
    Drawing setup, 12-7
    formatting drawing data, 12-21
    issues, 12-20
    layers and colors, 12-27
    Management, 12-4, 12-20
    networking, 12-28
    plotting, 12-23
    Project Work Book, 12-5
    reviewing drawing files, 12-23
    scale, 12-26
    sheet size, 12-26
    Short/Intermediate/Long term Goals, 12-4
    solutions, 12-21
    temporary drawing files, 12-23
    transferring data, 12-27
Project Drawing Layout
    purpose, 12-5
Project Log Sheet
    purpose, 12-6
Project Number Log
    purpose, 12-6

R

Reports
    CSI Specification Division Report, 11-1
    Framework III Report Forms, 11-17
    Generating a Specification Reference Report,
11-25
    Specification Reference Report Illustration, 11-25
Real world symbols, 2-1
RULESURF and EDGESURF, 5-29

S

Site Plan Design Drawing
    Adding Pavement, 4-41, 4-43
    Adding Setback Lines, 4-9
    Adding South Drive, Islands & Parking Area,
4-44
    Adding Streets and Walks, 4-39
    Adding Text, 4-56
    Adding Parking Striping, 4-47, 4-51
    Adding Trash and North Parking Area, 4-46
    Baseplan, 4-39
    Completed Floor Plan Design, 4-38
    Completed Site Plan Design, 4-56
    Defining Curb Cut Slopes, 4-50
    Defining Curbs, 4-48
    Defining Pavement Edges, 4-43
    Enhanced Building Lines, 4-53
    Inserting North Arrow, 4-54
    Laying Out the Property Lines, 4-7
    North Arrow, 4-54
    Pavement & Building Curbs, 4-48
    Placing the Base Plan, 4-38
    Sequence, 4-7
    Setback Area, 4-9
    Site Plan Areas Illustration, 4-41
    Site Plan Property Lines, 4-7
Site Plan Drawing
    Adding Construction Joints, 6-6
    Adding Dimensions to Site Plan, 6-18
    Adding Landscaping, 6-14
    Adding Site Plan Enhancements, 6-4
    Adding Text & North Arrow Symbol, 6-22
    Completed Site Plan, 6-22
    Dimension Variable Table, 6-18
    Dimensions for the Site Plan, 6-18
    Labeling Contours, 6-12
    Placing New Contours, 6-9
    SKETCHED Contour Lines, 6-9
    Tree Placement, 6-14
Specification Log Sheet
    purpose, 12-5
Structural Drawings
    Adding Beams, 9-11
    Adding Text and Titles, 9-13
    BASEPLAN, 9-6
    Beam Layout, 9-11
    Completed Roof Framing Plan, 9-13
    Defining Framing Members at Roof Openings,
    9-12
    Defining Joists and Bridging, 9-9
    Joist and Mansard Layout, 9-9
    Modifying the BASEPLAN, 9-6
    Roof Framing Plan, 9-2, 9-6
Symbol Libraries, 2-1
    1-unit by 1-unit, 2-3, 2-4, 2-7

Annotation symbols, 2-3, 2-4, 2-5
Annotations, 2-1
Arrow Symbol, EELEV, 2-16
Attributes and Symbols, 2-5
AutoLISP routines, 2-5, 2-6, 12-11
Base or Insertion Point, 2-5
blocks, 2-3
Completed Section Arrow Symbol (EELEV), 2-16
Constant Attributes, 2-6
Creating Detail Symbol, 2-14
Creating Door Annotation Symbol, 2-7
Creating Lavatory Symbol, 2-18
Creating Section Arrow, 2-16
Creating Symbols, 2-6
Creating Toilet Symbol, 2-21
Creating Window Annotation Symbol, 2-9
Defining Attributes, 2-5
Detail Symbol Drawing, 2-14
Detail Symbol, DETAIL, 2-14
documentation, 12-10
door annotation, 2-7
Door Annotation Symbol Drawing, 2-7
Door Symbol, DRSYM, 2-8
Invisible Attributes, 2-6
Lavatory Symbol Drawing, 2-18
Lavatory Symbol, LAVRT, 2-18
layer 0, 2-5
management, 12-8
Naming, 11-4, 12-11
Naming Symbols, 2-4
Real World symbol, 2-18
Real World Symbols, 2-1, 2-3, 2-4, 2-5
Room Annotation Symbol Drawing, 2-11, 2-12
Room Symbol, RMSYM, 2-12
Rules For Creating Symbols, 2-3
Rules for specifying Insertion Point, 2-5
Scale Factor, 2-4
Section Arrow Symbol Drawing, 2-16
solutions, 12-10
Symbol Types, 2-3
Symbols and Layering, 2-4
Toilet Symbol Drawing, 2-21
Toilet Symbol, TOILET, 2-21
updating, 12-11
USERI1, 2-6
Variable Attributes, 2-6
Visible Attributes, 2-6
AutoLISP Routine , 2-33
System Planning, 1-1
    AutoCAD and Manual Methods, 1-3

blocks, 1-4
combined methods, 1-5
Construction Document, 1-1
Creating and Using Prototype Drawing, 1-20
CSI, 1-21
Directory Table, 1-19
Drawing Name Table, 1-20
Drawing Number Log, 1-11
Drawing Number Log Form, 1-11
Drawing Specification Log, 1-12
Drawing Specification Log Form, 1-12
ENGR-PT, 1-21
entities, 1-4
Hardware, 1-14
Layer Colors and Line Weights Table, 1-21
Log Forms, 1-0
Naming Convention for the Project, 1-18
Naming Layers, 1-21
PLAN-PT, 1-20
Planning, 1-1
Production Concepts, 1-2
Production Decisions, 1-2
Production Flow Chart, 1-8
Production Flow Chart Diagram, 1-8
Production Issues, 1-5
Production Planning, 1-6
Production Variables, 1-4
Project Drawing Layout, 1-9
Project Naming Convention, 1-6, 1-19
Project Number Log, 1-10
Project Specification Log Form, 1-13
Project Work Book, 1-7
repeatable items, 1-4
SECT-PT, 1-20
Sheet Layout A-1 Drawing, 1-9
Sheet Layout A-2 Drawing, 1-9
SITE-PT, 1-20
System Development and Standards, 1-15
Third party menu, 1-5
Using Layer Colors, 1-21
working drawings, 1-4
System Software Components
    AutoCAD, 11-6
    Database and Spreadsheet Software, 11-6
    Framework III, 11-10
    Local Area Network, 11-6
    Microsoft Word, 11-10
    NOVELL's ELS II, 11-7
    Setting Up the Software Components, 11-10
    Software Components, 11-4

Word Processor, 11-6
System Standards
    Systems Standards Manual, 12-6

T

Technique
    3D Corrections, 5-10
    3D Curb, 5-19
    3D Curved Surface, 5-34
    3D Horizontal Roof Surfaces, 5-36
    3D Recessed Areas Drawing, 5-23
    3D Roof Curve, 5-27
    3D Vertical Ruled Surfaces, 5-31
    Adding Special Notes to Dimensions, 6-21
    AEBORDER Drawing, 1-28
    Adjusting the Areas, 4-20
    Base Drawing, 7-20
    BASEPLAN Drawing, 9-7
    BASEPLAN Plumbing Drawing, 9-23
    Calculating Areas, 4-13
    Ceiling Surface, 5-39
    Ceiling Grid System, 6-40
    Column, 5-15
    Construction Joint, 6-7
    Design Layout, 4-20
    Detail Base Drawing, 8-33
    Diagonal Striping, 4-52
    Dimensioning, 6-19
    Dimensions with Special Notes, 6-21
    Drawing New Contours, 6-9
    Enhancing & Correcting Exterior Entities, 5-10
    First Curb, 5-17
    Floor Plan Area Layout, 4-16
    Foundation, Floor Slab & Wall Components, 8-10
    Grade and Curb, 7-15
    Horizontal Frame Members, 7-34
    Joist Web Members, 8-24
    Labeling Contour Lines, 6-12
    Landscaping, 6-14
    Light Fixture Symbol Drawing, 6-45
    Major Component, 7-8
    Mullion, 4-34
    Other Half of the Section, 8-28
    Outlining 3D Roof Curves, 5-26
    New Contours Smoothing, 6-10
    Property Line, 4-7
    Prototype Drawings, 1-22

    Remaining 3D Curved Surfaces, 5-33
    Ruled Surface, 5-31
    Setback lines, 4-9
    Setback Block, 4-12
    Sloped Roof Lines, 7-11
    Store Front and Doors, 7-32
    Striping, 4-52
    Supply and Return Grilles, 9-16
    SURFTAB1 and SURFTAB2 variables, 5-31
    User-Defined Dimension Block, 6-20
    Vertical Storefront Member, 7-33
    Wall Poucheing, 6-33
    Wall Thicknesses, 4-22
    Window, 7-17
Title Blocks, 10-7
    Adding the Attributes, 10-12
    AETITLE Illustration, 10-8
    AETITLE Setup Table, 10-9
    Attributes, 10-12
    Completed AETITLE, 10-16
    Creating AETITLE Block & Completing
AEBORDER Drawing, 10-7
    Creating the Title Block, 10-8
    Defined Main Drawing Area, 10-9
    Defining Border Titles, 10-16
    Drawing Number & Company Attributes, 10-12
    Project Area Attributes, 10-13
    Revision Area Attribute, 10-15
    Sheet Title Area, 10-10
    Sheet Title SubAreas, 10-11

U

UCS
    Coordinate Systems, 5-2
    Plan View according to UCS, 5-3
Using
    AutoCAD and Manual, 12-3
    AutoCAD and pin registration , 12-3

W

WCS
    Coordinate Systems, 5-2

New Riders Library

AutoCAD for Architects and Engineers
A Practical Guide to Design, Presentation and Production
By John Albright and Elizabeth Schaeffer
544 pages, 150+ illustrations
ISBN 0-934035-53-9 **$29.95**

The only AutoCAD book specifically written for Architects and Engineers — Master you AEC project using high-powered design development with AutoCAD Release 10. Learn how to construct working drawings using techniques from real life AEC projects. The book shows you how to generate reports. and produce stunning computer presentations with AutoLISP, AutoShade and AutoFlix.

INSIDE AutoCAD   Over 250,000 sold
The Complete AutoCAD Guide   Fifth Edition — Release 10
D. Raker and H. Rice
864 pages, over 400 illustrations
ISBN 0-934035-49-0 **$29.95**

INSIDE AutoCAD, the best selling book on AutoCAD, is entirely new and rewritten for AutoCAD's 3D Release 10. This easy-to-understand book serves as both a tutorial and a lasting reference guide. Learn to use every single AutoCAD command as well as time saving drawing techniques and tips. Includes coverage of new 3D graphics features, AutoShade, and AutoLISP. This is the book that lets you keep up and stay in control with AutoCAD.

CUSTOMIZING AutoCAD   Second Edition — Release 10
A Complete Guide to AutoCAD Menus, Macros and More!
J. Smith and R. Gesner
480 Pages, 100 illustrations
ISBN 0-934035-45-8, **$27.95**

Uncover the hidden secrets of AutoCAD's 3D Release 10 in this all new edition. Discover the anatomy of an AutoCAD menu and build a custom menu from start to finish. Manipulate distance, angles, points, and hatches — ALL in 3D! Customize hatches, text fonts and dimensioning for increased productivity. Buy CUSTOMIZING AutoCAD today and start customizing AutoCAD tomorrow!

INSIDE AutoLISP   Release 10
The Complete Guide to Using AutoLISP for AutoCAD Applications
J. Smith and R. Gesner
736 pages, over 150 illustrations
ISBN: 0-934035-47-4, **$29.95**

Introducing the most comprehensive book on AutoLISP for AutoCAD Release 10. Learn AutoLISP commands and functions and write your own custom AutoLISP programs. Numerous tips and tricks for using AutoLISP for routine drawing tasks. Import and export critical drawing information to/from Lotus 1-2-3 and dBASE. Automate the creation of scripts for unattended drawing processing. *INSIDE AutoLISP* is the book that will give you the inside track to using AutoLISP.

STEPPING INTO AutoCAD   Fourth Edition—Release 10
A Guide to Technical Drafting Using AutoCAD
By Mark Merickel
380 pages, over 140 illustrations
ISBN: 0-934035-51-2, **$29.95**

This popular tutorial has been completely rewritten with new exercises for Release 10. The book is organized to lead you step by step from the basics to practical tips on customizing AutoCAD for technical drafting. Handy references provide quick set-up of the AutoCAD environment. Improve your drawing accuracy through AutoCAD's dimensioning commands. It also includes extensive support for ANSI Y14.5 level drafting.

AutoCAD Reference Guide
Everything You Want to Know About AutoCAD — *FAST!*
By Dorothy Kent
256 pages, over 50 illustrations
ISBN: 0-934035-57-1, **$11.95**

All essential AutoCAD functions and commands are arranged alphabetically and described in just a few paragraphs. Includes tips and warnings from experienced users for each command. Extensive cross-indexing make this the instant answer guide to AutoCAD.

INSIDE AutoSketch
A Guide to Productive Drawing Using AutoSketch
By Frank Lenk
240 pages, over 120 illustrations
ISBN: 0-934035-20-2, **$17.95**

INSIDE AutoSketch gives you real-life mechanical parts, drawing schematics, and architectural drawings. Start by learning to draw simple shapes such as points, lines and curves, then edit shapes by moving, copying, rotating, and distorting them. Explore higher-level features to complete technical drawing jobs using reference grids, snap, drawing layers and creating parts.

The Autodesk File
The Story of Autodesk, Inc., the Company Behind AutoCAD
Written and Edited by John Walker
532 pages
ISBN: 0-934035-63-6 **$24.95**

The unvarnished history of Autodesk, Inc., the company behind AutoCAD. Read the original memos, letters, and reports that trace the rise of Autodesk, from start-up to their present position as the number one CAD software company in the world. Learn the secrets of success behind Autodesk and AutoCAD. Must reading for any AutoCAD user or entrepreneur!

AutoCAD Software Solutions

New Riders AutoLISP Utilities
Disk 1 — Release 10
ISBN 0-934035-79-2 **$29.95**

This disk contains several valuable programs, utilities and subroutines. These are useful to any AutoCAD drawing application. Some of the tools are:

CATCH.LSP CATCH is great for selecting the new entities created by exploding blocks, polylines, 3D meshes, and dimensions.

HEX-INT.LSP is a set of hexadecimal arithmetic tools that make dealing with entity handles easier.

SHELL.LSP contains the SHELL function that executes and verifies multiple DOS commands with a single AutoCAD SHELL command execution (DOS only).

MERGE-V.LSP contains MERGE-V, which combines two files and verifies the copy procedure (DOS only).

GROUP.LSP contains functions to create and select *groups* of entities in AutoCAD drawings.

GRPT.LSP contains GRPT, a function that draws GRDRAW temporary points with any PDSIZE or PDMODE system variable setting.

XINSERT.LSP contains XINSERT, an external block extraction and insertion program.

STACK.LSP is a function loading program to minimize memory conflicts in using several moderate to large AutoLISP functions at once.

The AutoLISP programs and subroutines are not encrypted and are well documented by comments in the filename.LSP files. Watch for or call New Riders Publishing for information on additions to the New Riders Utilities Disk family.

Ketiv Technologies, Inc.

The AE-MENU routines used in this book are a small subset modified from the ARCH-T menu from Ketiv Technologies, Inc. The ARCH-T menu is a full-featured menu for architecture and engineering use. It supports AutoCAD Release 10. Ketiv Technologies, Inc. is an authorized AutoCAD developer. For more information about the ARCH-T menu, write or call Ketiv Technologies, Inc.

Ketiv Technolgies, Inc.
6645 N.E. 78th Court C-2
Portland, OR 97218
(503) 252-3230